Mathematics and Statistics for the Quantitative Sciences

Mathematics and Statistics for the Quantitative Sciences was born from a radical reimagining of first-year mathematics. While calculus is often seen as the foundational mathematics required for any scientist, this often leads to mathematics being seen as some, ultimately useless, hoop that needs to be jumped through in order to do what someone really wants to do. This sentiment is everywhere at every level of education. It even shows up in how people stereotype mathematics courses.

What this book aims to do, therefore, is serve as a foundational text in everyday mathematics in a way that is both engaging and practically useful. The book seeks to teach the mathematics needed to start to answer fundamental questions like "why" or "how". Why do we only need to take census data once every few years? How do we determine the optimal dosing of a new pharmaceutical without killing people in the process? Or, more generally, what does it even mean to be average? Or what does it mean for two things to actually be different? These questions require a different way of thinking — a quantitative intuition that goes beyond rote memorization and equips readers to meet the quantitative challenges inherent in any applied discipline.

Features

- Draws from a diverse range of fields to make the applications as inclusive as possible
- Would be ideal as a foundational mathematical and statistical textbook for any applied quantitative science course

Mathematics and Statistics for the Quantitative Sciences

Matthew Betti
Mount Allison University, Canada

CRC Press
Taylor & Francis Group
Boca Raton London New York

CRC Press is an imprint of the
Taylor & Francis Group, an **informa** business
A CHAPMAN & HALL BOOK

First edition published 2023

by CRC Press
6000 Broken Sound Parkway NW, Suite 300, Boca Raton, FL 33487-2742

and by CRC Press
4 Park Square, Milton Park, Abingdon, Oxon, OX14 4RN

CRC Press is an imprint of Taylor & Francis Group, LLC

ISBN: 978-1-032-20814-5 (hbk)
ISBN: 978-1-032-20826-8 (pbk)
ISBN: 978-1-003-26540-5 (ebk)

DOI: 10.1201/ 9781003265405

Typeset in Latin Modern font
by KnowledgeWorks Global Ltd.

Publisher's note: This book has been prepared from camera-ready copy provided by the authors.

For you, quite literally.
May you take what you learn here
and make the world a better place.

Contents

Preface

The intent of this book is to serve as a foundational text in everyday mathematics. I don't mean the everyday mathematics people think they need, like addition, subtraction, multiplication, and division. In fact, we are going to take those things for granted. These are the mathematics you need to start to answer questions like *why* or *how. Why do we only need to take census data once every few years?* Or, *how do we determine the optimal dosing of a new pharmaceutical without killing people in the process?* Or, more generally, *what does it even mean to be average?* Or *what does it mean for two things to actually be different?*

These questions require a different way of thinking to answer. They require a *quantitative intuition*; an ability to see equations and relationships in the world around us[1]. In this book, I hope that together we can build such an intuition by relying on shared experiences. While many people find math hard and confusing, it's really not supposed to be. It's supposed to make sense. It's a set of tools[2] that we use to describe the world around us that transcends language differences, geographic region, and, theoretically, even our planet.

This book was born from a radical reimagining of first-year mathematics[3]. While calculus is often seen as the foundational mathematics required for any scientist, this often leads to mathematics being seen as some, ultimately useless, hoop that needs to be jumped through in order to do what someone *really* wants to do. This sentiment is *everywhere* at *every level of education* [16, 26, 15, 23]. It even shows up in how people stereotype mathematics courses [8].

I don't think it's quite correct that calculus, algebra, and the like are *useless* unless you're a mathematician, I think the emphasis is, generally, on the wrong skills. Linear Algebra being the single most important

[1]You may have just had the thought *this isn't for me; I don't have that.* Keep with it.

[2]A deep, elegant, beautiful set of tools that, in the right light, can look more like a work of art.

[3]We at Mount Allison were not the first to reimagine, nor will we be the last, or best.

field of mathematics that any scientist can learn is a hill I will die on[4]. Concepts of independence, spanning, and organization of information into vectors, and manipulating that information using matrices are concepts that appear everywhere in our computer-driven world. Linear Algebra is well-structured, proofs are relatively simple, and it can help people build that quantitative intuition.

What is more important than learning all the derivative rules, or which integral substitution to use and when to use it, is to understand *why* these things came to be in the first place. What do these mean as descriptors of the real world? How do we use all the symbols and theorems and rules to say things about the real world?

This book is, if I may bastardize Nabokov, a love letter to mathematical thinking. The rules and theorems are, for the most part, secondary to the idea that, say, calculus is the study of change and change is the only thing we really have to make sense of the world around us[5]. Linear Algebra is the study of how linear things interact with one another. Probability and statistics is the study of counting possibilities and making sense of the randomness inherent in the world around us[6].

There are different paths through this book depending on what your needs are. Mathematics builds upon itself[7] and so some chapters of this book require other chapters either for the tools introduced or for context. The book works best as a whole, re-emphasizing the need for context, re-grounding ourselves in reality throughout, and always focusing on building intuition and reaffirming the value of specialized knowledge.

Other than reading the book cover-to-cover, figure 1 shows one possible path that will take you on a journey through what it means to *change* in a quantitative sense. This is as close to a traditional, one-semester introductory calculus course you will find embedded in this book. The biggest change is the introduction of linear algebra in Chapter 3 that aides multi-dimensional thinking and the introduction of differential equations in Chapter 5[8].

Figure 2 is the path used in my Introduction to Data Science course which introduces basic computational ideas, statistics, probability, and

[4]That linear algebra can serve any human no matter their life path is a hill I will be mortally wounded on.

[5]It's at least one perspective. How do we differentiate between wall and window? A change of material. Your experiences are just changes in your being caused by changes in your surroundings.

[6]Which is itself a form of change

[7]In the words of Peter Griffin: it insists upon itself.

[8]While I like to end a course with the "cliffhanger" of ODEs, it's also possible to skip right ahead to Chapter 6.

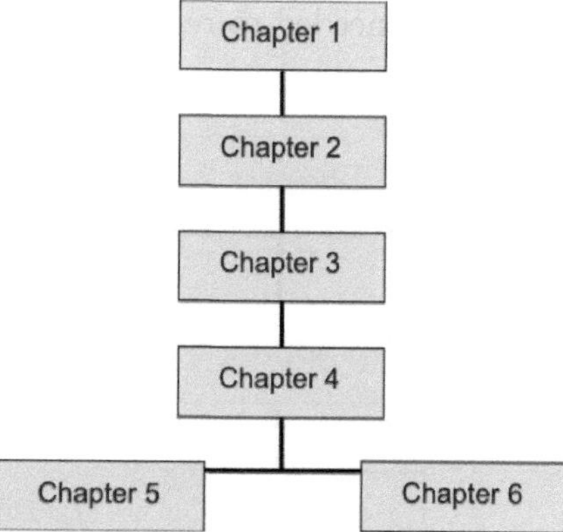

Figure 1: If you are interested in the study of change, this path of chapters will help develop your intuition.

communication. These chapters are written largely independent of Chapters 1–6, as the goal of Introduction to Data Science is open for all and has seen arts and humanities students succeed with minimal background.

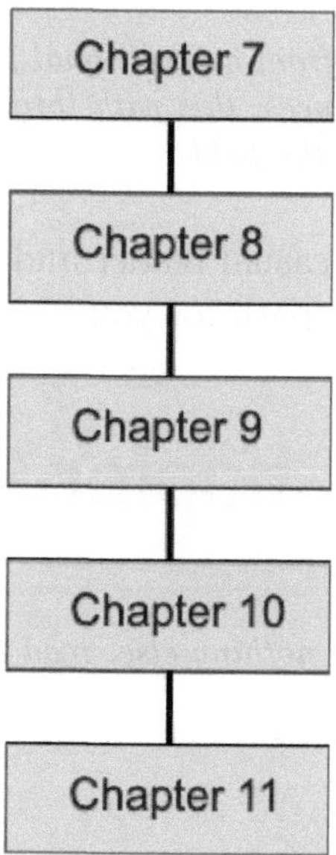

Figure 2: This set of chapters describes the foundational tools of probability, stats and communication which together form a foundation of Data Science.

Figure 3 can be used as a follow-up to a traditional first-term differential calculus course. This path is designed to give students in the life sciences specifically the tools needed to remain quantitatively literate as they continue their studies.

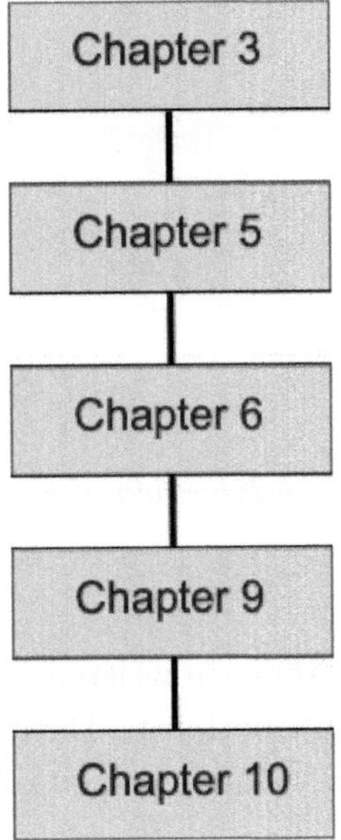

Figure 3: If you are coming from a traditional first-term calculus course and are interested in the life sciences, this path through the book is a guide to the tools which are most used in the field.

Of course, if you are a casual reader and only have time for a single chapter, then path 4 is the path for you.

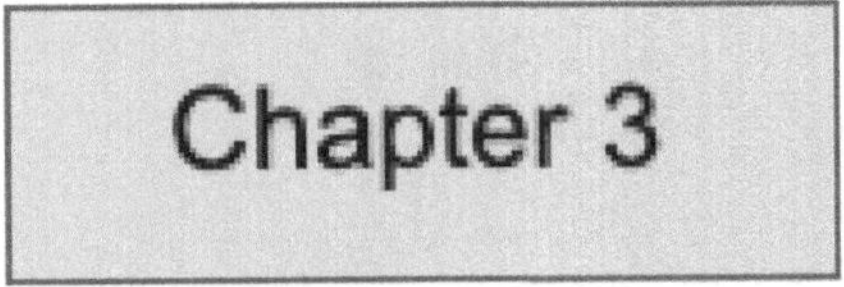

Figure 4: If nothing else, read this chapter.

Author Bio

Matthew Betti is an applied mathematician focusing on mathematical modelling of ecological and evolutionary problems, and disease spread. He is currently situated in Sackville, NB at Mount Allison University where he has developed and taught most first year courses in mathematics and computer science. Betti focuses on blending the rigorous with the intuitive to renew interesting in mathematics and statistics in students who see the subjects as a hurdle. Betti also incorporates discussions of ethics and social problems and the place of the mathematical sciences in this context.

Betti's research on disease spread and the ecology of honey bees has been published in numerous international journals. His approachable synthesis of complex material has been recognized by a number of presentation awards, and consultation work with governments at all levels.

I

Applied Mathematics

The Plot (so you don't lose it)

The first part of this book is structured like a story, because I like stories. I find them captivating. I want this course to be captivating. Like with some stories though, sometimes some things don't quite make sense until the end[9]. With some stories, you sometimes have to take a step back and think about them, maybe review certain scenes or flip back to remind yourself *oh yeah, that's* **that** *Mr Knightley*[10]. Sometimes, it helps to read the back of the book first, or watch a trailer for the movie so that you get a feel for what to expect[11]. This short introduction is supposed to be like the trailer. I'll introduce the main characters and the overarching conflict of our story, and then we'll dive in.

In case you don't know, most stories follow a very similar structure. This structure is outlined in the following figure.

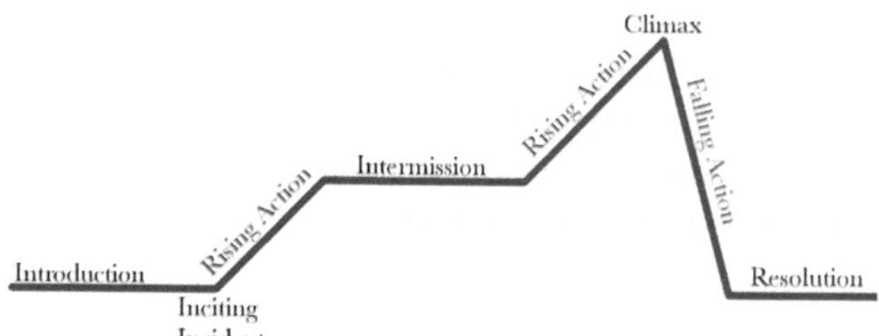

The **introduction** is where we set the stage. We will revisit functions, we will look at them in a new light – as a way to describe the world around us. We will focus on some functions of particular interest to the natural and social sciences. We will see how functions and data can interact to

[9] I'm talking about you, Memento.

[10] Why Ms Austen decided to name two characters identically I'll never know.

[11] Watching the trailer is at least 1% of the reason I refused to watch the movie *Cats*, for instance.

tell us about the world around us. We will learn how to translate our experiences into the mathematical language of functions.

Then, at the **inciting incident** we introduce the unlikely hero of our story: the derivative. They do not come from a vacuum; they are the descendant of a long line of mathematical concepts that ends with the *limit*. We see how the derivative can be used to describe the way the world is *changing* around us.

As we delve into the story of the derivative, we learn of their abilities[12]. We learn how to use the derivative to describe more and more of the world around us. We learn their rules, their limitations, and their strengths.

Then we take a brief **intermission**. We hit a roadblock of sorts. There are abilities of the derivative that are just out of our grasp; we don't have the tools to unlock them. We break and go off on a side-quest into the world of vectors and matrices. We learn what they are, what they do, and why we need them. Equipped with these tools, we are able to return.

When we return, we open up a whole new realm of potential for our lowly derivative. We learn that it doesn't just work on squiggles and lines, but on all kinds of weird shapes and surfaces. We have a set of tools now that can really bring insight into our world.

And so, we reach the **climax**. Everything starts to fall in place. We see how the derivative and all their abilities coupled with functional descriptions of the world and data are all just parts of a whole. We learn how to bring them together; we learn that math and the natural and social sciences are symbiotic.

Then, we fall into a **resolution**. All of our tools have found their purpose, all of the guns introduced in the first act have gone off[13]. Have we answered every single question there is to answer? Not even close. That is because this is just one course.

[12]Like Lara Croft or Megaman, each new lesson brings about a new ability.

[13]See Chekhov's gun

CHAPTER 1

Functions

1.1 ANATOMY OF A FUNCTION

What is a function?

We can think of a function as a mathematical machine. A function takes a thing[1] and transforms it into another thing[2]. The function is the rule (or set of rules) that tells us *how* to transform the input. In this way, a function is like a factory[3]; raw materials go in, the factory processes[4] the materials through a set of steps and spits out something new on the other side.

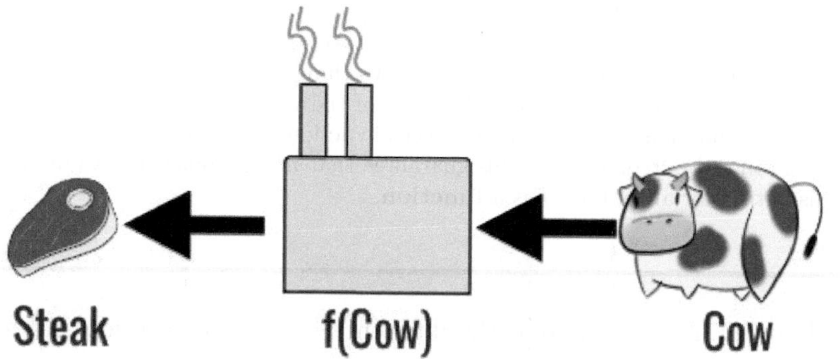

Figure 1.1: A factory and a function are incredibly similar. Within the factory (or function) is a set of manipulations of the input data which result in an output. In this case Steak = $f(Cow)$, where we would define f as the unspeakable things that happen to the cow within the factory walls.

[1]This is intentionally vague. An input is *usually* a number, but it doesn't have to be. Functions can take just about anything as an input as long as we can represent it mathematically.

[2]Again, intentionally vague

[3]Conveniently, f is for *f*unction and *f*actory.

[4]i.e. transforms

DOI: 10.1201/9781003265405-1

Mathematically, we write this as

$$y = f(x) \tag{1.1}$$

where y is our output, x is our input, and f is the name of the function, or the set of rules, by which we transform input x.

A factory would be pretty useless if every time we gave it the same raw materials, it processed them in a different way and gave us back a different product every time. If we give a sheet of blue denim and brown thread to a jeans factory, we should always get back blue jeans with brown stitching. If instead the colour changed randomly, or sometimes they made shirts instead of jeans, it would be pretty difficult to plan around our inventory[5]. Much like we would ask for some form of consistency in our factory, we want some form of consistency in our functions as well.

We require that any given input produces one and only one output. In math jargon, this is written as:

Definition 1.1.1. Consider a function f and two inputs x_1 and x_2. If

$$y_1 = f(x_1)$$
$$y_2 = f(x_2).$$

If $x_1 = x_2$, then $y_2 = y_1$. Basically, this is what we described above: if we put in two things that happen to be the same, we get two things out that are exactly the same. In a way, functions[6] are how we make sense of the world around us. Touch a stove, burn your hand; don't study, do poorly; certain actions in the real world have specific and guaranteed outcomes. It is this guarantee that makes something a function. If this is true for all x, then f is a **function**.

Figures 1.2 and 1.3 show the difference between a function and not a function.

Otherwise, an input, rule and output is called a **relation** because it relates inputs to outputs. A relation just relates two different things without the added condition that each input produces a unique output. A factory that takes a sheet of denim but sometimes produces shirts and sometimes produces jeans and sometimes produces jackets is not a function but a relation.

[5] But possibly justify the resurgence of the Canadian tuxedo

[6] Without the math

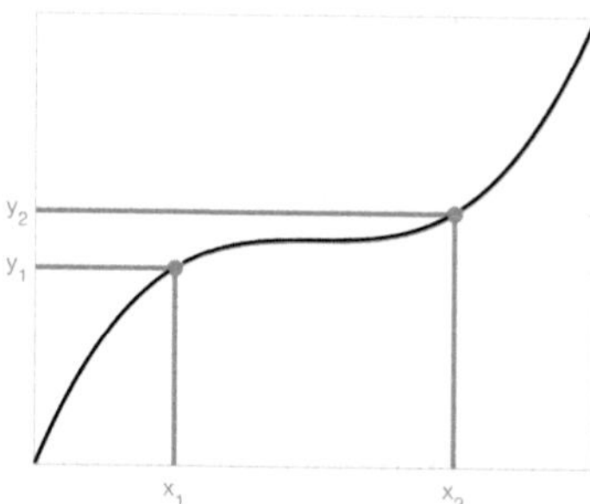

Figure 1.2: The black line is a function, showing the relationship between x and y for any input (like x_1 or x_2), we can follow the graph of the function to a unique output (y_1 and y_2, respectively).

Knowing the input[7] and a rule[8] does not guarantee an output[9]. If for any one input we could have an ambiguous output, the rule in question is not a function.

In the real world, most things are truly relations. We can never account for everything that causes a certain output, but that doesn't mean we can't get close. In some fields[10], we can often get closer than others[11] but one thing we can always do is **approximate** real-world behaviours[12] by functions[13].

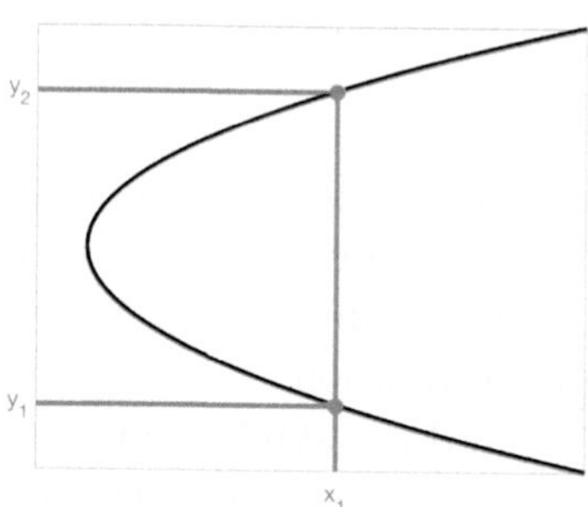

Figure 1.3: The black line is **not** *the graph of a function. For the value x_1, we could potentially end up at y_1 or y_2. One input produces two possible outputs.*

[7] Denim

[8] Say, cut the denim and stich it together.

[9] Jeans

[10] Like physics

[11] Like biology

[12] Inputs, outputs, and rules

[13] Inputs, outputs, and rules **that result in unique input–output pairs**

Functions need not be limited to one single input. If it is, it is called a **one-dimensional** function. Its graph existed in *two* dimensions[14]. A function can have as many inputs as needed, but we can only really visualize up to two inputs at a time[15]. Inputs are separated by commas in function notation,

$$z = f(x, y) \tag{1.2}$$

and if we have more inputs than letters, we usually use subscripts

$$y = f(x_1, x_2, x_3, \cdots, x_n).$$

Notice here that x_1 and x_2 in Definition 1.1.1 referred to specific points, but here they are referring to variables and y in one instance refers to an input and in another instance refers to an output.

IT'S WORTHWHILE TO GET THIS POINT OUT OF THE WAY EARLY: THESE LETTERS HAVE NO INHERENT SIGNIFICANCE. Most symbols in math are context-sensitive. You must *read carefully*[16] the context surrounding a symbol to determine what its significance is. In some cases, y may be an output as in equation (1.1), in some an input as in equation (1.2), and in others a number. A function need not always go by the name f, nor must f always refer to a function.

We could very well write

$$f = x(y)$$

where f is the output, x is the function, and y is the input. Nothing has changed here except the context of our symbols. Standardizing f to mean a function, x a generic input and y a generic output is nothing more than tradition and convention.

The ability to see beyond the symbols and into what they represent is a first step to abstract, generalized mathematical thinking. In this chapter, we are dealing with inputs, functions, and outputs. The symbols we use to represent these three abstract concepts may vary, but the principles will not.

---

[14] 1 input + 1 output = 2 dimensions

[15] This is a consequence of living in a world of three spatial dimensions with brains that can't process more than that.

[16] Why did I emphasize the word read? Sometimes people, myself included, think they're reading when really they're just internally pronouncing words while their mind is elsewhere. Reading, from this point forth, will be *actively* reading. It means pronouncing the words in your head, deriving meaning from them, and internalizing it. In this way, "read carefully" is redundant.

We haven't seen any explicit examples of functions yet and that is an intentional choice on my part. The idea of a function as a rule that takes an input to a unique output is, in my opinion, something that transcends examples. That being said, the following few non-mathematical examples may help you better conceptualize the idea of a function.

EXAMPLES OF FUNCTIONS

A **table of contents** is a function that takes a topic/section/chapter and returns a page number. Whereas an **index** at the back of a book is *not* a function. It takes a key word and points you to *multiple* page numbers.

A **set of student numbers** is a function that takes a unique set of digits as an input and returns a person. Provided we stay at one institution, student numbers are not repeated and no person has more than one student number. If you provide me a student number, in theory, I could give you back a person without any ambiguity.

EXAMPLES OF NOT-FUNCTIONS

An **index** of a book is not a function. An index takes a word that we might want to know more about and sometimes gives us several page numbers on which we could learn more about that word. For one input, we may have many different outputs to choose from. Each person who looks up a word[17] in the index may end up on a different page[18]. Does this mean that when there is only one page entry for a word in an index that those parts of the index *are* functions? No. Whether or not something is a function happens wholesale. If even one entry in the index has more than one output, then the *whole* index loses the ability to call itself a function[19].

One of the reasons we use student numbers is because **a set of names** is not a function. If we tried to use names to identify all people registered at an institution, we would quickly run into problems. In some cases, you may be able to give me a name[20] and I will be able to say *oh yeah, that person* and map them exactly to one unique individual. Let's say instead you give me the name James Smith, depending on the part of the world

[17]Input

[18]Output

[19]Although we could always enforce more rules to make it a function, we will talk more about this later.

[20]Like Guy Incognito

I will not be able to identify a single, unique individual based on this information[21].

Why should we care about functions?

As we've spent more than our fair share discussing already, functions give us a relationship between an input[22] and an output. They inform our understanding of cause and effect for different systems whether they be physical, biological, economical, social, or otherwise.

Once we have a handle on cause and effect, we can try to *predict* future outcomes based on current information. Sometimes these predictions are *qualitative*[23]; other times they are *quantitative*[24].

GRAPHS OF FUNCTIONS help us to understand qualitative relationships. In figure 1.4, the solid line represents known information and the dashed line represents one possible way of *extrapolating* to predict what happens for input values beyond 100. Of course, if all we have is a graph, we don't know how correct these predictions are, but the *trend* is probably correct, at least for values close to 100. This is a **qualitative** prediction: in figure 1.4, if the input increase, the output increases as well.

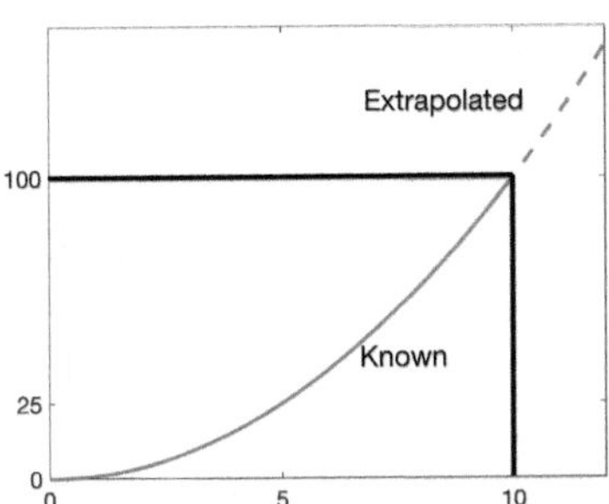

Figure 1.4: The solid line is "known" information; a relationship between cause and effect represented as a graph. Since each unique input appears to have a unique output, we know that this is a function. We can predict what might happen for input values > 10 *using the dashed line.*

[21] In the United States, James Smith is the most common combination of first and last names followed by Michael Smith, Robert Smith, and Maria Garcia.

[22] Or a set of inputs

[23] If you eat more without exercising more, your mass will increase.

[24] If you have a surplus of 3000 Calories, you will gain 0.454 kg of mass.

But *how much* does the output increase if the input increases by, say, 60 from 100? This would require a *quantitative* prediction and while the graph may be able to give us an estimate, a far better way to answer the question would be to have a **formula** for exactly how the output changes with the input.

The graph is figure 1.4 was generated using the simple formula

$$y = f(x) = x^2.$$

The output is y, the input is x, and the function, f, is the act of squaring x.

We can see this is indeed a function. If I set an input of $x = 1$, I get back $y = 1$. If I set $x = 2$, I get back $y = 4$. If I put in $x = -2$ I get back $y = 4$. This may give you pause but remember, our definition of a function says *nothing* about repeated outputs, only that an input should have at most *one* output. Two inputs can have the same output, but one input cannot have two different outputs.

Given this formula, it's very easy to answer the question

How much y do we have when $x = 160$?

Such a formula or set of formulas that allows for qualitative and/or quantitative predictions is called a **mathematical model**.

Let's instead consider the formula we used to generate figure 1.3

$$y = f(x) = \sqrt{x}.$$

The square root function is bf not a function because if we input $x = 4$ we have two possible outputs: $y = 2$ or $y = -2$. We have a choice to make, and we may not all choose the same thing. Without additional rules, this isn't a function. If we all agree on the additional rule to ignore the negative values,

$$y = f(x) = \sqrt{x} \qquad y \geq 0$$

then we can *make* this a function because we have eliminated the ambiguity.

Sometimes we are able to impose extra rules to turn not-functions into functions. For instance, with our example of a book index if we impose the *additional rule* that we select the first page value when looking up a word, then we have turned our index into a function. This comes at a cost though and that cost is ignoring the additional information we have.

Why might we want to turn relations into function? The simple[25] reason is that we have a much, much better understanding of functions than we do of relations. We have a hammer, our understanding of functions, and so we try turn everything into a function so we can use the powerhouse generational knowledge of mathematics[26].

1.2 MODELLING WITH MATHEMATICS

A MATHEMATICAL MODEL is a description of a particular aspect of the world in mathematical language.

WE USE MATHEMATICAL MODELS EVERYWHERE in everyday life[27] to make predictions.

IN ORDER TO MODEL CALORIES BURNED WITH GREAT ACCURACY FROM OUR ACTIVITY TRACKER, we would need a lot of individualized information. The number of Calories burned is correlated with the following:

- Your age
- Your current heart rate
- Your weight
- Your body composition
- Your height
- The ambient temperature
- Your activity level
- Your sex[28]
- Duration of exercise

[25] And if you tend towards the pessimistic, kind of sad

[26] This is a pattern you will see if you continue taking math courses. A lot of math work is getting a problem into a state where we can appeal to past mathematics for a solution. This is because mathematics is **powerful**. When something is proved true in mathematics, it cannot be undone. Mathematical truth is eternal, and so we can appeal to anything previously shown to be true when solving a problem.

[27] The translation of the data on your activity tracker (mainly heart rate) to calories burned is one example of a mathematical model.

[28] Biological, not frequency

along with a host of other things that may affect calories burned indirectly. Yet, all we have is fitness tracker measurements like steps and heart rate and some basic, approximate information about the person[29]. So, it would be in our best interest to build a mathematical model that calculates Calories burned using information we have at hand. Namely, age, weight, height, heart rate, and sex. Such a model might look like

$$C = f(a, w, h, H, s)$$

where a is age, w is weight, h is height, H is heart rate, and s is sex. The model is a function that takes these variables as inputs and gives us back Calories burned.

OUR MODEL IS CLEARLY INCOMPLETE; we have left out some important factors like genetics or how our weight is divided between muscle mass and fat mass. Mathematically, it would be impractical to account for every single thing that could affect our Calories burned, and often times increasing mathematical complexity leads to unpredictability and confounds any interpretation or insights that can be gleaned from a model.

On the other hand, if we do not include enough in our model, then we may miss some important relationships between inputs and output. The key is to try and strike a balance between mathematical complexity and interpretability. Simpler models are readily interpretable and easily understood by a wide breadth of people, but they may miss key factors. For instance, we could say that a human on average burns 96 Calories per hour, and then just multiply this number by the number of hours a fitness tracker is worn. This would give us a model with it time as the input:

$$C(t) = 96t$$

but that doesn't account for exercise, or periods of movement versus periods of rest or even the fact that we are all shaped and sized differently. It may be valid in general for a person who is just living their day-to-day life, but we might also assume that someone wearing a fitness tracker is trying to be *different* than average[30].

[29] Like height, weight, sex, and possibly age

[30] Since the average, at least in Canada, is fairly sedentary

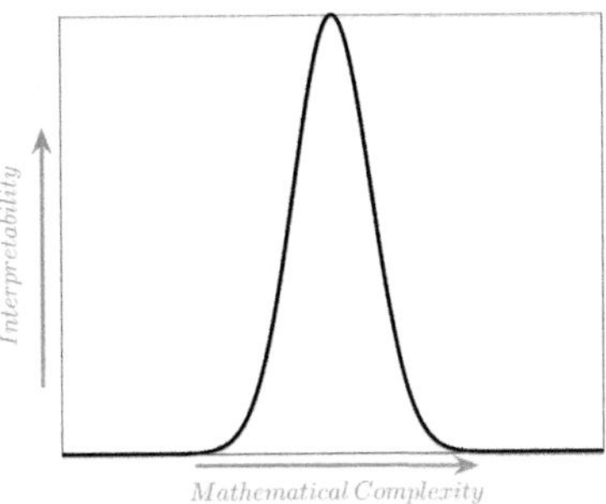

Figure 1.5: As a mathematical model becomes more complicated, it becomes harder to interpret. Often, when a model becomes overly complicated, it may be able to capture all possible behaviours, but we may not know why or how it is doing so. A model, therefore, should be as complicated as it needs to be and no more.

THIS COMPROMISE BETWEEN WHAT TO INCLUDE, HOW TO INCLUDE IT, AND WHEN TO INCLUDE IT IS FOUNDATIONAL TO MATHEMATICAL MODELLING. Some inputs are more important than others. For instance, your age has a much greater impact on your Calories burned than the ambient temperature outside. Knowing what to include and how to include it, and maybe most importantly, knowing how and when our conclusions are valid is what separates mathematical modellers from users of mathematical models.

Often times, and we will do this in this book, we name the output and the function the *same thing* for simplicity. For our Calories burned function above, we will often leave out the f and instead just write

$$C(a, w, h, H, s)$$

and understand that we are looking for a value C for given inputs a, w, h, H, and s.

SOME OTHER EXAMPLES OF MATHEMATICAL MODELS can be found in just about everything you use. **Weather forecasts** are generated by taking the current weather patterns and running them through multiple mathematical models to predict short-term changes[31]; we may include things like temperature, wind speed, weather in nearby areas, *etc.* **In retail**, mathematical models are used to determine when an item should

[31]In fact, the % chance of rain means that M/N models predicted rain.

go on sale and how much it should be discounted and whether these sales should differ between geographical regions, or even individual stores.

LET'S LOOK AT A VERY SIMPLE MATHEMATICAL MODEL which shows average battery life since time of last charge for a phone. Again, there are many, many things that can influence battery life[32], but **time** is the most influential thing on battery life. The rest can be "hidden" in the numbers that we incorporate in the model[33]. We can see a possible model for this in figure 1.6.

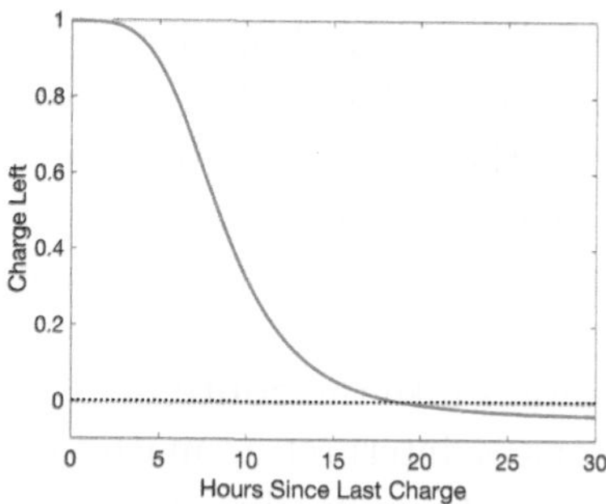

Figure 1.6: Battery Life is a function of time since last charge. This function can help us estimate how long a battery in a phone will last before it needs another charge.

This figure was generated using the formula

$$B(t) = 1 - 1.04\frac{t^4}{5440 + t^4}.$$

Again, all those other factors that might influence battery life are wrapped up in the numbers 1.04, 5440, and even the power 4[34]. If we change either of these numbers, we will change the function and the *quantitative* estimates of battery life will change. What *won't* change is the *qualitative* behaviour of the battery life.

The point where the *output* of the function is equal to zero is called the **x-intercept**. We will discuss intercepts more later, but in this instance it tells us the total number of hours until our battery is drained. If we know our current battery life, we can use this function to *predict* how many hours of battery we have left.

[32]Temperature, phone usage, screen brightness, and age of the battery to name a few

[33]Like A and B in our stylus model

[34]These specific numbers correspond to the average battery life of my iPhone XR, as tracked by me over a random assortment of days.

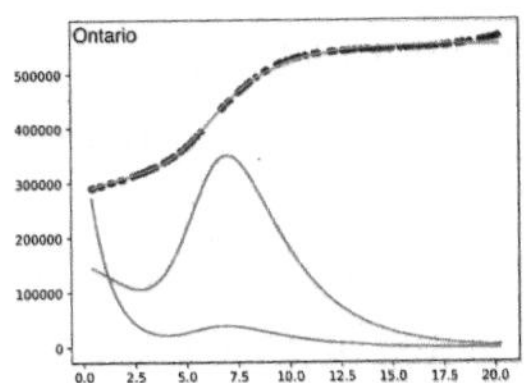

Figure 1.7: A model of the cumulative cases of an infectious disease in Ontario overlaid with the actual measured data.

Notice that beyond the intercept, the function is negative. What is the meaning of negative battery life? Spoiler alert: there is no meaning. This function *approximates* real life, it does not capture it perfectly. Our model isn't valid beyond the x-intercept, because it stops making sense. We could also argue that since our model doesn't take into account behaviours such as usage, screen brightness, and others, the model isn't even valid *near* zero.

It's important to always take the **context** of a function into account when making predictions or interpreting math in the real world. While something may be true mathematically, that does not mean it is true in the real world. All models have a sphere of influence for which they are valid. There is no model that is all encompassing. Knowing these limitations is necessary for proper interpretation of results and can act as a guide when creating new models.

As an example, let's look at two figures (figure 1.7 and figure 1.8) produced from models that I have developed for the spread of disease. You'll notice that the model for Ontario fits the data much better than the model for Yukon. One of the limitations of the model is that it is valid for large, dense populations that are relatively unchanging. The further we stray from condition, the worse the model will perform. As long as we know this limitation and clearly communicate it with users

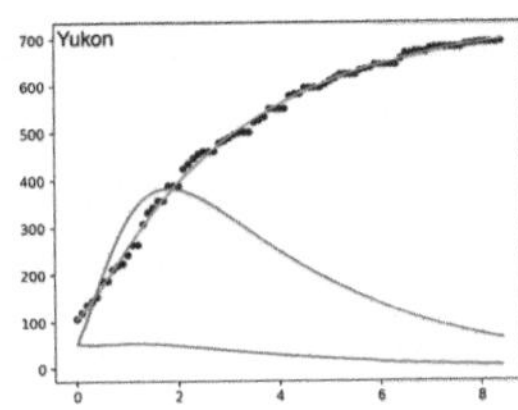

Figure 1.8: A model of the cumulative cases of an infectious disease in Yukon overlaid with the actual measured data.

of the model, we can separate valid model results from invalid model results.

THE REST OF THIS CHAPTER will deal with some important types of functions and the kinds of behaviour they model. We will use this information as a foundation to help us develop and interpret more complex mathematics later.

1.3 CONSTANTS AND LINEAR FUNCTIONS

THE SIMPLEST TYPE OF FUNCTION IS A CONSTANT FUNCTION:

$$f(x) = C$$

where C is a number[35].

A constant function represents something that is *unchanging* with respect to your input variable(s). Some examples include

- The colour of a pen
- The difference in age between you and your mother
- The number of legs on a particular horse

On their own, they are not all that interesting, but they creep into all other types of functions. When we are modelling behaviours in the real world, it is often not the case that an input of 0 causes an output of 0. For instance, looking at your mother's age (call it A), we would need her age at the time of your birth (let's call it A_M for age of mom), then your age (call it just a). A function that models your mother's age would look like

$$A(a) = A_M + a$$

Here A_M is the constant term. Your mother was a particular age when you were born and that age cannot be changed. A_M will equal a particular number for time immemorial. It is constant.

LINEAR FUNCTIONS are exactly as they seem: when graphed, they are straight lines. The general form of a one-dimensional linear function is

$$y = f(x) = mx + b$$

[35]Literally any number

where m and b are the **slope** and **y-intercept**, respectively. The term mx is the linear part of the function, and b is the constant part.

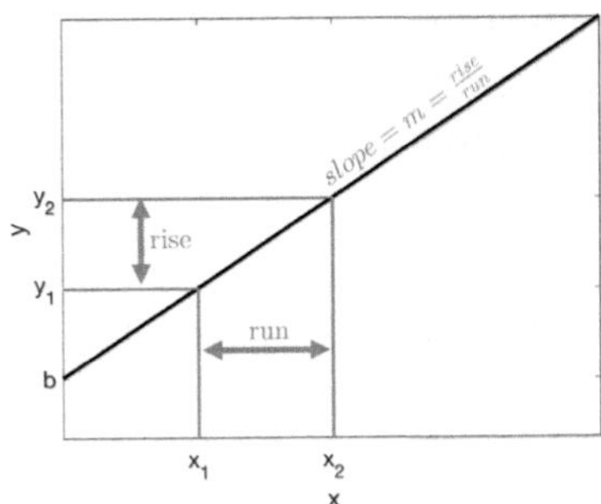

Figure 1.9: A linear function always has the same components; b is where the function meets the y-axis, and m is the rate at which the function grows. If $m > 0$, the function is increasing as x increase. If $m < 0$, the function is decreasing as x increases.

LINEAR FUNCTIONS ARE THE SIMPLEST WAY TO MODEL A REAL-WORLD SYSTEM. The world is a complicated place, and often the relationships between two things is not quite linear, but at least has a linear *component*. For **example**, in the function

$$y = f(x) = b + mx - ax^2 + ce^{-kx}$$

the blue terms make up the linear component of the function.

This linear component can help us make predictions about situations that are "close"[36] to a known situation. While there are few truly one-dimensional linear relationships in the world, many have a linear component and so linear functions are ubiquitous and extremely useful in mathematical modelling.

IN ORDER TO WRITE A 1-D LINEAR FUNCTION, you need *two* pieces of information. If we are given two points (x_1, y_1) and (x_2, y_2), we can solve the system of equations

$$y_1 = mx_1 + b$$
$$y_2 = mx_2 + b$$

[36]What is and is not close is fairly arbitrary and dependent on the context of the problem. Are Toronto, ON, and Chicago, IL close? Depends on whether you're used to global travel, or staying within your city limits.

for m and b. Doing so, gives us the standard equation for the slope of a line

$$m = \frac{rise}{run} = \frac{y_2 - y_1}{x_2 - x_1} = \frac{\Delta y}{\Delta x}.$$

This formula is also known as the **average rate of change** of y between x_1 and x_2.

THERE ARE THREE WAYS OF KNOWING IF A FUNCTION IS LINEAR[37]:

- When the function is in its simplest form, x^1 and x^0 are the only powers of x that appear[38].
- The graph of the function is a straight line.
- If the function is given as a table of values, the average rate of change between *any* two points is the same.

1.4 POLYNOMIALS

WHAT IF A FUNCTION IS NOT LINEAR? The possibilities are infinite in that case[39]. The next simplest class of functions[40] is the class of polynomials. In one dimension, these are given as sums of powers of x. You are probably familiar with quadratic, cubic, and maybe quartic polynomials. The general form of a polynomial is given as

$$y = f(x) = a_0 + a_1 x + a_2 x^2 + a_3 x^3 + a_4 x^4 + a_5 x^5 + \cdots + a_n x^n.$$

Any of these a_i, called **coefficients** can be zero, which is how we can use the above formula to get any polynomial we like. The highest power with a non-zero coefficient that appears in a polynomial is called the **order** of the polynomial[41].

ANOTHER MORE USEFUL FORM OF A GENERAL POLYNOMIAL[42] is given as

$$y = f(x) = (x - x_1)(x - x_2)(x - x_3)\ldots(x - x_n).$$

[37] In 1-D

[38] You may wonder if $y = mx + b$ actually fits this criteria. Since $x^1 = x$ and $x^0 = 1$, we may write $y = mx + b = mx^1 + b \cdot 1 = mx^1 + bx^0$.

[39] In fact, linear functions are so exceptional in the world of mathematics that most mathematical objects are classified as *linear* or *non-linear*.

[40] Arguably, of course

[41] As a quick example, linear functions are polynomials of order 1.

[42] Again, arguably

This is the factored form of a polynomial. In this form, we can identify all the points where $y = 0$[43]. If $x = x_1$ or $x = x_2$ or $x = x_3$ etc., the corresponding term in f is zero, thus the entire product is zero. Therefore, the points x_1, x_2, x_3, …, x_n are the x-**intercepts** of the function $f(x)$.

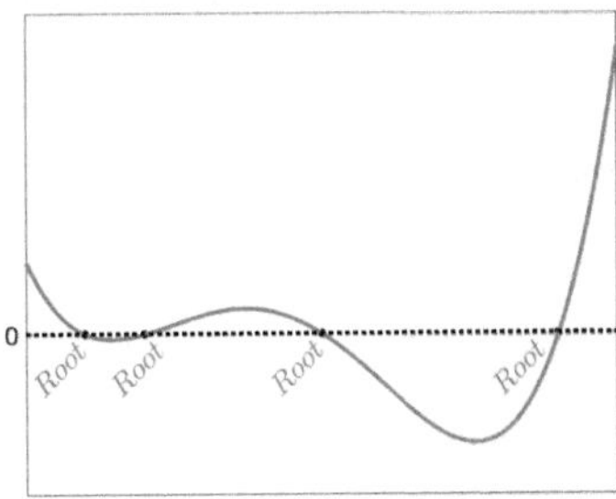

Figure 1.10: Graphically, the roots of a function are all the places where the graph of the function $f(x)$ crosses the x-axis.

This idea of where a polynomial crosses the x-axis can be generalized to *any* function.

Definition 1.4.1. A **root** of a function $f(x)$ is a point x^* where

$$y = f(x^*) = 0.$$

A function can have many roots, one root, or even no roots.

NOTICE HOW THIS DEFINITION IS NOT PARTICULAR TO POLYNOMIALS but instead is applicable to *all* functions. While generally roots of polynomials are easier to find[44], given the graphical interpretation of a root, there is no reason we need to limit this concept to polynomials.

Modelling with Polynomials

Let us consider the following set of data of age vs. systolic blood pressure:

When we graph the data, we get figure 1.11 and we start to see a trend.

[43] Why these are noteworthy will become clearer as we gain more tools for analysing functions.
[44] Even then, it's only really possible to analytically find roots for polynomials up to order 4.

Table 1.1: Table of blood pressure measurements for different individuals at different ages.

Age	Blood Pressure
41	110
45	110
48	130
57	150
64	128
69	140

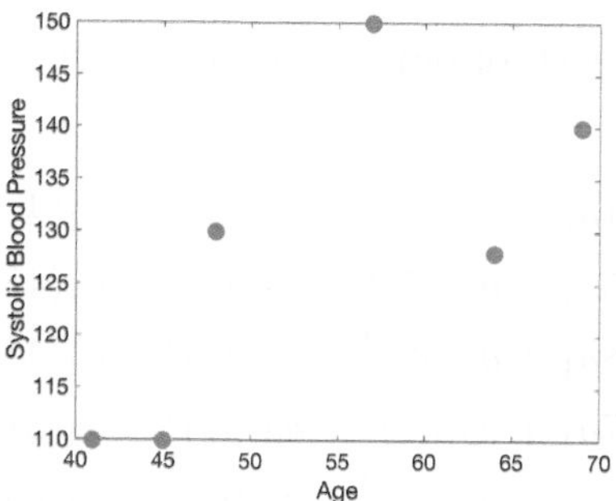

Figure 1.11: Plotted points from the data in the table.

NOW, WE CAN TRY TO FIND A FUNCTION THAT FITS THESE POINTS exactly. We are looking for a function of age that outputs blood pressure and must agree with our data at six key points. Let's write out the general form of what we want:

$$BP = f(A)$$

where BP is blood pressure and A is age.

With 6 x and y pairs, given in our table, we can solve for *up to* six coefficients, so let's look at a polynomial of order 5:

$$BP = f(A) = a_0 + a_1 A + a_2 A^2 + a_3 A^3 + a_4 A^4 + a_5 A^5$$

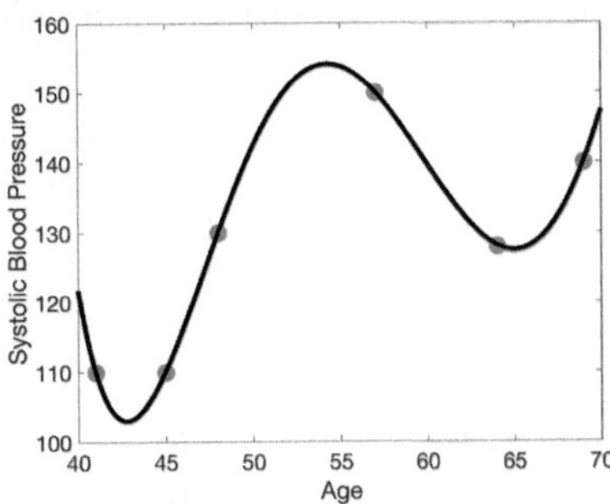

Figure 1.12: A function which fits our data points.

We can now plug in each of our six points to get six equations and six unknowns.

$$
\begin{aligned}
110 &= a_0 + 41a_1 + a_2(41)^2 + a_3(41)^3 + a_4(41)^4 + a_5(41)^5 \\
110 &= a_0 + 45a_1 + a_2(45)^2 + a_3(45)^3 + a_4(45)^4 + a_5(45)^5 \\
130 &= a_0 + 48a_1 + a_2(48)^2 + a_3(48)^3 + a_4(48)^4 + a_5(48)^5 \\
150 &= a_0 + 57a_1 + a_2(57)^2 + a_3(57)^3 + a_4(57)^4 + a_5(57)^5 \\
128 &= a_0 + 64a_1 + a_2(64)^2 + a_3(64)^3 + a_4(64)^4 + a_5(64)^5 \\
140 &= a_0 + 41a_1 + a_2(69)^2 + a_3(69)^3 + a_4(69)^4 + a_5(69)^5
\end{aligned}
$$

The function that we get after solving is

$$
\begin{aligned}
BP = f(A) = &- 0.000077490208A^5 + 0.023434649A^4 - 2.7864888A^3 \\
&+ 162.79298A^2 - 4671.04417A + 52772.40816
\end{aligned}
$$

Don't get too hung up on reproducing this function, or thinking about where it came from. We will later develop some of the tools needed to do this ourselves. Just know that it works and if we plot the function $f(A)$ with our points, we get figure we see we have a function that goes through all of our points. This is an amazing model because it agrees with our data everywhere!...Right[45]?

NOW, WE HAVE TO ASK OURSELVES IF THIS MAKES SENSE. Let's start by extending our function to ages 35 to 75 as seen in figure 1.13. We see that our model function makes very little sense. Do we really expect a 35 year old to have double the blood pressure of a 70 year old[46]?

[45] Not right

[46] The answer is no. No we don't.

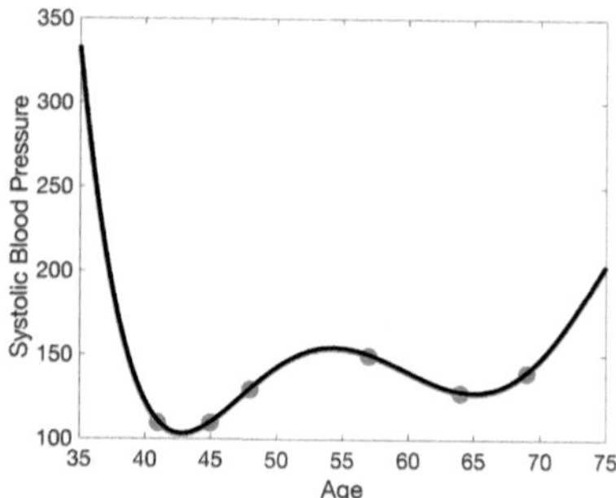

Figure 1.13: A function which fits our data points.

If a 37 year old was admitted to hospital with a blood pressure of 300, this model would tell us that everything is normal and we should send them home[47].

This is the inherent danger of trying to fit data *exactly*: measurements are hardly ever absolutely correct. If we step back from figure 1.11, and taking the points as *suggestion* instead of dogma, we can instead pull out the *trend* in the data: blood pressure increases with age, at the very least for middle-aged to senior adults[48]. It's important to remember that blood pressure depends on *more than just your age*[49]. Our model should not fit the data perfectly because age should not explain all the variation in the data. We have no reason to believe that it should. Perhaps we should stick to a simpler function to try and determine a *qualitative* relationship between age and blood pressure.

IT'S FAR MORE LIKELY THAT THE DATA SHOWS A LINEAR RELATIONSHIP, but with some "fuzziness" caused by imperfect measurement, or other factors that we don't have in our data. One line we might draw to fit this data is given in figure 1.14. In this case, we took our data, and drew a line through the first and last points to determine, roughly, the trend in the data: that blood pressure increases with age.

Figure 1.15 shows that this isn't the only way to pick a trendline. Determining the best way to fit data is more a problem for statistics. We will touch on this later, but for now the important part is understanding the difference between fitting data points and fitting trends in the data.

[47]For clarity, a blood pressure reading of 300 is beyond critical for a human, and only normal for a giraffe.

[48]120 is a normal number; we don't expect younger individuals to all have low blood pressure.

[49]Your height, weight, health, activity level, stress...

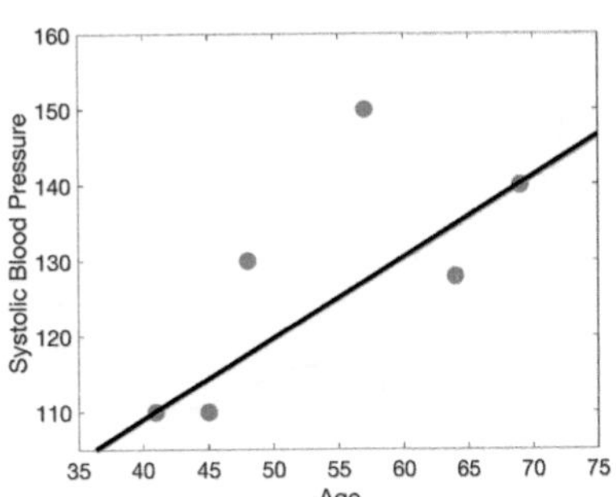

Figure 1.14: A line which fits the trend in our data.

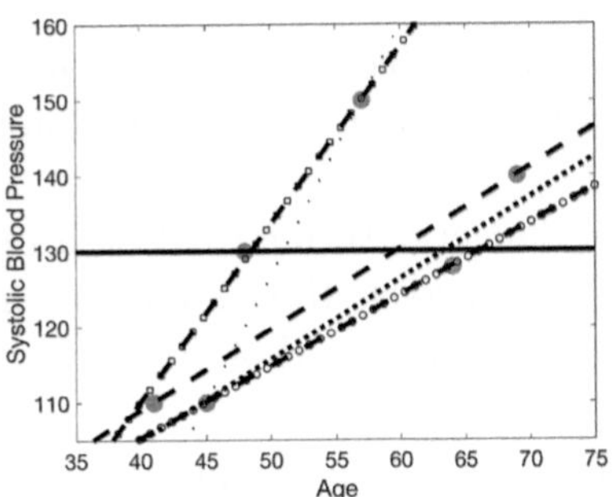

Figure 1.15: All these lines fit the trend in the data.

GENERALIZING FROM OUR EXAMPLE, we can think about how we might model blood pressure as a function of age as a thought experiment. Let's work in the *domain* $40 < A < 70$ since this is roughly where we know there's a trend (from the data).

First, we might ask, "Is there some minimum blood pressure?" And we might answer ourselves with, "Yes, indeed there is. Since 0 blood pressure would most certainly imply death." Therefore, at the very least, our function should contain this minimum value:

$$BP = f(A) = BP_{min}$$

Next, we might ask, "Do we expect blood pressure in increase if you're over 40?" We might this time answer, "Yes, because that's what the data seems to show." So we can modify our function so that

$$BP = f(A) = BP_{min} + mA.$$

Of course, we must also enforce that $m > 0$ here. If m were negative, we would be showing that blood pressure *decreases* with age.

Let's now ask, "Between 40 and 70, so we expect to find an age where our blood pressure is at a maximum or at a minimum?" The hard truth here is that between 40 and 70, your body is no longer optimal and so we would *not* expect to see a minimum or maximum in blood pressure. This doesn't actually modify our function, but for clarity we can show this *not* condition as

$$BP = f(A) = BP_{min} + mA + 0A^2.$$

Since the A^2 term is multiplied by 0, it is always 0 and contributes nothing.

We could continue to think about what we might expect to reasonably see in blood pressure as a function of age, by asking questions like, "Do we expect to find a maximal *and* minimal blood pressure between ages 40 and 70?" Again the answer should be "No" and thus we don't include an A^3 term in our model either.

Essentially, when using polynomials to model a phenomena, the order of the polynomial is related to how "wiggly" you expect the relationship to be. Often, you will find that scientists rarely model data with polynomials beyond linear or quadratic[50].

Without knowing anything about the physics of gravity[51], we could generate a model the height of a thrown ball over time. Assume the ground is height zero.

- Do we expect to throw the ball *from* the ground? No, we have a height, and so we probably start at some height above zero, call it h_0 for initial height[52].

$$h(t) = h_0$$

- Do we expect the height to change over time? Absolutely. If we let go of the ball, it won't just float. So let's add a linear term.

$$h(t) = h_0 + mt$$

- Do we expect the ball to reach some maximum height? Yes, if we throw a ball into the air, we eventually expect it to come back

[50]This is in part driven by Occam's razor: the simplest explanation is usually the best.

[51]Except from our personal experiences of living with gravity for our entire lives – astronauts excepted.

[52]Or height at time 0

down. There is some point in the sky that the ball will not go above. We have one max, so let's add a quadratic term:

$$h(t) = h_0 + mt + at^2$$

- The ball reaches a maximum, do we also expect it to reach a minimum? The answer is yes: the ground, but we will still not include a cubic term in the model because this is not a "natural" minimum. If the ball could, it would continue through the ground to the centre of the earth. The ground is more a barrier to the validity of our model than it is part of the *dynamics* of the thrown ball.

$$h(t) = h_0 + mt + at^2 + 0t^3$$

- We won't include any higher order terms because again, we can't justify them. Perhaps if it were a bouncy ball, we could argue we have many minima and maxima and they're all different. In which case, we might want higher order terms. Let's assume our ball is clay and doesn't bounce at all.

Our final model looks like

$$h(t) = h_0 + mt + at^2$$

If you look in any elementary physics text, you'll likely see the formula

$$h(t) = h_0 + v_0 t - 9.81t^2$$

These values have specific meanings and interpretations, but this isn't a physics book so I'll leave those explanations to the physicists. The key here is that the formula we came up with, and the physics one look the same. These are the kinds of ways we can translate the world around us into math. If you read up on the formula in a physics textbook, they will likely use different methods to get to the same formula, but because this is math for a general audience. We have appealed to intuition and personal experience and observation to arrive at the same thing.

1.5 EXPONENTIALS AND LOGARITHMS

THE OTHER LARGE, IMPORTANT CLASS OF FUNCTIONS are exponential functions and their inverses (logarithms). Exponentials are used as models when things depend on themselves[53].

[53] This is so very vague. Please read on.

Let's imagine a group of magical, immortal rabbits. If there are R rabbits, and every year each one of them has an offspring[54], then, after one year, we would expect to have all our original rabbits, R, plus one new rabbit for each rabbit we already had. In total, we should have $2R$ rabbits after one year.

Now, in year two, each one of *those* rabbits will *also* have an offspring, and each rabbit that we started with will have *another* offspring. So, we have our R original rabbits, plus another R rabbits from the first year, and then another $2R$ rabbits as offspring from all our rabbits. All in all, we should have $4R$ rabbits. In general, after t years, we will have

$$P(t) = R \cdot 2^t$$

rabbits.

This is an exponential function, and it arises because the population of rabbits is not growing by a fixed number of rabbits every year, but instead grown at a fixed *rate*. Our total population of rabbits doubles every year, or in other words increases *at a rate of* 100% each year.

THE GENERAL FORM OF AN EXPONENTIAL FUNCTION IS given by the formula

$$P(t) = P_0(1 + r)^t$$

where P_0 is the initial number of things you have and r is the rate at which P is increasing[55].

In the case of our rabbits, $P_0 = R$ and $r = 1$.

A lot of the natural world is modelled with exponentials. After we understand some of the mathematical properties of exponentials, we will look into some examples of how exponentials creep up in the world around us[56].

Linear Vs. Exponential

The main difference between a linear and an exponential function is the quantity that remains constant.

- If the **average rate of change** between any two points is constant, than the function is linear.

[54] This means that each pair of rabbits has two babies. Slightly unrealistic considering the reputation of rabbits.

[55] Or, if r is negative, the rate at which P is *decreasing*

[56] In more than just the reproductive habits of rabbits

- If the **relative rate of change**[57] between any two points is constant, than the function is exponential.

Definition 1.5.1. The **relative rate of change** between two points x_0 and x_1 is given by

$$R = \frac{f(x_1) - f(x_0)}{f(x_0)\,(x_1 - x_0)}$$

Another way to write this is the **average rate of change**, m, divided by the left most function value.

$$R = \frac{f(x_1) - f(x_0)}{x_1 - x_0} \frac{1}{f(x_0)} = \frac{m}{f(x_0)}$$

For exponential functions, R is constant[58].

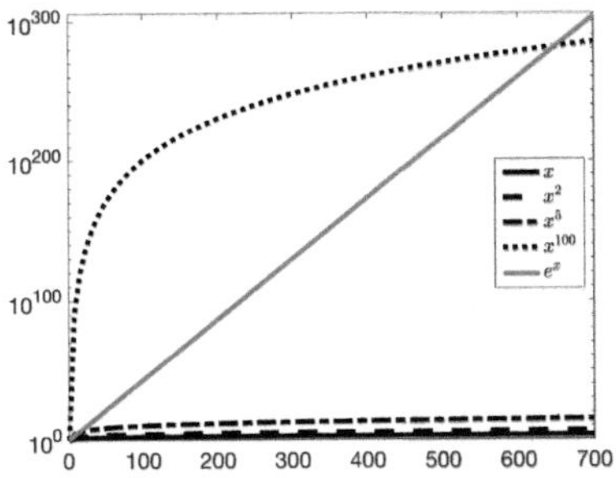

Figure 1.16: Comparing polynomial functions to the exponential function. Note that the y-axis is logarithmic; the distance between 1 *and* 10 *is the same as the distance from* 10 *to* 100*. Notice as we move right, the exponential function always dominates the polynomials.*

It is important to recognize exponential functions and know when they are used for modelling. One way to recognize them is as above: their *relative rate of change* is constant. Another way is that exponential functions always, eventually, grow faster than *any* polynomial, as seen in figure 1.16. Many biological processes, like our rabbits, are best modelled by exponential processes since life begets life[59].

---

[57] Before you panic and flip back through pages seeing if you've missed the definition of relative rate of change, this is the first time it has been mentioned.

[58] Yes, we used R above to mean rabbits. Remember, context. Symbols are just scratches on paper until we give them meaning.

[59] Put a bit more crudely, you need people to make people.

Logarithms

We can us the general form of an exponential to compute compounded interest over a period of time. If we start with M_0 dollars, and compound interest monthly, at a rate of 0.4% per month[60], we can calculate how much money we have in total after t months as

$$M(t) = M_0(1 + 0.004)^t.$$

What if instead we want to answer the *inverse* question? What if we want to know how much time it would take to triple our money? In other words for what value of t is $M(t) = 3M_0$? We may start by plugging this into our equation as:

$$\begin{aligned} 3M_0 &= M_0(1 + 0.004)^t \\ 3 &= (1 + 0.004)^t \qquad \text{Divide through by} M_0. \end{aligned}$$

How do we isolate t?

THE INVERSE OF AN EXPONENTIAL IS CALLED A **logarithm** and is defined as

---

Definition 1.5.2. Consider the equation

$$y = a^t$$

then

$$\log_a(y) = log_a(a^t) = t$$

We call $\log_a$ the **logarithm of base** a.

Another way to interpret the mathematical expression

$$\log_a(y)$$

is *for what value of t is $a^t = y$?*

There is no way to really compute a logarithm "by hand" except in very special circumstances. Here are a few that you can see without

[60]This amounts to 5% annual interest.

computation and will hopefully help solidify the concept of the logarithm in your head

$$\log_2(8) = 3 \quad \text{Since } 2^3 = 8$$
$$\log_3(9) = 2 \quad \text{Since } 3^2 = 9$$
$$\log_4(2) = \frac{1}{2} \quad \text{Since } 4^{\frac{1}{2}} = \sqrt{4} = 2$$

So, returning to our example, how long would it take to triple our money? We left off with

$$3 = (1.0004)^t$$

So, it seems we need the logarithm of base 1.0004 to isolate t.

$$3 = (1.0004)^t$$
$$\log_{1.0004}(3) = \log_{1.0004}((1.0004)^t) = t \qquad \text{Take the log of both sides}$$
$$2747.08 = t$$

It would take 2747.08 months[61] to triple our money at this interest rate[62]!

e and the natural log

If I had \$1, and I doubled it once a year[63], I would have \$2 after one year. Let's say I collected the interest twice a year, but kept that same interest rate of 100%. After 6 months, I would have \$1.50 and that \$1.50 would collect interest for the remaining 6 months, and I would have \$2.25 after one year. If I were to collect interest every 3 months, then at the end of the year, I would have \$2.44 since I increase the amount of money I am collecting interest 4 times a year. The more times I compound the interest, the more money I make.

The formula for compound interest is

$$P(t) = P_0 \left(1 + \frac{r}{n}\right)^{nt}.$$

This looks very similar to our canonical form of an exponential function, and it should since it is an exponential function. The difference is that we have added n which is the number of times per year we compound

[61] About 229 years

[62] Hold up, how did we get from $\log_{1.0004}(3)$ to 2747.08? A calculator; these things can't be done by hand.

[63] Say, because of a very kind bank or some kind of miracle

our interest, r. Usually when dealing with a bank or lender, r, n, and P_0 are fixed and the *variable* is how long you are collecting interest, t.

In our example, $P_0 = \$1$, $r = 1$, $t = 1$ and we are effectively looking at our money, P as a function of the number of times we compound our interest.

$$P(n) = \left(1 + \frac{1}{n}\right)^n \tag{1.3}$$

We are treating n as our variable instead, since we noticed that we make more money if we increase n.

The natural question to ask is, what if I compound interest continuously[64]? If we let n get so big that it is approaching infinity, do I end up with infinite money? Well, if we graph equation (1.3), we see that no we don't get infinite money. There is a limit to how much money we can make by doubling our money every year, but compounding it infinitely many times. The most we can make is approximately \$2.72.

Ok fine, maybe we can make infinite money if we *triple* the interest rate, and compound continuously.

$$P(n) = \left(1 + \frac{2}{n}\right)^n$$

Nope, we are capped out at \$7.39, which is a lot better than our \$1 that we started with, but is still a long way from infinity. Moreover, if we're clever, we can see that $\sqrt{7.39} \approx 2.72$! Obviously this number is special. The *exact* amount of money you will make is \$2.71828182846 $\cdots$. The number goes on forever.

Because of this property – the fact that even if you grow continuously, you are limited to a power of this number – this number has a special name. We call it e, named after Leonhard Euler[65]. This is the natural rate of continuous growth. All growth is a power of this number. It has some other fun properties that will pop up later on in our discussions.

Another way to think of this is the division of cells. To reproduce, cells go through a process called *cell division*. The process is just like the case where we compound our interest once per year[66]. A cell begins the division process and begins to create fractions of cells, but we don't

[64]This means at every conceivable moment, I collect whatever fraction of a penny I have "earned" and begin collecting interest on that as well.

[65]Pronounced "Oiler," not You-ler

[66]Except instead of years, cells usually only take one day to divide.

"count" those pieces until the cell is completely divided into two. If instead, as the cell was dividing, each piece also because dividing, we would not end up with infinite cells; we wouldn't even end up with three cells. We would end up with ≈ 2.72 cells.

ANY EXPONENTIAL FUNCTION CAN BE CONVERTED to use base e. Generally, if we have

$$y = f(x) = Ab^x$$

this can be rewritten as

$$y = f(x) = Ae^{ln(b)x}$$

Or, in our other form:

$$P(t) = P_0(1+r)^t$$

can be rewritten as

$$P(t) = P_0 e^{ln(1+r)t}.$$

SINCE e IS A SPECIAL NUMBER, IT HAS A SPECIAL INVERSE. The symbol ln[67] is shorthand for $\log_e$ (i.e. log with base e).

Remark. If $\ln(1+r)$ is positive, the exponential function is *growing*.
If $\ln(1+r)$ is negative, the exponential function is *decaying*.

WHY DO WE CARE ABOUT e? Why not just use "normal" numbers as bases? Because e is very convenient once we start talking about derivatives, so it is usually the base of choice. It has some very special properties that the other numbers don't have. If it didn't, we probably wouldn't even bother naming it.

I've waited until after we've defined e and ln to give a list of properties of exponentials and logarithms mostly because we will mostly stick to using e and ln from now on. Properties can be found in table 1.2.

Modelling with Exponentials

Just like we determined we sometimes want to fit messy data with lines, we sometimes want to fit messy data with exponentials. This is certainly the case when we know empirically or through reasoning that the data

[67] This comes from French, logarithme naturelle

Table 1.2: A list of common properties of exponentials and logarithms. While we use e *and* ln *here to demonstrate the properties, they are applicable to any base.*

$$
\begin{aligned}
\mathrm{e}^0 &= 1 \\
\frac{1}{\mathrm{e}^a} &= \mathrm{e}^{-a} \\
\mathrm{e}^a \mathrm{e}^b &= \mathrm{e}^{a+b} \\
\left(\mathrm{e}^a\right)^b &= \mathrm{e}^{ab} \\
\hline
\mathrm{e}^{\ln(a)} &= a \\
\ln\left(\mathrm{e}^a\right) &= a \\
\hline
\ln(ab) &= \ln(a) + \ln(b) \\
\ln\left(\frac{a}{b}\right) &= \ln(a) - \ln(b) \\
\ln\left(a^b\right) &= b\ln(a) \\
\ln(1) &= 0
\end{aligned}
$$

should change exponentially. In the same way that we fit lines to data, we can fit exponentials to data.

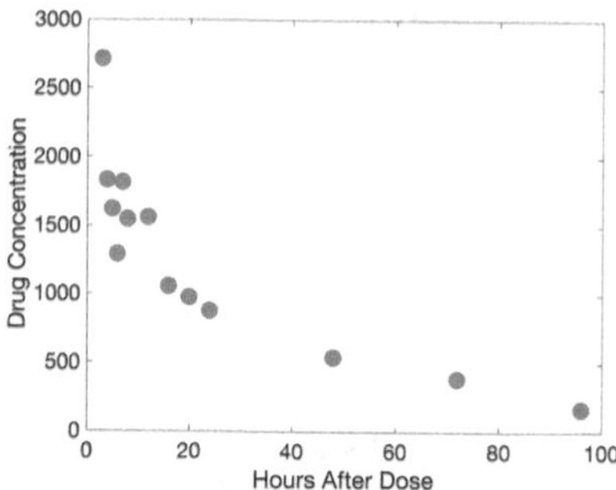

Figure 1.17: Blood drug concentration of efavirenz in Subject 12.

In figure 1.17, we isolate Subject 12 from a 2015 study [27] on the drug efavirenz – an antiretroviral drug used in the treatment of HIV. As you can see in the figure, just like in the linear case, there are different exponential functions that will fit the data. Just like in the case of a linear function, with two points we are able to determine an exponential function. But wait! With two points, we could draw a line through the data, why don't we just do that? We know, based on the way drugs behave in the body, that this data *should* be exponential. Moreover, if we take some of the data points and compute the *relative rate of change* we would see that it's roughly constant, a lot closer to constant than the average rate of change.

LET'S TAKE AS AN EXAMPLE the following two points from our data:

$$(t_1, y_1) = (4, 1838)$$
$$(t_2, y_2) = (48, 546.1)$$

If we know that we want an exponential function, we know it must have the form

$$y = Ae^{kt}$$

When we plug in both points, we get two equations with two unknowns that we must solve for

$$1838 = Ae^{4k}$$
$$546.1 = Ae^{48k}$$

We can solve the second equation for A,

$$A = \frac{546.1}{\mathrm{e}^{48k}} = 546.1\mathrm{e}^{-48k}$$

and plug into the first to get

$$1838 = 546.1e^{-48k}e^{4k}$$

Which we can solve

$$\frac{1838}{546.1} = e^{-44k}$$
$$\ln\left(\frac{1838}{546.1}\right) = -44k$$
$$k \approx -0.0275$$

We can then use this value to determine A:

$$A = 546.1\mathrm{e}^{-48*(-0.0275)} \approx 2052$$

This fit is shown in figure 1.18.

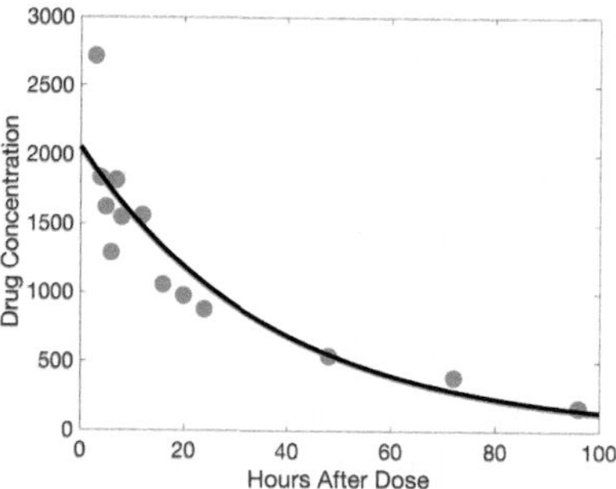

Figure 1.18: Blood drug concentration of efavirenz in Subject 12 and our calculated fit.

WHAT DO THESE CALCULATED VALUES TELL US? A tells us roughly the *dosage* of the drug at time $t = 0$. Our fit tells us that the patient ingested 2052mg of efavirenz. This isn't actually a bad estimate as the dosage was 2000mg! The fit also tells us that we should expect the rate of *decay*[68] in the bloodstream to be 0.0275, or in plain language we expect the concentration in the blood to go down by roughly 2% per hour. These values aren't so bad, but all this means is that we got lucky with the two points that we picked. If we had picked a different two points, we may have been *way* off. This is obviously not the ideal way to draw a function through data, but it's what we can do with the tools we currently have.

Half-life

WHAT'S THE POINT OF MODELLING THINGS LIKE DRUG CONCENTRATION? Experimentally, we can determine concentrations at which certain drugs are effective. Modelling drug concentration can help us determine things like dosing intervals and scheduling different drugs in the same system[69] that will prevent fatal interactions. Mathematics is cheap[70] and safe[71] compared to the other sciences. If we can draw conclusions mathematically, we can save money, lives, and potential consequences of running experiments. Mathematical models can help us be sure of what we are doing so that when we *do* experiment on people we are doing so ethically and with minimal risk.

[68]Decay because the value is negative
[69]i.e. body
[70]Financially
[71]In every sense

One particularly useful property of a drug is its half-life. This is the amount of time it takes the body to metabolize half of the drug in the system. A typical extra-strength Tylenol is 500mg. The half-life of Tylenol is the amount of time it takes from when you take the Tylenol until there is 250mg left in your system[72].

USING OUR MODEL OF EFAVIRENZ, we can find the half-life of the drug. In fact, all we need to know is parameter k in order to do this. No matter what the initial dose is, called it D_0 we want to know the *time*, t, when we have $0.5D_0$. So, our output for our function is $0.5D_0$ and we want to find the input that gives us this output.

$$0.5D_0 = D_0 e^{-0.0276t} \frac{0.5D_0}{D_0} \qquad = e^{-0.0276t}$$

You see that the initial concentration, D_0, cancels. The half-life is *independent of the initial dose.* Rearranging this equation we get

$$t_{HL} = \frac{\ln(0.5)}{-0.0276} \approx 25$$

for this model.

In general, the formula for half-life is

$$t_{HL} = -\frac{\ln(2)}{k}.$$

Obviously, this only works if the data is *decaying*[73]. If the data is growing, we should *never* expect the dose to be cut in half unless we can reverse time. If $k = 0$, well, the universe blows up.

1.6 FUNCTIONS IN HIGHER DIMENSIONS

UNFORTUNATELY, WE LIVE IN A COMPLICATED WORLD[74]. There is *rarely* a direct line from a cause to an effect; no one clear reason why things are the way they are. Often, what we see or feel or hear is the

[72] Approximately 2–3 hours, if you're curious. After 4–6 hours, there is approximately 125 mg left in your system, at which point you can take another dose. This will ensure that even if you take two pills at a time this will ensure that you remain under the critical 3000 mg threshold for acetaminophen.

[73] i.e. $k < 0$

[74] Fortunate for us in the business of describing that world; if things were simple, we wouldn't have jobs.

sum total of many small inputs. The more of these small inputs we can capture in a model, the more accurately we can describe the world around us and predict the effects of each small change on a possible outcome[75]. It's for this reason that we necessarily need to discuss functions in higher dimensions.

Definition 1.6.1. A **multivariate function** is a function with more than one input.

$$y = f(x_1, x_2, x_3, \cdots, x_n)$$

The function then exists in n-dimensions, and its graph in $n + 1$-dimensions.

This still must satisfy the basic principle that makes a function a function: the fact that a *unique* set of inputs produces a *unique* output. In math language, we write that for a function $f(x_1, x_2, x_3, \cdots, x_n)$, if

$$y_1 = z_1,\ y_2 = z_2,\ \cdots, y_n = z_n$$

then

$$f(y_1, y_2, y_3, \cdots, y_n) = f(z_1, z_2, z_3, \cdots, z_n).$$

This may not make complete sense the first time you read it, and that's okay. The beauty of printed word is that you can go back and read it until it does[76]. Try to pay close attention to the logic of the statement: *if* A *then* B. *IF* the inputs are exactly the same, *THEN* the outputs must be exactly the same. Think back to figures 1.2 and 1.3, to give you a clue to what the statement means. Just because there are more inputs, does not mean our underlying principles have drastically changed. I'd urge you to continually refer back to the one-dimensional case if things get too hairy in multiple dimensions. It will help strengthen your foundation and intuition.

TO CONCLUDE THIS RATHER NIHILISTIC INTRODUCTION to multivariate functions, I will tell you that imagining the graph of a function in anything

---

[75]But how many of these small changes are even measurable or quantifiable? How many *really* matter for a given purpose? These are questions mathematical modellers ask all the time.

[76]Don't worry, I'll wait.

higher than two dimensions becomes impossible[77] in a traditional sense. We live in a three-dimensional world, and so spatially we can only think and visualize in up to three dimensions. Higher dimensional functions exist as abstract objects. To analyse them, we fall back on rules and intuition developed from functions in dimensions we *can* picture. These rules and this intuition is the basis of calculus and allow us to speak to the properties of functions no matter the dimension.

Functions of two variables

WE WILL FOCUS ON FUNCTIONS OF TWO VARIABLES since we can graph these, but the concepts demonstrated apply to functions of *any* number of variables.

A function of two variables is represented symbolically as

$$z = f(x, y).$$

Again, I would like to point out that these are just symbols and have no inherent meaning other than their context. What we are saying with this group of symbols is that we have a factory that requires two resources from us, and from those resources we produce a single thing[78]. We may as well write

This is *still* a function because it takes the form of a function that we, as a society, have agreed upon and[79] satisfies the definition of a function. Speaking of the mathematical definition of a function in higher dimensions is exactly as it was for one-dimensional functions: for a given input, we expect one and only one output.

Definition 1.6.2. Consider a function f and two sets of inputs $(x_1, x_2, x_3, \cdots, x_n)$ and $(y_1, y_2, y_3, \cdots, y_n)$. If

$$z_1 = f(x_1, x_2, x_3, \cdots, x_n)$$
$$z_2 = f(y_1, y_2, y_3, \cdots, y_n).$$

[77] Anyone who says otherwise is probably being more than a little economical with the truth.
[78] Remember, I am intentionally vague with these descriptions.
[79] Hopefully

If

$$\begin{aligned} x_1 &= y_1 \\ x_2 &= y_2 \\ &\vdots \\ x_n &= y_n \end{aligned}$$

then $z_2 = z_1$.

Just like in the $1D$ case, if our inputs match, we should expect our outputs to match. There is only *one* possible output for a given input; the only difference is that our input is a set of things instead of a single thing.

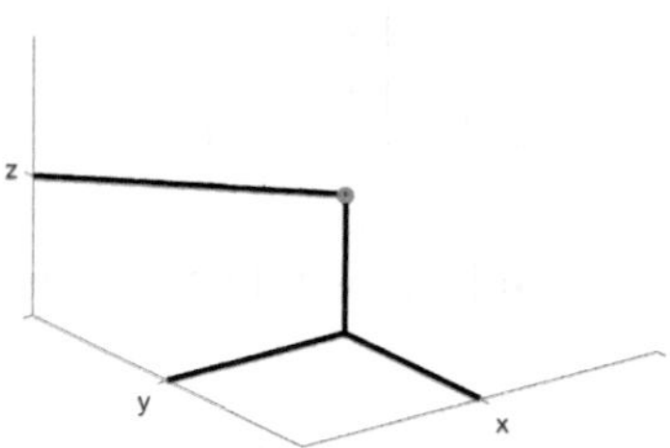

Figure 1.19: A 3D set of axes.

FUNCTIONS OF TWO VARIABLES ARE GRAPHED on a $3D$ set of Cartesian coordinates[80] (see figure 1.19) and create a surface when all points in a given range are plotted (figure 1.20).

FUNCTIONS OF TWO VARIABLES CAN ALSO BE GIVEN AS A TABLE, much like their one-dimensional counterparts. Consider the following example of this.

This table gives the wind chill temperature[81] in degrees Celsius. The wind chill temperature has two main contributors: the actual temperature

[80] Remember: 2 inputs + 1 output = $3D$

[81] The "feels like" temperature on your favourite weather app

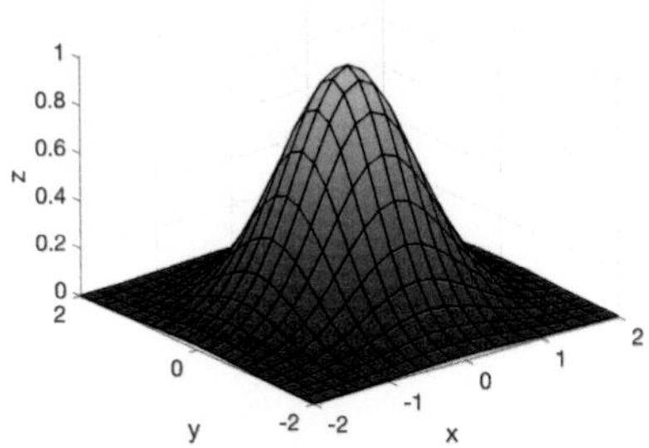

Figure 1.20: The graph of a 2D function is a surface in 3D

		Temperature				
		−10	−5	0	5	10
	0	−10	−5	0	5	10
	5	−13	−7	−2	4	10
Wind	10	−15	−9	−3	3	9
Speed	15	−17	−11	−4	2	8
	20	−18	−12	−5	1	7
	25	−19	−12	−6	1	7

(again, in degrees Celsius) and the wind speed (in km/h). We could write this as a function:

$$T_{WC} = f(T, W).$$

The values of the function are given by the interior values of the table. For instance,

$$f(-5, 10) = -9$$

and

$$f(0, 25) = -6.$$

Wind chill can also be estimated by a formula [9]:

$$T_{WC}(T, W) = 13.12 + 0.6215T - 11.37W^{0.16} + 0.3965TW^{0.16}$$

We can analyse the different terms in this formula by looking at certain points. By the formula,

$$T_{WC}(0, 0) = 13.12$$

We can interpret this as *if the temperature outside is* $0°C$ *and there is no wind, the wind chill temperature is* $13.12°C$. I'm going to ask you directly here to take pause. It's very easy to, when reading a textbook especially, just process the words without meaning. Go back and re-read that statement, consciously[82].

I hope that statement makes as little sense to you as it does to me. When there is no wind, there should be no wind chill. Moreover, if you have been outside on a day when the temperature is 0°C, and on a day when the temperature is 13°C, you would be pretty cognizant of the fact that they usually feel quite different[83]. Moreover, the formula doesn't match the table! The formula is an *estimate*, a model of the real world and quite clearly not applicable when there is no wind[84].

SO, LET'S GET BACK TO WHAT THESE TERMS MEAN instead of what they *don't* mean. Let's start with the term

$$0.6215T.$$

If this were the only term, this would mean that you feel (generally) warmer if the temperature is below 0, and cooler if the temperature is above 0[85].

The next term,

$$-11.37W^{0.16}$$

says that the wind speed *decreases* the perceived temperature. The wind speed is always positive, so this term is always taking away from the perceived temperature.

The last term is

$$0.3965TW^{0.16}.$$

Here, temperature and wind speed are multiplied together. This is because these two variables interact with one another. A warmer temperature, $T > 0$, means a warmer wind, which will have less of an effect on your perceived temperature than the same wind on a very cold, $T < 0$, day[86].

---

[82]If you already did the first time, good job!

[83]As different as a day in late fall, and a day in mid-spring

[84]You can compute $T_{WC}(T, 0)$ for different values of T to see this is indeed the case.

[85]This has almost everything to do with the fact that you are a living, breathing, energy-expanding, and heat-producing organism with a personal temperature regulator.

[86]To get a clearer picture in your head: think of a 10 km/h wind in the dead of winter, and a similar wind on a very hot day in the summer. Only one of these is usually described as pleasant.

Of course, all of these changes must be added or subtracted from *something*, and that's where the 13.12 come in. It is a translation in the formula that is found by fitting experimental data. It exists so that the experimental and theoretical values for wind chill agree to be within 1°C on the appropriate domain[87].

ALL OF THE FUNCTION TYPES WE SPOKE ABOUT IN ONE DIMENSION HAVE MULTIVARIATE COUNTER-PARTS. The wind chill formula is an example of a non-linear, multivariate function. Notice it doesn't fit nicely into any of the function families we talked about earlier because of the power 0.16. We will briefly discuss the same function families in a multivariate context, but you will see things quickly get out of hand. This is where the distinction between linear and non-linear becomes even more important!

Multivariate linear functions

A multivariate linear function is one that looks flat when graphed. The highest power of x or y is 1, and there is **no multiplications of x and y**. The general form of a multivariate linear function is

$$f(x_1, x_2, x_3, \cdots, x_n) = m_1 x_1 + m_2 x_2 + m_3 x_3 + \cdots + m_n x_n + b$$

Notice this just like a bunch of $1 - D$ lines added together, and that's because it essentially is. b is still the constant part, and the rest is the linear part.

We can compute the slope in any one direction by holding all the other values constant and just changing our variable in a certain direction. For instance, the average rate of change between points a and b in the x_1-direction is

$$m_1 = \frac{f(b, x_2, x_3, \cdots, x_n) - f(a, x_2, x_3, \cdots, x_n)}{b - a}$$

Notice that the terms for x_2 through to x_n will all cancel!

Just like in $1D$, if each of the average rates of change in our different directions is constant no matter which two points we take, then the function is linear. The difference is that this must be true for **all** the m's: m_1, m_2, m_3, all the way to m_n. If even one direction is not constant, then the function is not linear.

[87]The appropriate domain for this function is $W > 5$ and $T < 10$.

Non-linear multivariate functions

As soon as we move away from linear functions, the world gets particularly complicated. A natural conclusion, given the way we defined a multivariate linear function, is that a multivariate polynomial of order k might look like

$$\begin{aligned} f(x_1, x_2, x_3, cdots, x_n) =& a_{1\,k}x_1^k + a_{1\,k-1}x_1^{k-1} + \cdots + a_{1\,2}x_1^2 + a_{1\,1}x_1 + \\ & a_{2\,k}x_2^k + a_{2\,k-1}x_2^{k-1} + \cdots + a_{2\,2}x_2^2 + a_{2\,1}x_2 + \\ & a_{3\,k}x_3^k + a_{3\,k-1}x_3^{k-1} + \cdots + a_{1\,2}x_3^2 + a_{3\,1}x_3 + \\ & \vdots \\ & a_{n\,k}x_n^k + a_{n\,k-1}x_n^{k-1} + \cdots + a_{n\,2}x_n^2 + a_{n\,1}x_n \\ & + b \end{aligned}$$

which is just an addition of the $1D$ polynomials. It's a decent[88] first thought, but it's missing *so many possibilities.*

Since a function with n input variables and k powers in each of those variables can be really hard to read and manage, let's reduce our example to two input variables, and order 3. Our first pass would say that this polynomial should look like

$$f(x, y) = a_1x + a_2x^2 + a_3x^3 + b_1y + b_2y^2 + b_3y^3 + d$$

but what about a term like a_4xy? Is this valid? The answer is yes, it absolutely is and this is one of the possibilities missing from our first pass above. We could have *any* multiplication of powers of x and y and still get a valid polynomial. The canonical form of a polynomial of order 3 in $2D$ is

$$\begin{aligned} f(x, y) =& a_9x^3 + a_8y^3 + a_7x^2y + a_6xy^2 + a_5y^2 + a_4x^2 + a_3xy \\ & + a_2y + a_1x + a_0 \end{aligned}$$

In the case of multivariate polynomials, we calculate order a little differently as well.

Definition 1.6.3. The **order** of a multivariate polynomial is the largest exponent present. In the case where two variables are multiplied together, we **add the exponents**.

---

[88] And again, totally natural

Notice that even if $a_9 = a_8 = 0$ in the above polynomial of order 3, it would *still be a polynomial of order* 3. This is clear if we explicitly write in our exponents

$$\begin{array}{lcccccccc} & g(x,y) = a_7x^2y^1 & + a_6x^1y^2 & + a_5y^2 & + a_4x^2 & + a_3x^1y^1 & + a_2y^1 & + a_1x^1 & + a_0 \\ \texttt{Order of Term:} & 3 & 3 & 2 & 2 & 2 & 1 & 1 & 0 \end{array}$$

Now imagine increasing the order to 4, or increasing the number of input variables to 3; things get out of hand very quickly!

THINGS DON'T GET MUCH BETTER WHEN WE LOOK AT EXPONENTIAL FUNCTIONS IN MULTIPLE DIMENSIONS. In each dimension, we have a term like $A_n e^{k_n x_n}$ and we can combine these terms in many different ways. We can multiply them, add them, and multiply variables in an exponent. Again, much too many possibilities to effectively ground ourselves. We are better off developing general techniques that can be used on functions of any type and abandon trying to pigeonhole things as we were able to do in $1D$.

1.7 CONTOUR DIAGRAMS

Sometimes surfaces can be hard to visualize. Sometimes it's much easier to interpret things in two dimensions[89]. A **contour diagram** is a representation of a three-dimensional surface on the plane[90].

Consider the surface

$$z = x^2 + y^2$$

This is a relatively easy surface to draw, but sometimes we can glean information from a two-dimensional projection. We do this by setting $z = C$ for different values of C and plotting

$$C = x^2 + y^2.$$

For instance, we can plot

$$0.25 = x^2 + y^2.$$

This looks suspiciously like the equation of a circle with radius 0.5, and that's because it is a circle[91]. What this means is that at height

[89]This is mostly because we communicate mostly through a two-dimensional medium.

[90]"The plane" is your standard, two-dimensional xy axes.

[91]If it looks like a circle, quacks like a circle, and has the form $r^2 = x^2 + y^2$, then it's a circle.

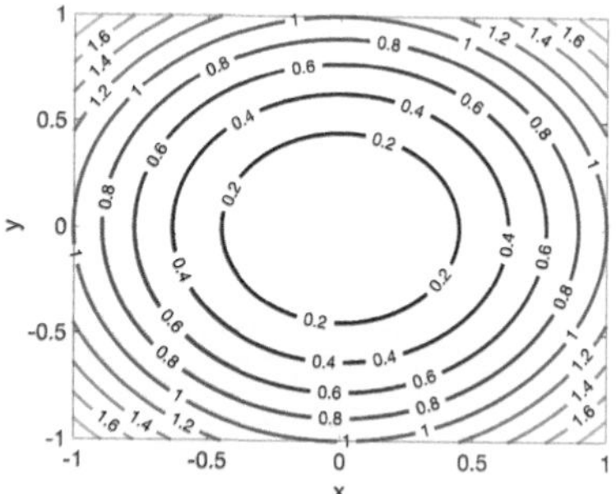

Figure 1.21: Contour plot of $z = x^2 + y^2$. The numbers on the curves are the heights of the various circles if we were to stretch this picture into the third dimension (out of the page).

$z = 0.25$, the *cross-section* of our surface is a circle of radius 0.5, which can be plotted on the xy-plane. If we do this with many different values of C, we can get a **contour diagram** of our surface as seen in figure 1.21. Each circle represents a different slice of the surface at a different height. Figures 1.22 and 1.23 show the relationship between a contour plot and the corresponding surface.

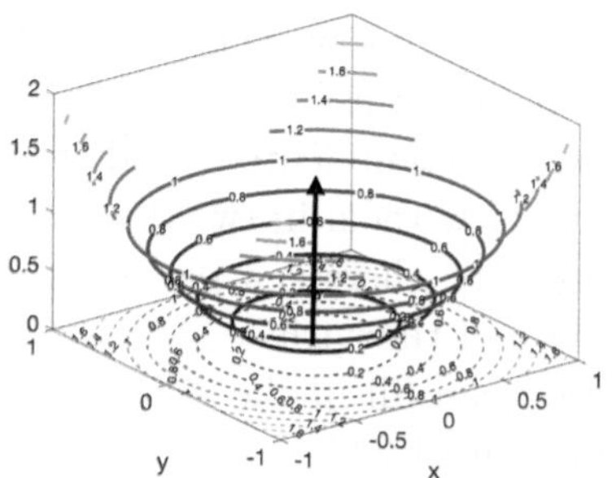

Figure 1.22: Contour plot of $z = x^2 + y^2$ being stretched into a surface.

You may be asking yourself, "how do I pick the C values?" or "How many C values do I pick?" or "If I pick *all* the C values[92], wouldn't I just end up with a totally coloured square?"[93]. The answer to the last question is yes. The answer to the second question is "pick a reasonable number – 4 or 5 is usually good; if it isn't then add more," and the

[92] For some reasonable definition of all

[93] If these questions seem ridiculous to you, then I guess teenage Matt was ridiculous.

answer to the first question is "the integers or other nice numbers are usually a good place to start"[94].

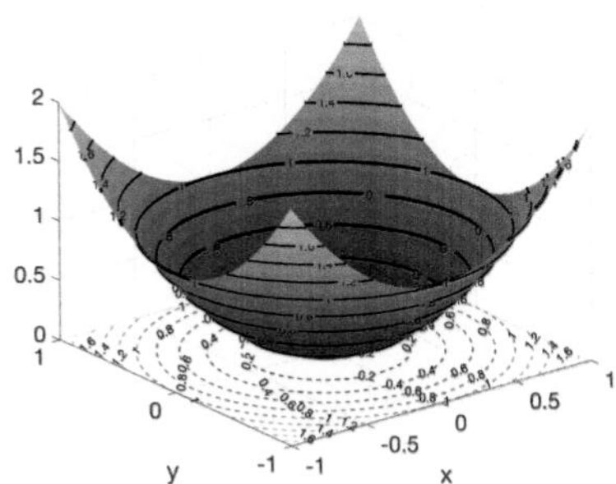

Figure 1.23: Contour plot of $z = x^2 + y^2$ and the corresponding surface.

YOU MAY BE ASKING WHY. We have surfaces, why in the world do we need another way to represent functions in two variables? Why must I learn yet another thing that I will never ever use? What will I learn that I will use? What job will I use those things in? What am I even going to do with my life? Why am I even here? These are all valid questions, and they should all be considered over the course of your life, but in order to avoid an existential spiral I think we will stick to the very first question: why do we need contour diagrams?

Well, surfaces can get complicated, and what makes them more complicated is trying to view them in two dimensions. Even for our very simple surface in figure 1.23, there is a piece of the surface that's being masked by another piece of the surface. Now, in that case it's fine; it's really simple. Our brain can fill in the missing bit just fine. But what about the surface[95] in figure 1.24?

This surface representation of the function misses some key details. Namely, it's very hard to see the sunken valley in the middle, and impossible to see the spikes protruding from the valley. If this were a trap set by your worst enemy and only have the surface plot, you may very well meet a grim end on the spikes. The contour plot in figure 1.25 shows all this detail and in our hypothetical situation may just save your life.

[94]These are the ridiculous answers I wish someone had given me had I been confident enough to ask my ridiculous questions.

[95]Yes, this is a surface that can be represented by a function, but it's a very, very complicated one.

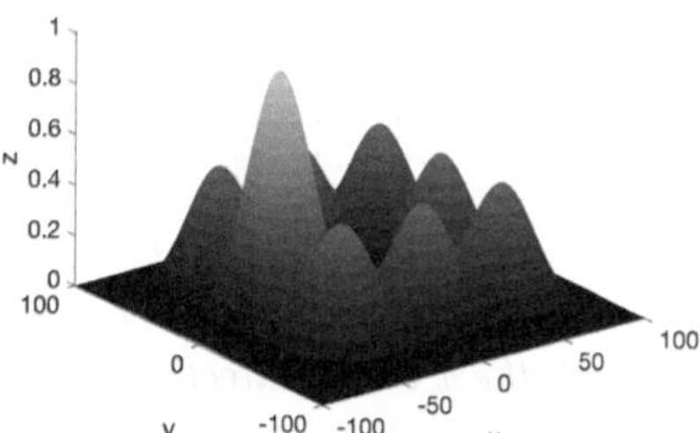

Figure 1.24: A complicated surface of hills and valleys. This two-dimensional representation of a three-dimensional surface hides some of its features, where as the corresponding contour diagram doesn't.

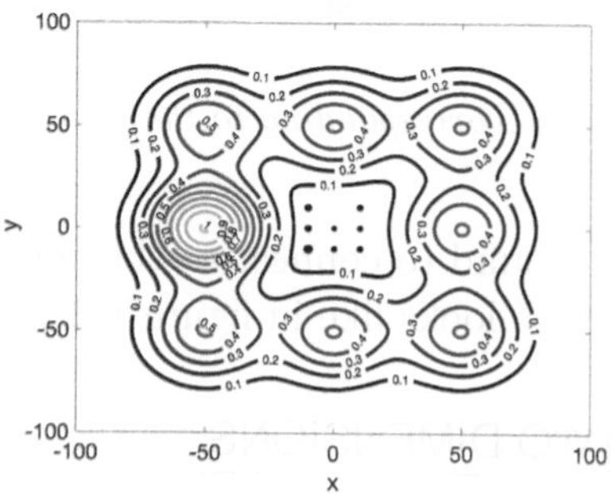

Figure 1.25: A complicated surface of hills and valleys. This two-dimensional representation of a three-dimensional surface hides some of its features, where as the corresponding contour diagram doesn't.

While our example above might seem artificial, the idea behind it is very natural: for complicated surfaces, a contour diagram holds more information than the corresponding surface plot. A very natural example of this is the use of contours on maps.

MOREOVER, CONTOUR DIAGRAMS MAKE IT VERY EASY TO DETERMINE WHAT IS AND ISN'T A FUNCTION. **The contour diagram of a function will never have two or more contour lines that cross each other.** If the contour lines do cross, we only have a relation.

So far, the examples of contours we have seen have all been **closed**. A closed contour is one that forms a loop; there is a clear *inside* and a clear *outside* of the contour. Not all contours need be closed though as in figure 1.26. We see the contours extend off to infinity in both directions

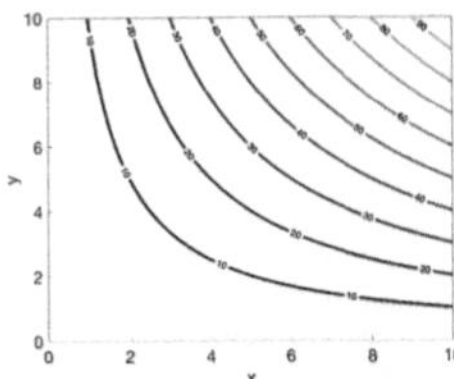

Figure 1.26: Not all contours need to connect as closed curves. These contours extend off to infinity in both the x and y directions.

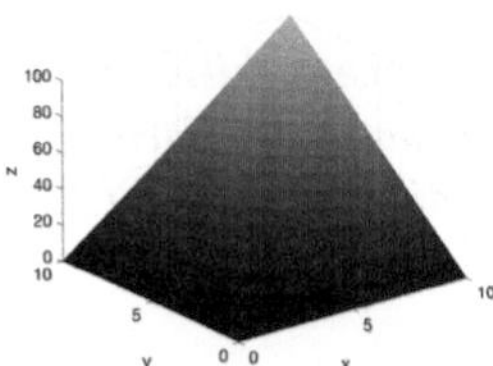

Figure 1.27: A surface corresponding to the contours in figure 2.12.

and never come back around to connect to the side[96]. We can see what the corresponding surface looks like in figure 1.27.

1.8 MODELS IN TWO DIMENSIONS

WE'VE ALREADY SEEN A MODEL IN TWO DIMENSIONS with the model of wind chill. That was an *empirical* model, meaning that the coefficients and powers are determined through experimentation. We could, in contrast, build a *theoretical* model based on intuition and what we expect to happen in a given situation.

Let's take a look at a model of credit limit as a function of age and annual income. Let's call this function

$$c(a, I)$$

where c is for credit limit, a is for age, and I is for annual Income.

When building a model, the first question to ask is, *how might these things be related based on our previous knowledge?* The first question we might ask is what do we expect to happen to a credit limit as our age increases? Generally, as we get older we become less "risky" to creditors[97].

[96]If they did, they would at some point have to cross the *other* contours that are going off to infinity, which would leave this not a function.

[97]Ageism at its finest

The less of a risk you are, the more likely people are to lend you larger amounts of money. We might then posit that as we get older we will be more likely to be given more credit. So, the first term we may include in our model is

$$c(a, I) = Ma$$

where M is the average amount of new credit we expect to receive for each year older we get. If we expect credit to increase with age, we would want to state that $M > 0$.

We also might be able to argue that the more money you make, the more trusting you are perceived to be and thus the more credit you will be allowed[98]. We might also expect your credit limit to increase with your annual income, so we add the term

$$c(a, I) = Ma + nI$$

and since we expect an *increase* we state that $n > 0$.

Do age and income have any dependence on one another? If you've ever watched the news, or met with the executives at any company[99], then you would know that the answer is a resounding yes. Perhaps then we would want to then also include the term

$$c(a, I) = Ma + nI + kaI$$

For good measure, we should also include a constant term since there is generally a minimum amount of credit a bank or a credit card company is willing to lend you[100]. So we add the term

$$c(a, I) = Ma + nI + kaI + b$$

This constant term will act as a baseline for credit upon which the other terms will act.

Our next step when modelling is to make sure that the model we built actually agrees with the real world, in order to do that we need data[101]. This is where we, as mathematical modellers, rely on the other sciences heavily. We depend on scientists, social scientists, and humanities

[98]We could also argue this is a form of classism meant to keep down the proletariat and benefit the aristocracy, but this is after all just a math book.

[99]Whatever institution you're studying at is likely *not* the exception; if it is hurrah!

[100]For instance, you can't get a credit card with a five limit.

[101]Precious, precious data

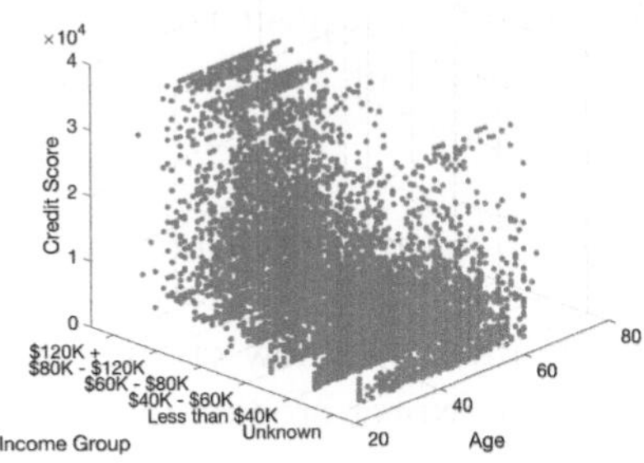

Figure 1.28: Credit score as a function of age and income.

scholars to collect data that can help validate our models. With validated models, we can be confident in the insights gained from our analysis. Of course, those insights then need to be tested and verified. If our model still agrees with the world around us, great, we continue with it. If not, we don't abandon it and hang our heads in shame; we ask *what are we missing* and we refine and improve until we can explain what was missing.

Luckily, this example was chosen because there is some data available [13] and it's plotted in $3D$ in figure 1.28. Right now, we don't have the tools to prescribe numbers to M, n, k and b but we could approximate them in much the same way we did with our $1D$ functions.

1.9 VARIABLES VS. PARAMETERS

While we use letters and symbols to represent many things in mathematics, there are two fundamentally different categories of symbols: variables and parameters. Unfortunately we use letters[102] to represent *both* categories. This can lead to some confusion, especially when we are just learning. It's important to keep in mind *context*; it is the best way to distinguish between the two categories.

The first category is the one we are most used to: **variables**.

VARIABLES REPRESENT NUMBERS THAT WE WISH TO VARY over a **domain**. They have no inherent value, but can take a *range* of values. They are often distinguished from parameters by their inclusion in the inputs of a function. This means they are things that we *intend* or *expect* to change.

[102] Greek or Latin, usually

PARAMETERS, IN CONTRAST, ARE SYMBOLS THAT REPRESENT A SPECIFIC VALUE. While we can often treat them as variables as well, context tells us that parameters are meant to be known quantities, or knowable quantities. They may differ, depending on the context, but generally a parameter is a fixed value or that we can expect to be fixed.

Let's take a look at a basic equation for disease transmission in a population[103]. The number of infected individuals on day d is given by the following:

$$I_d(I_{d-1}, S_{d-1}) = I_{d-1} + \beta I_{d-1} S_{d-1} - \mu I_{d-1}$$

We read this in the following way. The number of infected people, I, on day d, is the number of infected individuals on the previous day, $d-1$; *plus* all susceptible individuals, S, who *interacted* with infected individuals, and subsequently caught the infection with probability β; *minus* the number of infected individuals who recovered from illness with probability μ.

In this model, β and μ are **parameters**. We can figure out through experiments and observations the probability of infection and the probability of recovery for a specific disease. Once we know these numbers, we can use the above equation to try and predict how many infections we will have on a particular day. This type of equation is used every day by public health to prevent and mitigate epidemics[104].

AS ANOTHER EXAMPLE, let's use a general form of the wind chill model from earlier:

$$T_{WC}(T, W) = T_0 + aT - bW^k + cTW^k$$

In this model, the inputs tell us what our variables are and it is *assumed* that the other values are parameters. In this model, temperature, T, and wind speed, W, are our variables because they appear as inputs to our function. These parameters are set through experimentation, and largely depend on the surface body temperature of a human and other very human qualities. The *values* of the parameters given earlier are average values for the average human, but we all experience temperature and heat loss differently. We could determine parameters for individual people

[103] Of humans, but could be any living thing

[104] This is almost mostly a lie. We actually use differential equations for this. We will revisit the dynamics of diseases later when we talk about simple differential equations, but we have **a lot** of ground to cover before we can get there.

if we so wanted to and it's likely that they would be quite different for different people. For instance, taller people tend to retain more heat than shorter people and thus will likely feel wind chill differently [10].

Sometimes we write

$$T_{WC}(T, W; T_0, a, b, c, k)$$

the semi-colon separates variables from parameters in the model. This notation starts to blur the line between the two.

THAT'S NOT TO SAY WE COULDN'T MAKE ANY OF THE PARAMETERS INTO VARIABLES. We could just as easily write

$$T_{WC}(T, W, k) = T_0 + aT - bW^k + cTW^k$$

Perhaps k changes with species, and we wish to look at wind chill across many animals and choose to use k to represent that. k is now a variable instead of a parameter. We assume it is something varying as opposed to knowable.

THIS ALL MIGHT SOUND A LITTLE VAGUE AND STRANGE AND POINTLESS but being able to tell the difference will be pretty important in our upcoming study of calculus. The key takeaway is that variables are provided as inputs to functions and are *expected to change*, whereas parameters as *assumed constant* and do not appear as an input to a function[105].

[105] Or if they do, they appear *after* a semicolon.

PRACTICE PROBLEMS

Functions & Models

1. For each of the following, denote whether it is a function or not a function.

   (a) $f(x) = 5x + 2$

   (b) $f(x) = e^{\sqrt{x}}$

   (c) $f(x) = \sqrt{x}$

   (d) A store inventory system where you input a product code and it returns to you the current price.

   (e) A store inventory system where you input a specific price in order to find a specific item; it returns multiple items that all have that price.

   (f) In the original Super Mario Bros. video game, there were only six buttons: Up, Down, Left, Right, A and B. A made Mario[106] jump, as did B; Left made Mario move left, Right made Mario move right, Down made him crouch, and Up didn't really do anything. If the buttons are inputs, are the controls of the game a function?

   (g) Mario's actions are to move left, move right, crouch, and jump. If these are inputs and the outputs are the buttons that need to be pressed to perform the actions, is this a function?

2. You are tasked with modelling the size of an algal bloom in the Bay of Fundy. What types of inputs might you think to include?

3. Consider the graph in figure 1.29. Is this a function? Why or why not?

4. Consider the function

   $$P = f(a, b, c)$$

   If P is probability of a forest fire, what might a, b, and c be?

5. The graph in figure 1.30 shows historical stock data for Sony Corporation. What do you think may have caused the spike around the year 2000? There are a few things that were in the works for Sony at the time[107].

---

[106] Or Luigi if you happened to be the younger sibling, or more timid friend

[107] Data taken from https://finance.yahoo.com/quote/SNE/

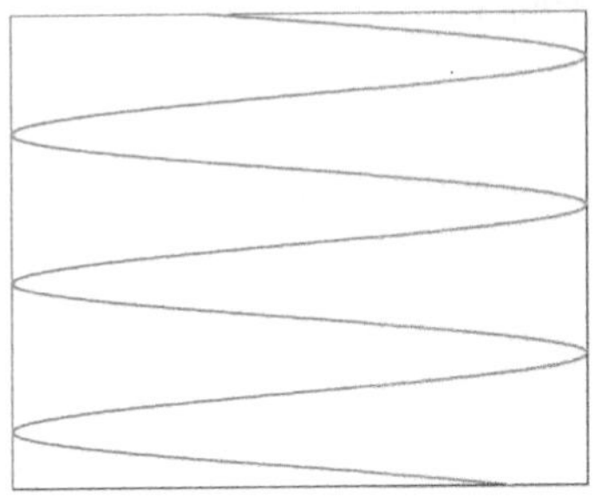

Figure 1.29: Figure for question 2.

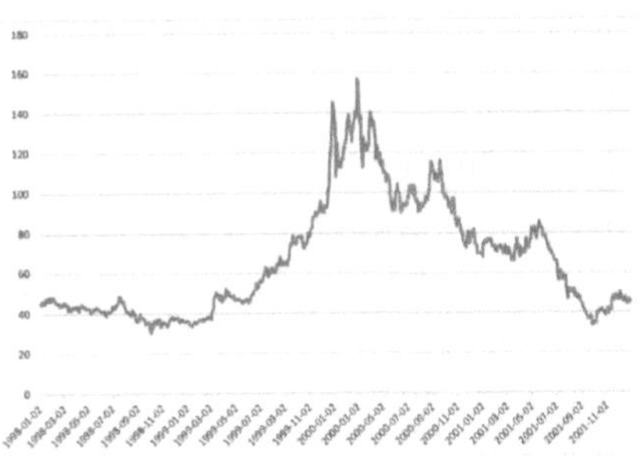

Figure 1.30: Figure for question 4.

6. Let's pretend we have a model of the number of mosquitos in a region as a function of the number of bats[108].

$$M(B)$$

What do you think $M(0) = 10000$ means? What about $M(-10)$? How would you interpret $M(B) = 0$?

Linear & Polynomial Models

1. Consider the function

$$f(x) = 5x + 2$$

Identify the slope and y-intercept.

[108] Fun fact: bats are a primary predator of mosquitos.

2. Consider the function

$$f(x) = \frac{8x - 72}{4}$$

Identify the slope and y-intercept.

3. Consider the function

$$y - 2 + x - 5 = 4$$

If y is the output, identify the slope and y-intercept.

4. Consider the function

$$D(g) = -5g + 50$$

This function represents the loss of genetic *D*iversity in a population after a mass extinction[109], g represents the generation since the bottleneck event, D represents the genetic diversity on a scale of 0 to 100.

   (a) Identify the slope. What might it represent?

   (b) Identify the y-intercept. What might this represent?

   (c) When might this model lose validity?

5. Consider the function

$$V =$$

6. Figure 1.31 shows annual average land temperatures for the globe. There are 166 data points from 1850 to 2015[110].

   (a) Develop a linear model for the average global land temperature as a function of year.

   (b) Develop a quadratic model for the average global land temperature as a function of year.

   (c) Which one might be a better fit? Why?

7. Once again, referring to figure 1.31, do you see more than one possible trend in the data? Where do the dynamics seem to shift dramatically? What happened around this time in history?

[109] Typically called a bottleneck event

[110] Temperature data actually goes back further than this, but it's far less reliable.

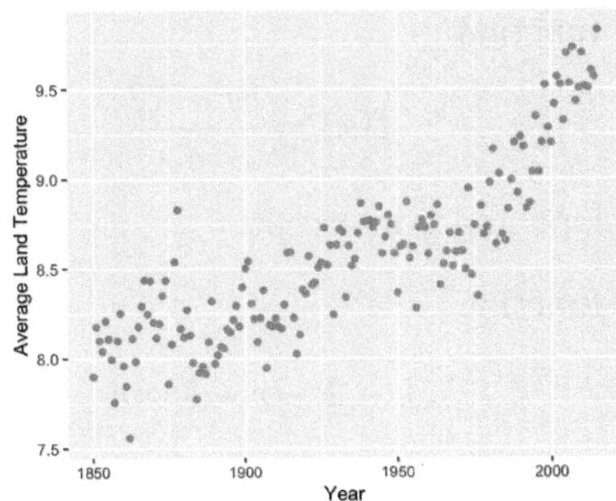

Figure 1.31: Figure for question 1. I hope that you are not reading this book on a hostile, hot planet.

8. Here is some anonymized car dealership data (figure 1.32) for the year 2016. Each point shows the number of trucks sold in that month. This data is purposefully sparse and ambiguous.

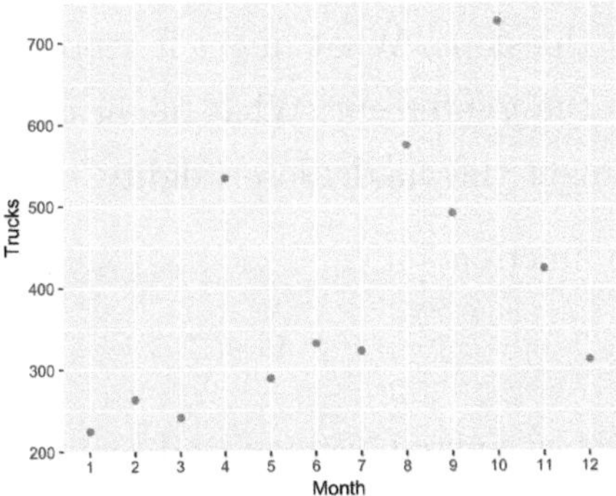

Figure 1.32: Figure for question 3.

(a) Given the context of the data, would you expect there to be a maximum or minimum? Do you think a particular time of year may lead to more/fewer truck sales?

(b) Think about how after December comes January. While that January is in a different year, typically something like retail sales follows a similar pattern year to year.

(c) How would you model this data? Give it a go.

Exponentials and Logarithms

1. Take a look at figure 1.31 again. Could you fit an exponential to this curve? Give an estimate of the doubling time of the global land temperature.

2. Now, there is some debate as to whether the average global land temperature is increasing linearly, or exponentially. Look at figure 1.33. Can you fit an exponential to this? CO2 emissions and the average global temperature are related. What do you think that relationship might be? Can you write temperature, T, as a function of CO2 emissions?

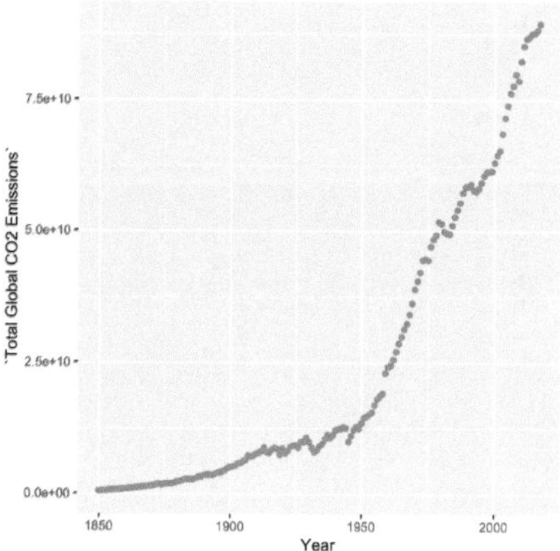

Figure 1.33: Figure for question 2.

3. Figures 1.34 and 1.35 show the full available data set for the drug efavirenz split into male and female participants, respectively. Develop an exponential model for drug decay in females and in males.

   (a) Does the rate of decay differ between the sexes? Would you say this difference is significant?

   (b) Using your models, see if the half-life is different between the two sexes.

   (c) What does each model predict about the initial dose? Why might this value be different between the two sexes?

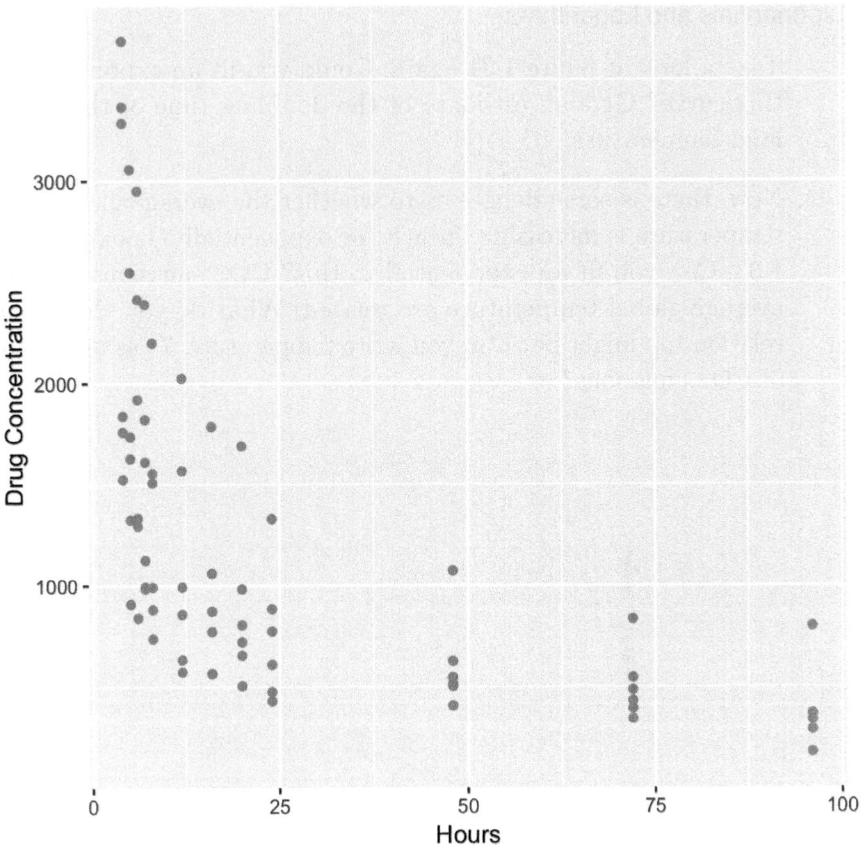

Figure 1.34: Figure for question 3.

4. A very common type of function used in biology is the **logistic function**. It looks like this:

$$P(t) = \frac{L}{1 + \mathrm{e}^{-k(x - x_0)}}$$

Pick a few different values for the parameters L, k, and x_0 and plot the logistic function. What might this be used to model? What might these parameters represent?

Functions of two variables

1. A complete picture of drug concentration in the bloodstream is given as a 2D function. A drug's concentration first increases as

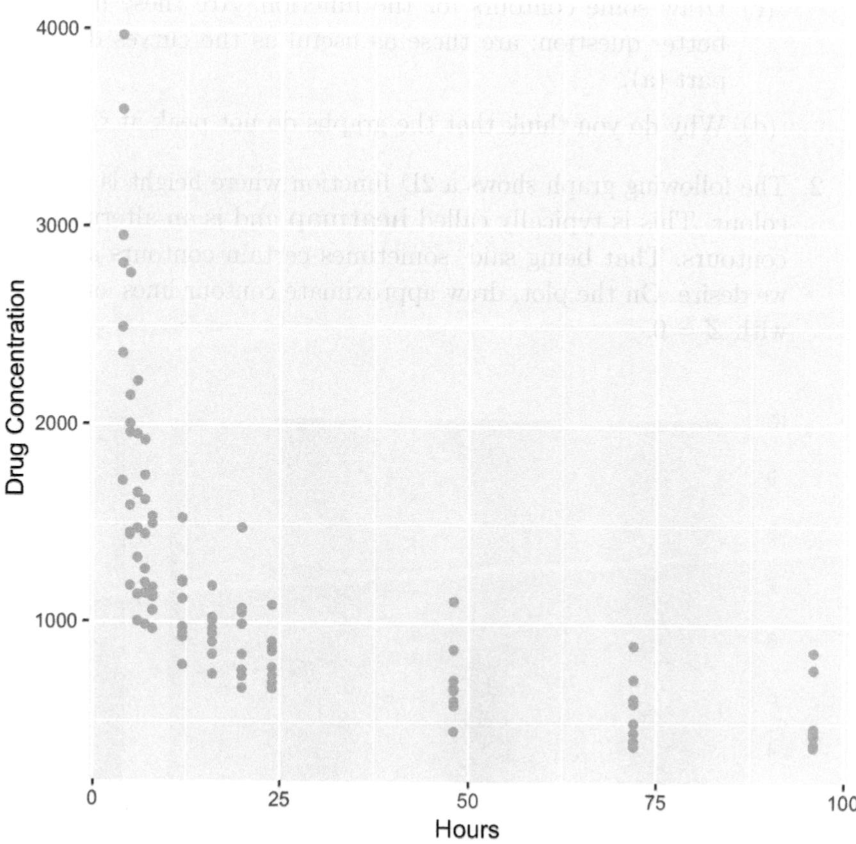

Figure 1.35: Figure for question 3.

it is absorbed into the bloodstream, and then it is metabolized by body. This process depends both on time and the initial does.

$$C(d,t) = te^{-rtd}$$

where d is the dose, t is the time since dose, and r is the rate at which the drug is metabolized.

(a) Set $d = A$, where A is a few different integer values. Plot the **one-dimensional** functions $C(A, t)$ all on the same set of axes.

(b) There are three "letters" in the function: r, t, and d. Which are variables? Which are parameters? How do you know?

(c) Draw some contours for the function. Are these useful? A better question: are these as useful as the curves drawn in part (a)?

(d) Why do you think that the graphs do not peak at the dose?

2. The following graph shows a 2D function where height is given by colour. This is typically called **heatmap** and is an alternative to contours. That being said, sometimes certain contours are what we desire. On the plot, draw approximate contour lines associated with $Z = 0$.

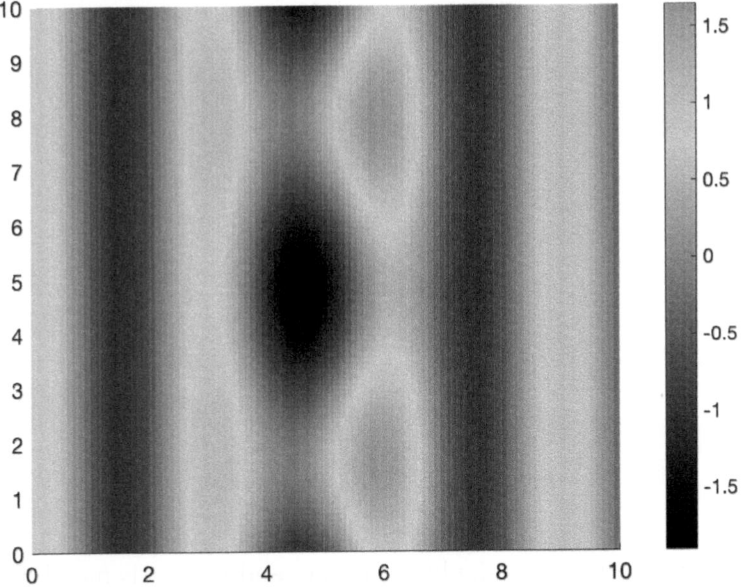

This figure is a heat map of altitudes for a Minecraft world where negative altitude is the beds of water (rivers, lakes, etc.). The contour $Z = 0$ shows the shorelines.

3. Car insurance premiums (the amount you pay per year for insurance) use multi-variable functions to determine how much you should pay. Generally, insurance premiums start at some fixed amount, they increase with the value of your car, and depend on your age. Younger drivers and older drivers are both more expensive to insure because of their increased risk of being involved in

collisions. As well, young people driving expensive cars tend to cause more collisions than young people in low-valued cars.

Build a basic model for insurance premiums as a function of age, A, and car value, V.

4. Match the following surfaces with their respective contour plots.

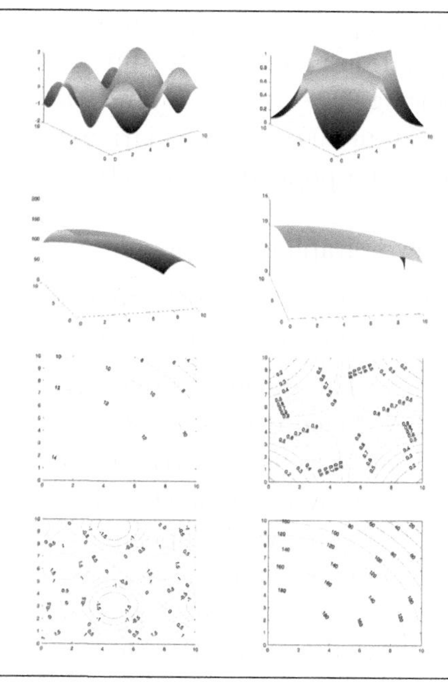

5. Human populations tend to establish near bodies of water. If we have a shoreline at $x = 0$, and start with a population of $100,000$ individuals, we can model both population growth and the sprawl outward away from the waterfront.

   We know that populations tend to grow exponentially, so let's say our population is growing at a rate r.

   Sprawl also tends to be exponential. Let's say that population decreases at rate k for every kilometre away you are from the water.

   (a) Write down a model for the population $P(x, t)$. Give this some thought. Do we expect the exponentials to add? Multiply? Why?

   (b) How long until the population at $x = 50$ is $100,000$?

(c) Let $r = 0.01$ and $k = 0.08$. Plot some contours. How do you interpret this graph?

Real Thinkers

1. The population of Sackville in 1901 was 1,444. In 1911, the population had increased to 2,309. By 1941, the population was 2,489. In 1951, there were 2873 people in Sackville. Over the next 10 years, the population increased steadily to 4,612 and increased again in the following 20 years to 5635. In 1996, the population was reported to be 5393, in 2001 the population was 5361 and the 2016 census data showed a population of 5331.

   (a) Draw a graph of the population changes in Sackville.

   (b) For much of its history, Sackville had a healthy foundry industry and was a central railway hub:

   > The National Policy of Prime Minister Sir John A. Macdonald's administration in the 1870s–1880s saw various industries cluster along the Intercolonial Railway in Amherst and Sackville. Sackville became home to two independent foundries; the Enterprise Foundry, and the Fawcett Foundry. Both produced stoves and related products with both businesses operating for more than a century. These competitors eventually merged and the Fawcett Foundry was closed and the foundry demolished in the **19??**s.
   >
   > Sackville grew in importance as a railway junction after Canadian National Railways established a dedicated railcar ferry service at Cape Tormentine in 1917. The Sackville railway yard and station were constantly busy until the opening of publicly funded highways following World War II started a slow decline. The abandonment of the Prince Edward Island Railway in **19??** saw the line to Cape Tormentine removed at the same time as the Trans-Canada Highway was being expanded to a four-lane freeway. As the railway consolidated to a single mainline running through town, businesses left, including offices of Atlantic Wholesalers.

Turns out, the closing of the foundry and the abandonment of the Prince Edward Island Railway happened at roughly the same time. Based on your graph, when do you think this was? Give a reason for your answer.

2. Metro supermarket is a Montreal-based grocery store. Here are some stock prices for Metro over a week:

Date	Price (CAD)
Aug 23, 2019	54.18
Aug 26, 2019	54.51
Aug 27, 2019	55.05
Aug 28, 2019	55.53
Aug 29, 2019	56.19

   (a) **Assuming the data is linear**[111], develop a formula for the stock price as a function of time. (Graphing it first might help)

   (b) Use your formula to predict the price on 30 August 2019.

   (c) The actual stock prices on 30 August 2019 was 56.19. How far off are your predictions?

3. Now, here is some longer term data for Metro stocks:

Date	Price
Aug 30, 2015	35.52
Aug 30, 2016	44.66
Aug 30, 2017	40.82
Aug 30, 2018	40.86

   (a) **Assuming the data is linear**[112], develop a formula for the stock price as a function of time (per year) (graphing it first might help).

   (b) Use your formula to predict the price on 30 August 2019.

   (c) The actual stock prices on 30 August 2019 was 56.19. How far off are your predictions?

4. What can you say about predicting stock prices one day in the future versus predicting stock prices one year in the future?

---

[111] It's not linear.

[112] Again, not linear

5. The growth of an algae bloom over space and time can be modelled by

$$b(x,t) = \mathrm{e}^{-\dfrac{(x-5)^2}{(t+1)^2}}$$

where B is density of the bloom, x is 1D space, and t is time in days.

   (a) Draw some contours for this function.

   (b) The REAL function (which is too complicated to draw contours for) is

$$B(x,t) = \frac{1}{t+1}\mathrm{e}^{-\dfrac{(x-5)^2}{(t+1)^2}}$$

   On the same axes, draw $B(x,0)$, $B(x,2)$, and $B(x,6)$.

   (c) What is happening to the algae bloom over time?

6. Stevens' power law says that perception, P, is proportional to the intensity of the stimulus (call it I) to the power n, where n depends on the type of stimulus (e.g. brightness, pain, electric shock, etc.). We can write an equation for Stevens' Power Law as

$$P(I) = kI^n$$

where k is some scaling factor.

Take the natural log of both sides. Use the properties of logarithms to expand the right-hand side of the equation as much as you can. What kind of equation is this? What are the values of the slope and y-intercept? What might they represent?

7. On average, dogs stop growing within 1 year. When Link (my dog) was little, I took monthly measurements of his weight **(t=0 is when I got him, so he was already 2 months old)**:

month	0	1	2	3	4
weight(kg)	0.91	1.2	1.6	2.2	2.9
weight (pounds)	2	2.6	3.6	4.8	6.5

Use kg or pounds for the following, which ever you're more comfortable with.

(a) Calculate the average rate of change of Link's weight between months 0 and 1; 1 and 2; and 3 and 4.

(b) Calculate the relative rate of change of Link's weight between months 0 and 1; 1 and 2; and 3 and 4.

(c) Why type of function would be more appropriate to model Link's growth? Why?

(d) Use the data to build a model of Link's growth.

(e) How much does the model say Link weighed on his first birthday?

(f) This is Link:

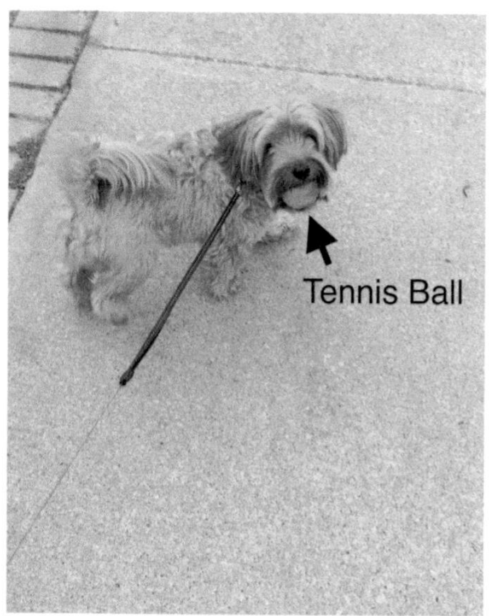

Does your answer in part (f) seem reasonable? What may have gone wrong?

CHAPTER 2

Derivatives

CALCULUS AS A WHOLE IS THE STUDY OF CHANGE. This study is split into two parts, the study of **instantaneous change, or differentiation,** and the study of **accumulated change, or integration**. The former studies how things move in infinitely small, imperceptible periods of time. The latter studies how infinitely many small changes add up to a noticeable, macroscopic change.

WE CAN THINK OF DIFFERENTIATION IN A MORE PRACTICAL SENSE if we first agree that every measurement we ever take, and every model we ever make is always wrong[1] and we can never be 100% certain of anything we do. The best we can do is get arbitrarily close to what is correct.

As an example, we may measure the air temperature as, say, 24.1°C, but the actual air temperature may be 24.15°C. Even if we find ways to be far more precise in our measurements, there is always a limit to our precision. If we measure the air temperature to 100 decimal places, it is still just an approximation[2].

That covers why our measurements are always wrong, what about our models? If we even look at something as simple as a falling ball and want to model its speed as we drop it from a roof, there are myriad things that affect this. There is wind, air temperature, air pressure, drag, imperfections in the shape of the ball, and possible microscopic particles floating in the atmosphere. The biggest influencer though is gravity, and so often we will just model gravity to get a decent estimate. Usually this

[1]But some are useful, according to George Box

[2]You may ask, *who cares? I can't feel the difference between* 24° *C and* 24.1° *C*. You can't, but weather systems can be incredibly dependent on the smallest changes. In fact, most of the reason we can't predict weather very well is because our measurements can never be precise enough.

DOI: 10.1201/9781003265405-2

is good enough[3], but how much would or should the neglected phenomena *change* the outcome?

So, we have imperfect models that use imperfect measurements to try and make perfect predictions. Why does this work? Or, a better question is, *when* does this work? Differentiation can be thought to be a way to measure sensitivity of functions[4] to small imperfections in inputs.

Let's say we want to use a function to predict the number of malarial infections from the number of mosquitos present, $p(M)$, where p is people infected with malaria and M is the number of mosquitos. We can probably never count the exact number of mosquitos in an area, but we can probably estimate the number decently. The question then is, if our input is close to the true value, is the output from our function close to the true number of people infected with malaria?

To answer this question, we have to study how functions *change*. We can think of change and sensitivity as analogous for the sake of this definition. This is all differentiation is, the study of how functions change.

Definition 2.0.1. The amount of change of a function, $f(x)$, at a point x_1 is called the **derivative** of $f(x)$ at x_1. Mathematically, the derivative can be represented in two different[5] ways:

$$f'(x_1) \qquad \text{or} \qquad \left.\frac{\mathrm{d}f}{\mathrm{d}x}\right|_{x=x_1}$$

2.1 THE TANGENT LINE

WE SAW IN CHAPTER 1 THAT THE AVERAGE RATE OF CHANGE between two points on a function, f, is given by the formula

$$\frac{\Delta f}{\Delta x} = \frac{f(x_2) - f(x_1)}{x_2 - x_1} \tag{2.1}$$

This is sufficient if we only care about where we start and where we end. We lose all information about what happens between x_1 and x_2, like in figure 2.1.

---

[3] But if we're looking for perfection, "good enough" is still wrong.

[4] And thus, mathematical models

[5] But equivalent

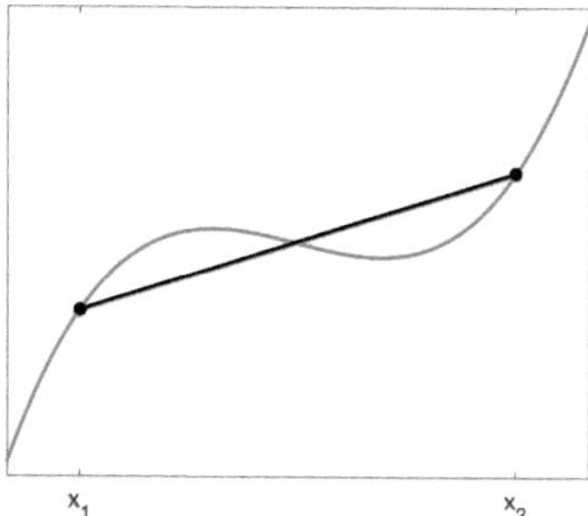

Figure 2.1: While the average rate of change gives us an idea of what is happening, we miss a lot of information about what is happening between our two measured points.

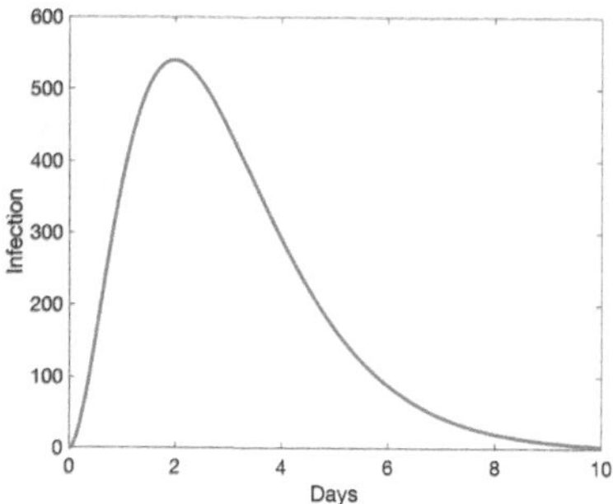

Figure 2.2: The number of infections on any given day for an outbreak of a disease.

Consider the plot in figure 2.2, if we calculate the average rate of change in the number of infected people over 10 days, we would see 0. We would miss the entire outbreak. The only information that we gain from the average rate of change is that the outbreak is localized and did not persist. In fact, if we take the average rate of change between the starting point and any day greater than 2, we will always underestimate the outbreak.

So, maybe, the natural[6] question to ask is, *is there a way to calculate the rate of change at a single point?* Another way to think of this is, given three photographs of triceratops with timestamps and distance markers, as in figure 2.3, you can use formula (2.1) to estimate the speed of the

[6]Natural for me, anyway

triceratops between picture 1 and picture 2,

$$speed_{p1-p2} = \frac{2-0}{10:25-10:17} = \frac{2}{8} = 0.25 m/min$$

and between picture 2 and picture 3

$$speed_{p2-p3} = \frac{3-2}{10:44-10:25} = \frac{1}{19} m/min \approx 0.053 m/min$$

Figure 2.3: We take three snapshots of a triceratops movement with timestamps. This allows us to calculate an average speed between these times, but it is impossible from this information to get the exact speed at, say, 10 : 24.

What we don't necessarily know is the *exact speed* of the triceratops when picture 2 was taken; we might be able to estimate that the triceratops was travelling roughly $0.152 m/min$ by averaging the two speeds we can calculate, but this is just an estimate. How might we find the true value of speed in picture 2? In other words, the speed of the triceratops at *exactly* 10 : 25? We'd need more pictures over a smaller time period; if we had pictures of the position of the triceratops at, say 10 : 24 and 10 : 25 as in figure 2.4 we can calculate

$$speed_{P1-P2} = \frac{2-1.91}{10:25-10:24} \approx 0.09 m/min$$

and

$$speed_{P2-P3} = \frac{2.08 - 2}{10:26 - 10:25} = 0.08 ^{m}/_{min}$$

Figure 2.4: As our snapshots get closer and closer together, we get a better idea of what is happening exactly at 10 : 25.

As our measurements get closer and closer together, we get a better and better idea of what is happening at exactly 10 : 25, this set of pictures would tell us the speed of the triceratops at 10 : 24 is roughly $0.085 ^{m}/_{min}$ if we average the two speeds we can calculate. If all three pictures happened within 1 minute of each other, we would get an even better estimate. The big idea here is that if we push things close enough together, we can get a value for the **rate of change** of a function at a **single point**.

We can see how this happens by analysing the graph of a function and pushing x_2 towards x_1 and drawing a line between them[7], like in figure 2.5. As x_2 gets closer and closer to x_1, we see the lines approach a very specific line. This line is called the **tangent**.

[7]Because, remember, the average rate of change between two points is the *slope* of the line connecting the points.

Definition 2.1.1. The **tangent line** at $x = a$ is a line that touches a function $f(x)$ at the *single point* $x = a$. The slope of this line is called the **derivative** of $f(x)$ at a. It is the *instantaneous rate of change* of f at a.

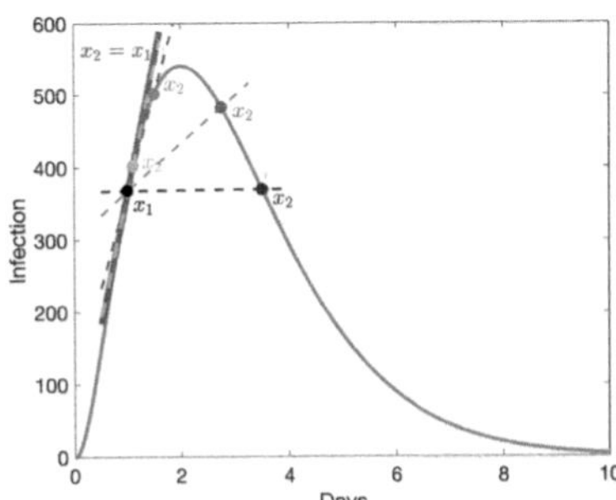

Figure 2.5: As x_2 approaches x_1 (or x_3 approaches x_4, the line approaches the tangent line at x_1 (or x_4). The slope of the tangent line is the derivative of the function at x_1 (or x_4).

Another way to think of this is that any function[8] is just a series of connected straight lines when we zoom in to imperceptibly small scales[9]. You can think of this almost as the opposite of the infinite. We treat *infinity* as something imperceptibly big; the biggest you can possibly thing, then bigger. The infinitesimal – or the scale on which functions are basically linear – is imperceptibly small. The slope of the tangent is how much our function value would change if we change x by an infinitesimally small amount, a value we will hence forth call ε[10].

The figure 2.6 shows, I hope, more clearly the difference between a tangent and not a tangent line. More specifically, notice the difference between the left-most red and violet lines. The red just barely touches the curve at a single point, where the purple touches more than one point in the local area[11]. The violet line (if you look closely) crosses the curve near the point the red line is tangent to.

---

[8]That satisfies some properties: the most important of which is that the instantaneous rate of change **exists** at the point.

[9]Remember, we are dealing with a and b being infinitesimally close together.

[10]So let it be written, so let it be done – Yul Brynner.

[11]*Local* is a math word that means "nearby." There are ways to make this more precise, but essentially if you draw a line that touches another point *near* the point of interest, then it is not a tangent.

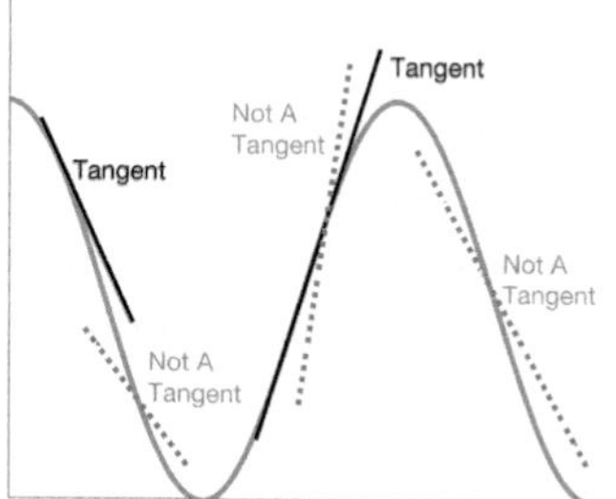

Figure 2.6: Sometimes the best way to see what a tangent (red lines) is to compare it to what isn't a tangent (violet lines).

2.2 APPROXIMATING DERIVATIVES OF FUNCTIONS

Let's say we have function values for two points a and b, the derivative at point a can be approximated by

$$f'(a) \approx \frac{f(b) - f(a)}{b - a}$$

The closer together a and b are, the better the approximation of the derivative. As a and b get closer together, the right-hand side gets closer to $^0/_0$, but as we saw in our triceratops example[12] we expect a *number* as b gets infinitely close to a. How do we make sense of this? If you've learned any math ever, you've probably learned that dividing by zero causes the universe to blow up. That's only true most of the time, but to figure out when it's not true we need a new tool called **the limit**.

2.3 LIMITS

Formally, the derivative is what we call the **limit** of the average rate of change. A limit is a value that we should logically expect, given things we already know.

When discussing limits, we will discuss *approaching* certain input values of functions usually without explicitly reaching the input value.

A very classical example of a limit is the dichotomy paradox of Zeno. In it, we are told of Atalanta, the fastest of all the Greek heroes. Let's say Atalanta wants to traverse the distance between Mount Parthenion and Calydon. In order to do so, she must first travel half the distance.

[12] Or when drawing lines that converge upon the tangent line.

From her point half-way between Mount Parthenion and Calydon, she must then traverse half of *that* distance. No matter where she is on her path towards Calydon, there is always a halfway point she must pass through. According to Zeno, she can never truly reach Calydon as she must always travel half the distance between her current position and Calydon and that will take some finite amount of time.

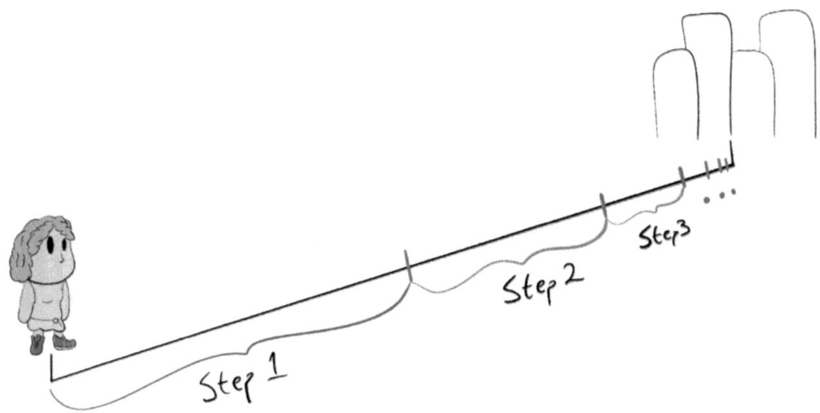

Figure 2.7: Even though, according to Zeno, Atalanta can never truly reach Calydon, she can get as close as she likes in a finite amount of time. For all intents and purposes, this is just as good as being there.

While there have been many proposed solutions to Zeno's dichotomy paradox, calculus helps us reframe the problem to something manageable. Yes, Atalanta needs to take infinitely many steps in order to get to Calydon but, since each step is necessarily smaller than the one before, as in figure 2.7 we can get *as close as we like* in a finite amount of time. The fact that we never truly reach the point in space called Sparta is irrelevant because we can get to a point that's good enough. In this sense, Calydon is **the limit** of Atalanta's journey. She may never reach it, but we can get as close to it as we like.

The straight line distance from Mount Parthenion to Calydon[13] was $\approx 200km$. By Zeno's claim we can demarcate the journey as After 20 steps, we have travelled $199.99981km$. If we need to get closer to Calydon, we can just take one more step. Still not close enough? Take another[14]

[13]Calydon doesn't actually exist anymore

[14]Very, very small

Step	Distance to Calydon
0	200 km
1	100 Atlantic Wholesalers km
2	50 Atlantic Wholesalers km
3	25 Atlantic Wholesalers km
4	12.5 Atlantic Wholesalers km
5	6.25 Atlantic Wholesalers km
6	3.125 Atlantic Wholesalers km
7	1.5625 Atlantic Wholesalers km
8	0.78125 Atlantic Wholesalers km
9	0.390625 Atlantic Wholesalers km
10	0.1953125

step. Limits give us a tool for dealing with things that we can never really reach, but since we can get arbitrarily close it *doesn't matter.*

Practically speaking, we know that we are able to travel and occupy a point a space. You can say *I'm going to walk over there* and you can walk to the exact spot you were referring to[15]. This also has an analogy in terms of limits, we can take limits towards values where things actually do exist, but the power of limits is in their ability to give value to things that would traditionally not have them.

Limits tell us that the difference between being *exactly somewhere* and arbitrarily close to somewhere is negligible.

Let's do a few mathematical examples. Consider the function

$$f(x) = x - 2.$$

We can ask the question, *what value do we expect the function to have as x approaches* 2? If we start plugging in values that get ever closer to 2, we start to see a pattern

$$\begin{aligned} f(1) &= -1 \\ f(1.5) &= -0.5 \\ f(1.75) &= -0.25 \\ f(1.9) &= -0.1 \\ f(1.9999999) &= -0.0000001 \end{aligned}$$

[15]It's this contradiction between Zeno's logical argument and reality that makes the dichotomy paradox a, well, paradox.

We can also approach from the other side and see the same[16] pattern

$$\begin{aligned} f(3) &= 1 \\ f(2.5) &= 0.5 \\ f(2.25) &= 0.25 \\ f(2.1) &= 0.1 \\ f(2.0000001) &= 0.0000001 \end{aligned}$$

No matter how we approach $x = 2$, the function always seems to approach 0. As our x values get closer to 2, we expect the function value to get closer to 0. We say that the limit of $f(x) = x - 2$ as x approaches 2 is 0. Mathematically we write

$$\lim_{x \to 2} x - 2 = 0$$

Now, when we verify this by checking $f(2) = 0$, we can say that $f(x)$ is **continuous**[17] **at** $x = 2$ since the limit matches the value of the function at that point.

Limits are all about *expectations*; what do we expect to happen when we get near a particular input value of a function? There are some tricks to setting expectations properly. For instance, we may ask

$$\lim_{x \to \infty} g(x) = \frac{x}{x + 2}$$

In this case, we don't have a number which we can check around since ∞ is a *concept* and not a *number*. That being said, we know that ∞ represents things that are unimaginably big, so we can see what happens when we put in big numbers.

$$\begin{aligned} g(100) &= 0.98039215686 \\ g(1000) &= 0.99800399201 \\ g(1000000) &= 0.999998 \\ g(1000000000) &= 0.999999998 \end{aligned}$$

shows us that as the value of x gets bigger and bigger, the value of $g(x)$ gets closer and closer to 1. Therefore, the limit as x approaches ∞ of $g(x)$ is 1, or

$$\lim_{x \to \infty} g(x) = 1$$

[16]But opposite

[17]You may have heard a definition of continuity as *any function we can draw without lifting our pen from the paper.*

Can we ever reach 1? No, in the same way that Atalanta can never reach Calydon. But we can get as close to 1 as we need to for our purposes. So close that we might as well be *at* 1.

Was there an easier way to get to this limit? For sure, if we know something about how big numbers work. When we are working with big numbers, especially numbers that are infinitely big, then eventually we get to the point where adding a number to a value approaching infinity is meaningless; $\infty + 1 = \infty$[18].

For our example, when we add 2 to 1000000000 we get 1000000002 which is *basically* still 1000000000. So as we approach infinity, we can say that

$$\lim_{x\to\infty} \frac{x}{x+2} \approx \frac{x}{x} = 1$$

and arrive at the same limit in a different way.

What if we had

$$g(x) = \frac{x}{x + 1000000000}$$

Then 1000000000 + 1000000000 = 2000000000 which isn't basically 1000000000 anymore. Does the trick still work? Yes! Because when we are heading towards infinity, there is *always* a bigger number; there is *always* a number so big that it will make all numbers before it seem inconsequential.

This is interesting because it allows us to define a limit for an entire class of functions all at once! You can pick *any* number $L > 0$ to make the function

$$g_L(x) = \frac{x}{x+L}$$

and

$$lim_{x\to\infty} g_L(x) = 1$$

every single time.

AGAIN, DO WE EVER ACTUALLY *reach* A VALUE OF 1? No, because we can never get to infinity[19]. We can get as close as we like if we pick x correctly. So, the *value* is never achievable, but we approach it gradually[20].

[18] In common parlance, it's like urinating in a lake.

[19] Infinity is more of an idea than a concrete number.

[20] Or quickly, depending on the function

We can look at a similar but contrasting problem as well. Let's think about the function

$$f_K(x) = \frac{K^x - 1}{x}$$

What happens if we start moving along the function towards $x = 0$? Clearly, if we just try to plug in $x = 0$, we will end up with a problem; we get

$$f_K(0) = \frac{K^0 - 1}{0} = \frac{0}{0}$$

and the whole universe blows up[21].

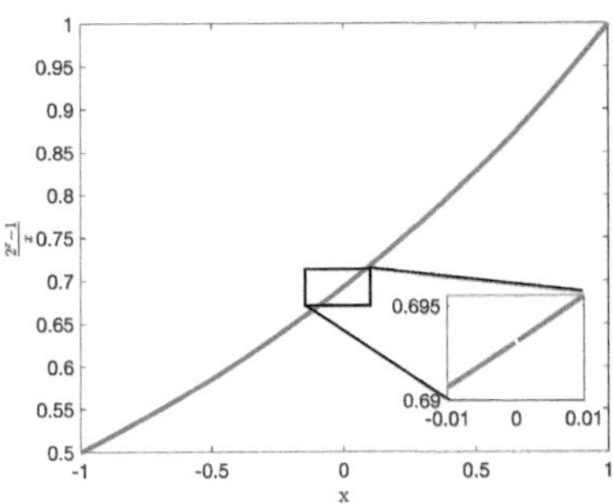

Figure 2.8: The function $f(x) = {}^{2^x-1}/_x$ doesn't actually exist at $x = 0$ exactly, as we can see by the inset, but we can get so close to $x = 0$ while the function still exists that we can effectively imagine what $f(0)$ should be in a perfect world.

So, right at zero we have a huge problem, but this is a very small problem[22] compared to *all the numbers everywhere.* In fact, if we plot this function using some fancy plotting software like in figure 2.8, if we don't plug in $x = 0$ exactly, it's almost impossible to see that there's a problem at all!

This is because the *limit* of f_K at $x = 0$ still exists. In fact,

$$\lim_{x \to 0} \frac{K^x - 1}{x} = 0$$

[21]Don't panic

[22]Despite the universe-ending thing

This means that we can pick a number as close as we like to 0, and there is an x that is close to $x = 0$ that will return that as a function value. We just have to be careful not to plug in $x = 0$ directly[23].

We can, however, make a function with no holes in it by redefining our function as a piecewise function

$$F_K(x) = \begin{cases} \dfrac{K^x - 1}{x} & \texttt{if } x \neq 0 \\ 1 & \texttt{if } x = 0 \end{cases}$$

Now we have an actual value at 0 and, more importantly, it matches the limit so that the function can be drawn without lifting our pen from the page.

2.4 LIMITS AND DERIVATES

GREAT, WHAT DOES THIS HAVE TO DO WITH TANGENT LINES AND DERIVATIVES AND CALCULUS? The limit is the basis of calculus. The whole basis of the derivative is bringing two points so close together that we get the fraction 0/0. While this doesn't exist in the technical sense, we definitely do *approach* a specific value, as we saw when we drew lines looking for the tangent line.

The formal definition of the derivative is

Definition 2.4.1. The derivative of a function $f(x)$ at x_1 is the limit of the average rate of change between x_1 and x_2 as x_2 gets closer to x_1. We write this as

$$f'(x_1) = \left.\frac{\mathrm{d}f}{\mathrm{d}x}\right|_{x=x_1} = \lim_{x_2 \to x_1} \frac{f(x_2) - f(x_1)}{x_2 - x_1}$$

We could also define the derivative at a point x_1 in terms of the distance, $h = x_2 - x_1$, between x_1 and x_2:

$$f'(x_1) = \left.\frac{\mathrm{d}f}{\mathrm{d}x}\right|_{x=x_1} = \lim_{h \to 0} \frac{f(x_1 + h) - f(x_1)}{h}$$

It's important to note here that the average rate of change at the point x_1 doesn't technically exist, but we know the value that we are

[23] As this will blow up the universe, again

arbitrarily close to. So we *define* the derivative at x_1 as the value of the limit much in the same way we *filled in* the little hole in $f_K(x)$ above with the value of the limit.

USING THIS AND OUR KNOWLEDGE OF LIMITS, WE CAN COMPUTE DERIVATIVES. For example, back to our triceratops example, the function used for the triceratops position, x, as a function of time, t, was

$$x(t) = \frac{t^2}{160}$$

where x is the triceratops's distance from the tree, and t is the time in minutes since $10:17$. So what is the derivative of $x(t)$ at $t = 8$?

$$\begin{aligned} x'(8) &= \lim_{h \to 0} \frac{x(8+h) - x(8)}{h} \\ &= \lim_{h \to 0} \frac{{}^{(8+h)^2}/_{160} - {}^{8^2}/_{160}}{h} \\ &= \lim_{h \to 0} \frac{64 + 16h + h^2 - 64}{160h} \qquad = \lim_{h \to 0} \frac{16h + h^2}{160h} \end{aligned}$$

From this point, we can cancel an h from the numerator and denominator

$$x'(8) = \lim_{h \to 0} \frac{16\not{h} + h^{\not{2}}}{160\not{h}} \qquad = \lim_{h \to 0} \frac{16 + h}{160}$$

Notice now that plugging in h directly into the formula doesn't cause any problems! We don't end up with any infinities or divisions by zero, so everything works out fine. This leaves us with

$$x'(8) = \lim_{h \to 0} \frac{16 + h}{160} = \frac{16}{160} = 0.1$$

So $x'(8) = 0.1$[24].

2.5 DERIVATIVE FORMULAS

LET'S SAY WE'RE GIVEN A FUNCTION, $f(x)$, as an equation. Do we really have to go through the trouble of plotting it, drawing a tangent line and then determining the slope of the tangent line in order to find the derivative at a point? Do we have to plug a single number into the

[24] Notice our estimate above got pretty close!

limit definition and determine the derivative that way for each point we need the rate of change at? Of course not, that would be silly.

We can determine formulas for a general x if we use the definition of the limit on a family of functions. For instance, what is the derivative of $f(x) = x^2$ for *any* x? Well, let's put it into our definition:

$$\begin{aligned} f'(x) &= \lim_{h \to 0} \frac{f(x+h) - f(x)}{h} \\ &= \lim_{h \to 0} \frac{(x+h)^2 - x^2}{h} \\ &= \lim h \to 0 \frac{x^2 + 2hx + h^2 - x^2}{h} \\ &= \lim h \to 0 \frac{\cancel{x^2} + 2hx + h^2 - \cancel{x^2}}{h} \\ &= \lim h \to 0 \frac{2\cancel{h}x + h^{\cancel{2}}}{\cancel{h}} \\ &= \lim h \to 0 2x + h \\ &= 2x \end{aligned}$$

So, if $f(X) = x^2$, then $f'(x) = 2x$ for *any* x. We could play this game with any function we please, although with some functions these games get more difficult to play.

IN MATHEMATICS, the idea is *not* to memorize anything intentionally[25]. The process, the reasoning, and the first principles tend to be the important parts of any mathematical pursuit because if you know those, then you need not memorize *any* formulas. Now, that being said, we are not prepared to derive these derivative formulas from first principles for all functions[26]. So, I will present to you the results, and tell you that the best way to let these sink in is to do practice problems.

You can see that if we plug $n = 2$ into the second rule, we have you see we get back the exact derivative formula we derived above, let this be evidence that you can trust these formulas.

The third row in our table answers the question from Chapter 1 of *who cares about e?* It is its own derivative! It's basically the free square in the middle of a bingo card[27]. By having exponentials with base e,

[25]We often memorize things just because we use them so often we can't help it.

[26]You could, if you're curious. It gets pretty involved and tricky for some functions.

[27]Or to use some references that will end up dated: the free battlepass in Fortnite or Call of Duty: Warzone, or that time in Squid Game when that woman got to skip a game because she didn't have a partner

Table 2.1: Table of derivatives for common functions.

$f(x)$	$\frac{\mathrm{d}f}{\mathrm{d}x} = f'(x)$
C	0
x^n	nx^{n-1}
e^x	e^x
e^{kx}	$k\mathrm{e}^{kx}$
$\sin(x)$	$\cos(x)$
$\cos(x)$	$-\sin(x)$
$\ln(x)$	$\frac{1}{x}$

Table 2.2: Using base e versus using base a.

$f(x)$	$\frac{\mathrm{d}f}{\mathrm{d}x} = f'(x)$
e^x	e^x
a^x	$\ln(a)a^x$

we can simplify a lot of formulas. Particularly when we're dealing with mathematical models and trying to analyse and interpret them, we want our formulas and equations to remain as simple as possible. We can compare the derivative of an exponential with base e to that with base a: we see that if we use base a, we have to now make sure to keep track of an extra $\ln(a)$. While not that much more complicated in this very simple case, once we start looking at more complex functions, the base e becomes much more convenient.

DERIVATIVES ALSO HAVE SOME OTHER SPECIAL PROPERTIES. If a function is the sum of two other functions, like

$$f(t) = g(t) + h(t)$$

Then the derivative is the sum of the individual derivatives!

$$f'(t) = \frac{\mathrm{d}f}{\mathrm{d}t} = \frac{\mathrm{d}g}{\mathrm{d}t} + \frac{\mathrm{d}h}{\mathrm{d}t} = g'(t) + h'(t) \tag{2.2}$$

IF A FUNCTION IS MULTIPLIED BY A CONSTANT, then the derivative is also multiplied by that constant[28]!

$$\frac{\mathrm{d}}{\mathrm{dx}}(c \cdot f(x)) = c\frac{\mathrm{d}f}{\mathrm{d}x} \tag{2.3}$$

THESE LAST TWO RULES ALLOW US TO FIND THE DERIVATIVE of many combinations of functions. For example, consider the function

$$f(x) = \sin(x) + 5x^2 + A\ln(x)$$

Equation (2.2) tells us we can take the derivative of each term separately and then add them all back together.

From our table directly, we see

$$\frac{\mathrm{d}}{\mathrm{d}x}(\sin(x)) = \cos(x)$$

and noticing that $5x^2$ is the second entry in our table multiplied by a constant, we use equation (2.3) and our table to get

$$\frac{\mathrm{d}}{\mathrm{d}x}\left(5x^2\right) = 5\frac{\mathrm{d}}{\mathrm{d}x}\left(x^2\right) = 5(2x^1) = 10x$$

The same holds for the last term, even though A is a letter and not a number, when we say we are taking *taking the derivative of a function* and that function is one–dimensional, it is assumed that we are taking the derivative *with respect to the input* unless otherwise stated. Since A is not an input[29], we treat it as a constant.

$$\frac{\mathrm{d}}{\mathrm{d}x}(A\ln(x)) = A\frac{\mathrm{d}}{\mathrm{d}x}(\ln(x)) = A \cdot \frac{1}{x} = \frac{A}{x}$$

Then we add everything back up in the right way!

$$\frac{\mathrm{d}f}{\mathrm{d}x} = \cos(x) + 10x + \frac{A}{x}$$

AS ANOTHER EXAMPLE, let's take the derivative of the function

$$f(x) = \mathrm{e}^R - bx$$

[28]Too many exclamation marks? I feel like they're not used enough in math texts.

[29]In case it needs clarifying: our input is x.

Again, this is a one-dimensional function, so it is assumed we are taking the derivative *with respect to the input*[30]. If we are looking at the function with x as the variable, that means it is assumed we know[31] R. Which means e^R is just a number (i.e. a constant). The derivative of a constant is zero, therefore, we get

$$\frac{df}{dx} = f'(x) = -b$$

THIS IS NOT TO SAY THAT THE FUNCTION ABOVE COULDN'T BE A FUNCTION OF R. Just by changing our notation slightly, we can go from

$$f(x) = e^R - bx$$

to

$$f(R) = e^R - bx$$

Given the overall context, one way to interpret the derivative of $f(R)$ is *how sensitive is our model to changes in* R? In other words, if R is a parameter that we measure, and we make a slight error in it, will the error in our predictions be small, or large? Of course, this interpretation holds for x as well. *If we make small changes in* x*, how large are our changes in* $f(x)$[32]? Once again, we see that context matters. The notation for functions serves many purposes and they all relate back to differentiating our variable from our parameters. When x is the input, we assume x is variable and R and b are parameters. If R is our input, we assume x is a parameter instead and R is variable.

2.6 THE PRODUCT RULE

We've seen that we can multiply functions by a constant number, and that we can add functions and we are still able to take a derivative. It is natural to ask then, what happens in the case that we *multiply* two elementary functions.

As an example, consider the function:

$$f(x) = Rxe^{-kx}$$

[30]In this case, x

[31]Or *could* know

[32]This notion of sensitivity is one of the defining characteristics of mathematical chaos.

Naively, we might think to do this:

$$\frac{\mathrm{d}f}{\mathrm{d}x} = \frac{\mathrm{d}}{\mathrm{d}x}(Rx)\frac{\mathrm{d}}{\mathrm{d}x}\left(\mathrm{e}^{-kx}\right) = -Rk\mathrm{e}^{-kx}$$

but this would be **WRONG**. Notice that if we take the derivative this way, then the derivative can never be zero. If we look at the graph in figure 2.9, we can see that there is clearly a point where the derivative is zero.

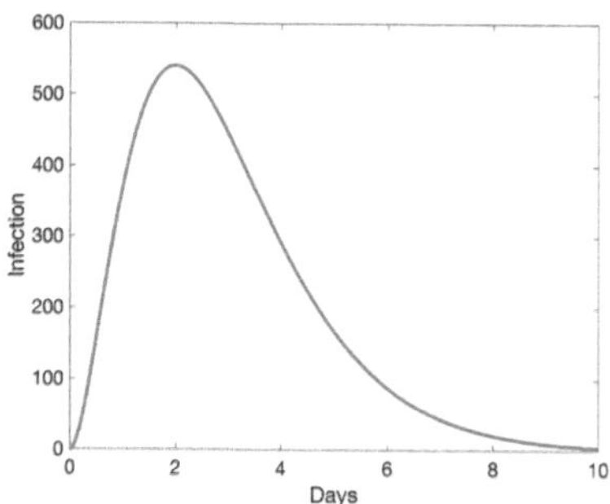

Figure 2.9: A typical infection within a population can be modelled by $f(x) = Rx\mathrm{e}^{-kx}$

This means we need a different strategy for taking derivatives of products. Without going into detail, we can state the rule which you can then use at your leisure.

Definition 2.6.1. Let $f(x)$ and $g(x)$ be two function and $h(x) = f(x)g(x)$. Then

$$\frac{\mathrm{d}h}{\mathrm{d}x} = \frac{\mathrm{d}}{\mathrm{d}x}(f(x)g(x)) = \frac{\mathrm{d}f}{\mathrm{d}x}g(x) + f(x)\frac{\mathrm{d}g}{\mathrm{d}x} = f'(x)g(x) + f(x)g'(x)$$

This is typically known as the **product rule** for derivatives.

The hardest part of the product rule is identifying *when* to use it and how to split if a function up. Generally, we try to split functions until each sub-function can be dealt with using table 2.1.

For our example function above:

$$f(x) = Rx\mathrm{e}^{-kx}$$

We notice that this can be split into the product of two functions

$$f(x) = Rx\mathrm{e}^{-kx}$$

We should give each of these pieces a name, let's call them

$$g(x) = Rx$$
$$h(x) = \mathrm{e}^{-kx}$$

Which are simple enough that we can use table 2.1 to calculate the derivatives:

$$g'(x) = \frac{\mathrm{d}g}{\mathrm{d}x} = R$$
$$h'(x) = \frac{\mathrm{d}h}{\mathrm{d}x} = -k\mathrm{e}^{-kx}$$

Then we can use the product rule:

$$\begin{aligned}\frac{\mathrm{d}f}{\mathrm{d}x} = \frac{\mathrm{d}}{\mathrm{d}x}\left(g(x)h(x)\right) &= g'(x)h(x) + g(x)h'(x)\\ &= R\mathrm{e}^{-kx} + Rx\left(-k\mathrm{e}^{-kx}\right)\\ &= R\mathrm{e}^{-kx}\left(1 - kx\right)\end{aligned}$$

THE PRODUCT RULE ALSO WORKS RECURSIVELY. If we have multiple functions multiplying each other, we use continued applications of the product rule until we have simple, elementary functions. As an example, let's consider the function

$$K(t) = t^2 \sin(t) \cos(t)$$

This can be split into two pieces:

$$K(t) = t^2 \sin(t)\cos(t),$$

which we can call

$$f(t) = t^2 \sin(t)$$
$$g(t) = \cos(t).$$

When we attempt to take the derivative of $f(t)$, we see that it itself is a product of two functions and thus must be split,

$$f(t) = t^2\sin(t)$$

and these pieces must be given names.

$$x(t) = t^2$$
$$y(t) = \sin(t)$$

Then we can compute

$$\frac{\mathrm{d}f}{\mathrm{d}t} = 2t\sin(t) + t^2\cos(t)$$

by the product rule, and

$$\frac{\mathrm{d}g}{\mathrm{d}t} = -\sin(t).$$

Finally,

$$\begin{aligned}\frac{\mathrm{d}K}{\mathrm{d}t} &= \frac{\mathrm{d}f}{\mathrm{d}t}g(t) + f(t)\frac{\mathrm{d}g}{\mathrm{d}t} \\ &= 2t\sin(t) + t^2\cos(t) + t^2\sin(t)\left(-\sin(t)\right) \\ &= t^2\cos(t) + 2t\sin(t) - t^2\sin^2(t).\end{aligned}$$

2.7 THE CHAIN RULE

ADDING AND MULTIPLYING FUNCTIONS ARE NOT THE ONLY WAYS we can combine elementary functions to model more complex dynamics. Sometimes we have **composite** functions, and composite functions have their own derivative rule called the **chain rule**.

AGAIN, WE WILL PROCEED BY EXAMPLE. Consider the function[33]

$$f(x) = \mathrm{e}^{-x^2}$$

We know[34] how to take the derivative of e^{-x} and we know how to take the derivative of x^2, but when we combine them in this way, the result clearly does not look like either of the elementary functions involved, like in figure 2.10.

It helps us to write $f(x)$ as a composition of the functions that we know, namely

$$g(y) = \mathrm{e}^{-y}$$
$$y(x) = x^2$$

[33]This is the infamous bell curve that most students believe they should be graded against. In reality, most grade distributions follow a bimodal bell curve.

[34]In theory

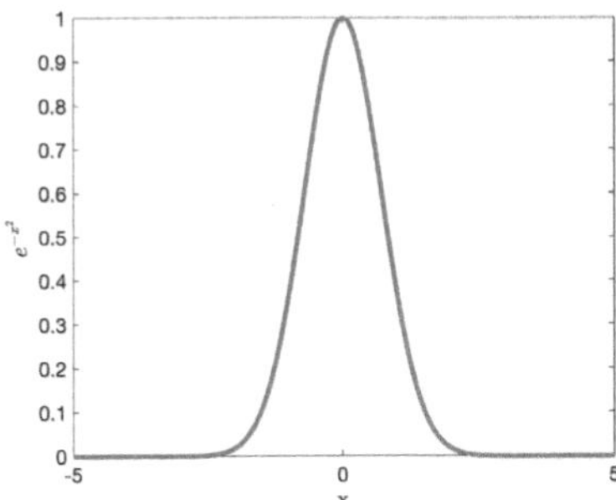

Figure 2.10: The canonical bell curve function, also known as a Gaussian or normal.

Then we can say that

$$f(x) = g(y(x))$$

or in plain words f *is equal to* g *with* y *as an input.*

WHEN WE HAVE FUNCTIONS THAT ARE COMPOSITIONS OF SMALLER FUNCTIONS LIKE THIS, we employ the chain rule to take derivatives.

Definition 2.7.1. Let $f(x) = g(y(x))$ then

$$\frac{\mathrm{d}f}{\mathrm{d}x} = \frac{\mathrm{d}}{\mathrm{d}x}\left(g(y(x))\right) = \frac{\mathrm{d}g}{\mathrm{d}y}\frac{\mathrm{d}y}{\mathrm{d}x}$$

This is the **chain rule** of differentiation.

For our example, this leads to

$$\begin{aligned}\frac{\mathrm{d}}{\mathrm{d}x}\left(\mathrm{e}^{-x^2}\right) &= \frac{\mathrm{d}}{\mathrm{d}y}\left(\mathrm{e}^{-y}\right)\frac{\mathrm{d}}{\mathrm{d}x}\left(x^2\right) \\ &= -\mathrm{e}^{-y}\left(2x\right)\end{aligned}$$

but our function is only supposed to be a function of x! Now we have these y's floating around! Remember that we *defined* $y = x^2$ above. So we need to now replace all our y's with the appropriate function of x that we defined above. This leads to

$$\begin{aligned}\frac{\mathrm{d}f}{\mathrm{d}x} &= -\mathrm{e}^{-y}(2x)\\ &= -2x\mathrm{e}^{-x^2}\end{aligned}$$

These compositions are not always easy to see, and it will take practice to be able to employ the chain rule fluidly. Practice, as with many things in life, is the only way to get better.

Let's look at a more complicated example.

$$\theta(L) = r_{max}\left(1 + \left(\frac{K}{L}\right)^n\right)^{-1}$$

This is the **Hill equation** from biochemistry. It is typically used to fit dose–response relationships where L is a dose of a chemical or drug and θ is the response.

For instance, when we eat sugary foods our bodies will produce a compound called insulin to remove sugar from our bloodstream. The concentration of insulin, L, is related to the *rate* at which glucose[35] is removed from the bloodstream (also called the glucose uptake rate), θ, through a Hill equation.

A question we could ask[36] is, *when is the glucose uptake rate changing the fastest?* In other words, when does increasing the concentration of insulin have the most effect?

To answer this, we might look for when we see the greatest change in the glucose uptake rate with respect to insulin concentration; i.e. we take the derivative of $\theta(L)$. This function has a lot of nested pieces, and it may be hard to see the answer from just looking at the problem, but that's no reason to give up. We should appeal to our fundamentals and work through the problem slowly.

First, we see that most of our function is contained within a power of -1. We know how to take the derivative of powers, but only when the power is x^n, not necessarily stuff^n. So let's give the inside of this power a new name

$$\theta(x(L)) = r_{max}\,(x(L))^{-1}$$

[35] i.e. sugar

[36] And answer with a derivative, because that's the one tool we have

where

$$x(L) = 1 + \left(\frac{K}{L}\right)^n$$

Can we take the derivative of this piece? Almost. We know that 1 is a constant, and its derivative is 0 so we needn't worry about the 1. The second term looks almost like L^n, for which we could take a derivative, but we have the pesky problem of $1/L^n$. Luckily, we can appeal to one of our exponent rules to fix this: $1/L^n = L^{-n}$. Therefore,

$$x(L) = 1 + \left(K^n L^{-n}\right)$$

Remembering that the derivative of a sum is the sum of derivatives we get

$$\frac{\mathrm{d}}{\mathrm{d}L}(1) = 0$$

and since **multiplicative constants do not affect derivatives** we have

$$\frac{\mathrm{d}}{\mathrm{d}L}\left(K^n L^{-n}\right) = K^n \frac{\mathrm{d}}{\mathrm{d}L}\left(L^{-n}\right) = -nK^n L^{-n-1}$$

Therefore

$$\frac{\mathrm{d}x}{\mathrm{d}L} = 0 - nK^n L^{-n-1}$$

That takes care of the inside! What about the outside? Remember we have

$$\begin{aligned} \theta(x) &= r_{max} x^{-1} \\ \frac{\mathrm{d}\theta}{\mathrm{d}x} &= -r_{max} x^{-2} \end{aligned}$$

So by the chain rule we have

$$\begin{aligned} \frac{\mathrm{d}\theta}{\mathrm{d}L} &= \frac{\mathrm{d}\theta}{\mathrm{d}x}\frac{\mathrm{d}x}{\mathrm{d}L} \\ &= \left(-r_{max} x^{-2}\right)\left(-nK^n L^{-n-1}\right) \\ &= r_{max} n K^n \frac{1}{L^{n+1}\left(1 + K^n L^{-n}\right)^2} \end{aligned}$$

This derivative function, $\theta'(L)$, tells us how much our glucose rate increases for 1 unit increase in insulin. We can plug in some numbers for our parameters so that we can get a more concrete idea of interpretations. For instance, if we were dairy cows [7], we could parameterize our model with the following values

$$\begin{aligned} r_{max} &= 18.51 \\ K &= 76.41 \\ n &= 1 \end{aligned}$$

and the rate at which our glucose response is changing is then

$$\frac{\mathrm{d}\theta}{\mathrm{d}L} = 18.51 \cdot (76.41) \frac{1}{L^2 \left(1 + {}^{76.41}\!/_{L}\right)^2}$$

We can plug in different values for L to determine the rate at which our glucose uptake response is changing.

$$\begin{aligned} \theta'(0.1) &= 0.248 \\ \theta'(1) &= 0.236 \\ \theta'(10) &= 0.189 \\ \theta'(100) &= 0.0453 \\ \theta'(1000) &= 0.0012 \end{aligned}$$

We can see just from these four values that our glucose response to insulin changes fastest when we are near $L = 0$[37]. As insulin concentration gets higher and higher, we see less and less of a change in glucose uptake. This is evident when we plot our function $\theta(L)$ as in figure 2.11.

Why is $n = 1$? This has to do with the fact that one insulin molecule binding to one glucose molecule does *not* affect the ability of other insulin molecules from binding to other glucose molecules. The individual processes are independent of one another. In cases like haemoglobin in your red blood cells, picking up some oxygen makes them better at picking up *more* oxygen[38]. In this case, when binding is cooperative,

[37] In fact, while $L = 0$ is undefined, the limit exists. What is it?

[38] This is extremely; simplified; for more information visit your local biochemistry department.

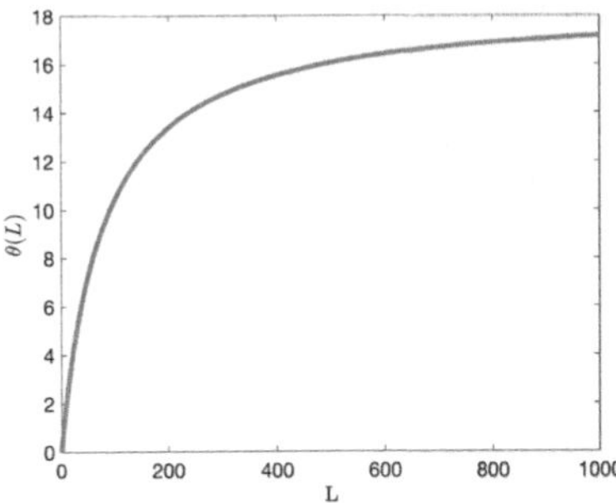

Figure 2.11: Response curve of the rate of glucose removal from the bloodstream as a function of insulin in the bloodstream. While it is indeed true that more insulin means more glucose removal, higher concentrations of insulin are less efficient at increasing the rate of insulin removal than low levels of insulin.

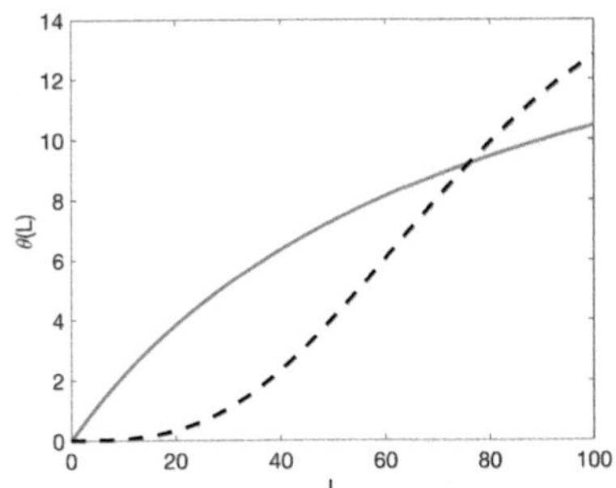

Figure 2.12: The difference between $n = 1$ and $n = 3$ in our function $\theta(L)$. The solid line is $n = 1$ and the dashed line is $n = 3$. We can see that cooperative binding is slower to get started but eventually outpaces independent binding.

$n > 1$. In fact, in his original paper, Hill estimated $n = 3.5$, you can see the difference between independent and cooperative binding in figures 2.11 and 2.12, respectively. A higher n leads to a ramping up of binding. If we look at the derivative at different points with $n = 3.5$[39], we see qualitatively different behaviour.

[39] We will keep the same model parameters to highlight the comparison even though our independent process was insulin in cows and our cooperative process is oxygen in human blood.

$$\theta'(L) = 18.51\frac{3.5(76.41)^{3.5}}{L^{4.5}\left(1+K^{3.5}L^{-3.5}\right)^2}$$
$$\theta'(0.1) = 5.25 \times 10^{-8}$$
$$\theta'(1) = 1.66 \times 10^{-5}$$
$$\theta'(10) = 5.22 \times 10^{-3}$$
$$\theta'(100) = 0.131$$
$$\theta'(1000) = 7.99 \times 10^{-6}$$

In this case, we see that there is specific concentration of oxygen in the blood when the rate at which oxygen is attaching to haemoglobin is changing the fastest. The value is when L is near 100. We don't have the tools yet to figure out *exactly* when this is changing the fastest; we will get there, but we need to be sure of ourselves when it comes to working with derivatives first.

2.8 MIXING RULES

The world wouldn't be nearly as interesting as it is if we only even encountered one derivative rule at a time. Knowing how and when to use each rule is a cornerstone for analysing complex functions and models and extracting relevant information from the analysis. The thing to keep in mind is that everything is hard until you have practiced it enough. Enough is different for everyone, and that's ok. You can't rush brilliance; but here are some base rules that might help it along:

- Identify any products of functions.
- Identify any composite[40] functions.
- Isolate any term in a sum and work on that.
- Each individual derivative you take should be from the table above[41].
- Put everything back according to the sum/chain/product rules.

These guiding principles aren't meant to be perfectly understood just upon reading. In fact, writing them down was nigh impossible because it's me trying to translate what happens in my head when taking derivatives

[40] i.e. nested

[41] At least as far as the scope of this book is concerned

Table 2.3: A table of the most commonly used derivative rules.

Function, $f(x)$	Derivative, $f'(x)$
C	0
x^n	nx^{n-1}
$\sin(x)$	$\cos(x)$
e^{kx}	ke^{kx}
$\ln(x)$	$1/x$
$g(x)h(x)$	$g'(x)h(x) + g(x)h'(x)$
$g(h(x))$	$g'(h(x))h'(x)$

into words. It's a system that was developed from years and years of practice taking derivatives and developed into a sort of intuition. There was[42] a lot of struggle to build said intuition, I'm hoping with some guiding principles you can avoid many of the missteps and struggles.

Let's work through an example that uses multiple rules all at once. Consider the function

$$P(t) = \frac{KP_0e^{rt}}{K + P_0\left(e^{rt} - 1\right)}$$

This is called a **logistic function**. It models the size of a population that starts with P_0 members, a maximum reproduction rate, r, and is limited in its growth by a resource K. For instance, the population of a classroom is physically limited by the number of seats in the room, usually. If you decide to overfill the room, you are still limited by the fact that every living, breathing human being takes up some physical space in the world which cannot be usurped by another[43]. There is still a limit to the number of individuals you can fit in a room. The parameter K is the limiting resource either way and is open to interpretation. It could represent food required to sustain a population, physical space as we've already mentioned, or something more abstract like social structures and hierarchy.

If we measure time, t, in years then the derivative of $P(t)$ gives us the growth rate of the population at a given time. In order to find this, we should use our principles above to try and simplify the derivative problem.

First, we will rewrite the function so that it looks like a product:

$$P(t) = KP_0e^{rt}\left(K + P_0\left(e^{rt} - 1\right)\right)^{-1}$$

[42] Also, is

[43] Or, I guess, it also applies to non-living, non-breathing humans as well.

then we should identify the two factors in our product and give them names

$$P(t) = f(t)g(t)$$

where

$$\begin{aligned} f(t) &= KP_0\mathrm{e}^{rt} \\ g(t) &= \left(K + P_0\left(\mathrm{e}^{rt} - 1\right)\right)^{-1} \end{aligned}$$

In doing this, we know that

$$P'(t) = f'(t)g(t) + f(t)g'(t)$$

So we only need to find the derivatives of $f(t)$ and $g(t)$. The first derivative is easy as it fits one of our tabulated rules in table 2.1. We note that K and P_0 are constant with respect to time and are multiplying a function of t; this means that they also multiply the derivative.

$$f'(t) = \frac{\mathrm{d}}{\mathrm{d}t}KP_0\mathrm{e}^{rt} = KP_0\frac{\mathrm{d}}{\mathrm{d}t}\mathrm{e}^{rt} = KP_0r\mathrm{e}^{rt}$$

The second derivative is a little more complicated as it needs the chain rule since we have something more than just t in our power. In this case, the inner function can be written as

$$y(t) = K + P_0\left(\mathrm{e}^{rt} - 1\right)$$

and the outer function is

$$h(y) = y^{-1}$$

and then

$$g(t) = h(y(t))$$

and we can use the chain rule to determine

$$\frac{\mathrm{d}g}{\mathrm{d}t} = \frac{\mathrm{d}h}{\mathrm{d}y}\frac{\mathrm{d}y}{\mathrm{d}t}.$$

Since $h(y)$ and $y(t)$ are simple enough, we can use table 2.1 to get

$$h'(y) = -y^{-2}.$$

For $y(t)$, we can expand the bracket to get $y(t) = K - P_0 + P_0 e^{rt}$. We do this because then it's very clear that we can ignore the K and the $-P_0$ since they are constant terms and have no t terms multiplying them. What's left is just an exponential!

$$y'(t) = P_0 r e^{rt}$$

Now that we have all our pieces, we need to put them back together in the same way that we took them apart. First we need to put $h'(y)$ and $y'(t)$ together to get $g'(t)$

$$g'(t) = -y^{-2} P_0 r e^{rt}$$

WAIT! g is only a function of t; what's with this y? We need to get rid of it using our previously made definition for $y(t)$.

$$g'(t) = -P_0 r e^{rt} \left(K + P_0 \left(e^{rt} - 1\right)\right)^{-2}$$

The whole reason we defined f and g to begin with was to use the product rule, so now we fill in f, f', g, and g' in our equation for $P'(t)$

$$P'(t) = K P_0 r e^{rt} \left(\left(K + P_0 \left(e^{rt} - 1\right)\right)^{-1}\right) - K P_0^2 r e^{2rt} \left(K + P_0 \left(e^{rt} - 1\right)\right)^{-2}$$

and we're done!

The quotient rule

The product rule and chain rule are sufficient to handle taking derivatives for any way we can combine elementary functions, but we can also create a rule for dividing two functions. It comes from one application of the chain rule and one application of the product rule[44].

Note that any division can be written as a product with a negative power:

$$F(x) = \frac{f(x)}{g(x)} = f(x)(g(x))^{-1}$$

Now, this is just the product rule, but $g(x)$ is embedded inside another function; let's give it another name

$$F(x) = f(x)h(g)$$

[44] And was used in the previous example in a way

where $h(g) = g^{-1}$. We will apply the product rule and chain rule.

$$\frac{\mathrm{d}F}{\mathrm{d}x} = \frac{\mathrm{d}f}{\mathrm{d}x}h(g) + f(x)\frac{\mathrm{d}h}{\mathrm{d}g}\frac{\mathrm{d}g}{\mathrm{d}x}$$

We can't say much about $f(x)$ and $g(x)$ since we haven't defined these functions, but we can say something about $h(g)$ since we know the form of it.

$$diffhg = \frac{\mathrm{d}}{\mathrm{d}g}(g)^{-1} = -(g)^{-2}$$

Plugging this back in to our derivative for $F(x)$, we get

$$\frac{\mathrm{d}F}{\mathrm{d}x} = f'(x)\left(g(x)\right)^{-1} - f(x)g'(x)\left(g(x)\right)^{-2}$$

or, putting everything over a common denominator,

$$F'(x) = \frac{\mathrm{d}}{\mathrm{d}x}\left(\frac{f(x)}{g(x)}\right) = \frac{f'(x)g(x) - f(x)g'(x)}{g(x)^2}$$

This is the **quotient rule.**

2.9 CRITICAL VALUES

Maxima and Minima

These are called **local maxima/minima** because they are higher or lower than function values that are nearby. A derivative of zero does not guarantee that it is the largest or smallest function value across a large domain. This is because the derivative is effectively a *local* measurement and can only carry information about *local* behaviour. Figure 2.13 shows this phenomena. Even though the second maximum is truly the highest point, the derivative test for a maximum or minimum identifies three points.

LET'S WORK THROUGH AND EXAMPLE OF FINDING LOCAL MAXIMA/MINIMA. Consider our favourite function from pharmacokinetics

$$C(t) = t^n \mathrm{e}^{-\frac{t}{m}}$$

where $n > 0$ and $m > 0$.

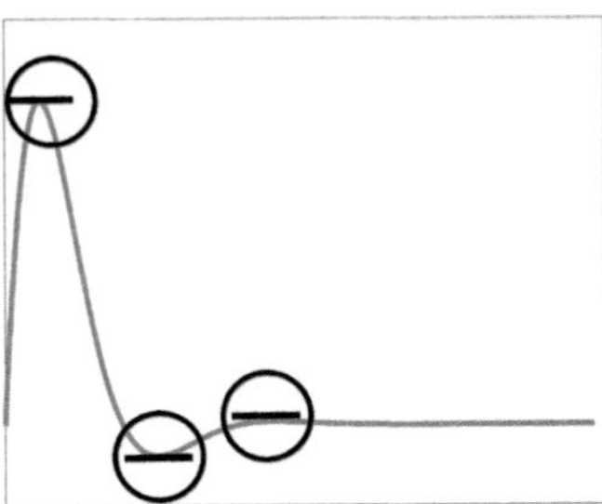

Figure 2.13: The derivative is a local quantity and thus using it to detect maxima and minima can only detect whether or not a point is the largest/smallest of the points nearby.

We may want to know when our drug concentration is highest. So first we will take a derivative

$$C'(t) = nt^{n-1}e^{-\frac{t}{m}} - \frac{t^n}{m}e^{-\frac{t}{m}}$$

If we are looking for a critical point[45], we want to set this expression equal to zero.

$$C'(t) = nt^{n-1}e^{-\frac{t}{m}} - \frac{t^n}{m}e^{-\frac{t}{m}} = 0$$

If we then solve for t, we can determine the time at which drug concentration is either maximized or minimized.

$$\begin{aligned} nt^{n-1}e^{-\frac{t}{m}} - \frac{t^n}{m}e^{-\frac{t}{m}} &= 0 \\ t^{n-1}e^{-\frac{t}{m}}\left(n - \frac{t}{m}\right) &= 0 \end{aligned}$$

Since e^{kx} is *never* zero, we see that the only possible solutions are

$$\begin{aligned} t_1 &= 0 \\ t_2 &= nm \end{aligned}$$

THE NEXT QUESTION WE SHOULD ASK is, *which one is the maximum and which is the minimum?*

[45] Instead of writing max/min everywhere, we often call points where the derivative is zero **critical points**.

We have all the information we need to figure this out. Let's use figure 2.13 as a guide. If we look to the *left* of a maximum, we see that the function is increasing, and to the *right* of a maximum the function is decreasing. This is true for all local maxima[46]. With that in mind, let's check a point that is close to our critical value, but a little to the left. This means we want to set $t = nm - \varepsilon$.

NB: the symbol $\varepsilon > 0$ is often used in math to represent *a little bit*[47]. How little is a little bit? At little as we need, but always greater than zero.

The derivative evaluated at this point gives us

$$\begin{aligned} C'(nm - \varepsilon) &= (nm - \varepsilon)^{n-1} \mathrm{e}^{-\frac{nm-\varepsilon}{m}} \left(n - \frac{nm - \varepsilon}{m}\right) \\ &= (nm - \varepsilon)^{n-1} \, \mathrm{e}^{-\frac{nm-\varepsilon}{m}} \left(\frac{\varepsilon}{m}\right) \end{aligned}$$

From here, we look at the individual terms and see if we can say anything useful about this quantity.

Since ε is just a little bit, i.e. a very small positive number, we can say for certain that if $nm > 0$ then $nm - \varepsilon > 0$[48], as well $\mathrm{e}^{\text{anything}}$ is always positive, and finally if $\varepsilon > 0$ and $m > 0$ then $\dfrac{\varepsilon}{m} > 0$. If we multiply three positive numbers, the product also must be positive. So we can say for certain that

$$C'(nm - \varepsilon) > 0$$

So the function $C(t)$ is *increasing* to the *left* of the critical point.

Now, we have to check a little bit over to the *right* of the critical point. We do this by looking at the critical point plus a little bit more. In other words, we set $t = nm + \varepsilon$.

$$C'(nm + \varepsilon) = (nm + \varepsilon)^{n-1} \, \mathrm{e}^{-\frac{nm+\varepsilon}{m}} \left(-\frac{\varepsilon}{m}\right)$$

Again, the first two terms are positive, but this time the third term is negative. If I multiply two positive numbers and a negative number, the result is negative. Therefore,

$$C'(nm + \varepsilon) < 0$$

---

[46] If you think about this for a little bit, you may come to find that we couldn't even call it a local maximum if this wasn't the case.

[47] As with everything though, context is important. We only have so many symbols, so sometimes ε means other things!

[48] If it's not, we just make ε an even smaller number.

which means our function is decreasing to the *right* of the critical point. Therefore, we can conclude it is a maximum.

Concavity and the Second Derivative

You may notice that local maxima always have this shape ∩ and local minima are always ∪ shaped[49]. The difference between these two shapes is the **concavity** of the function. **Maxima** are **concave down** and *minima* are *concave up*.

GRAPHICALLY THERE IS A FAIRLY STRAIGHT-FORWARD WAY TO DETERMINE A SECTION A FUNCTION. If we draw a straight line between two points on a curve and then draw an arrow from the curve to the straight line, if the arrow points up then our function is concave up. If our arrow points down, then our function is concave down, see figure 2.14.

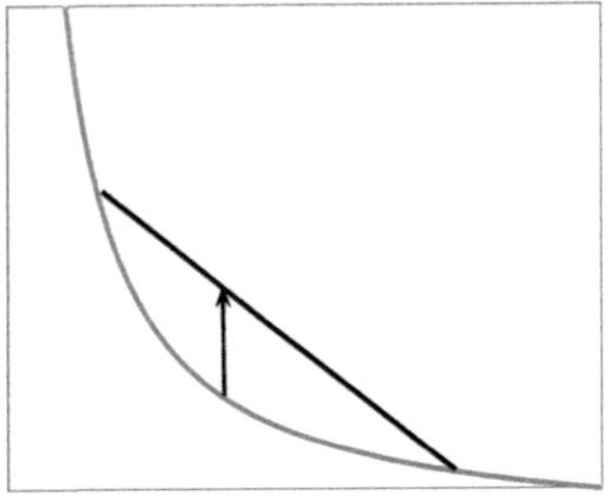

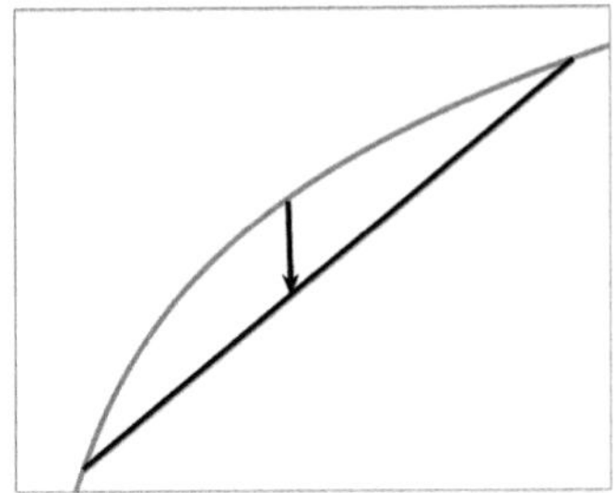

Figure 2.14: The left function is concave up; the right is concave down.

WE CAN ALSO RELATE CONCAVITY TO THE SECOND DERIVATIVE OF A FUNCTION. For a function $f(x)$, if $f''(x) > 0$ then the function is concave up, and if $f''(x) < 0$ the function is concave down.

How can we relate this back to our study of maxima and minima? Well, quite simply: if $f(x)$ has a critical point at $x = p$ so that $f'(p) = 0$ then:

- If $f''(p) < 0$, then $f(p)$ is a **maximum**.
- If $f''(p) > 0$, then $f(p)$ is a **minimum**.

[49] You may ask, could a point be a local maximum if it isn't ∩ shaped? No, no it cannot.

We can use this rule to check our *other* critical point in our drug concentration example, $t = 0$. If we take a second derivative of our function we get.

$$C''(t) = t^{n-2}e^{-\frac{t}{m}}\left(n(n-1) - 2n\frac{t}{m} + t^2\right)$$

if we plug in $t = nm$ we see that

$$C''(nm) = -(nm)^{n-2}e^{-n} < 0$$

which confirms that $t = nm$ is a maximum. If we plug in $t = 0$, we see that

$$C''(0) = 0$$

IF YOU'RE OBSERVANT, YOU MAY HAVE NOTICED we specifically left out the case when $f''(p) = 0$. When this happens, we could be at a maximum, or a minimum, or what is called an **inflection point**.

An inflection point may have $f'(p) = 0$ but additionally has $f''(p) = 0$ **and** is neither a max or a min. An inflection point is a point in a function where the concavity changes.

The simplest example of an inflection point is $f(x) = x^3$. If we look for any critical points, we will find

$$f'(x) = 3x^2 = 0$$

which means that the only possible critical point is $x = 0$. If we plot this function, as in figure 2.15, we see that $x = 0$ is neither a max nor a min. Instead, what is happening is that the concavity of the function is

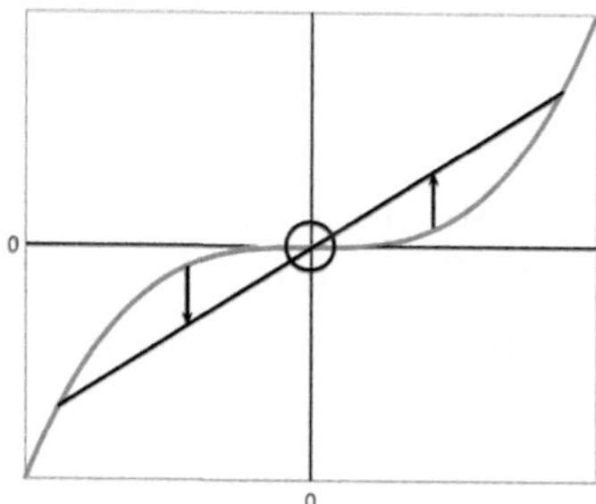

Figure 2.15: The point $x = 0$ on this graph looks like it would be a max or min, but instead is an inflection point marking a change in concavity.

changing at $x = 0$. We must always take care that our critical point does indeed represent a minimum or maximum and not an inflection point[50].

Following up with our example is $C(0)$ an inflection point? The answer is *it depends.* Remember, an inflection point means concavity changes at the critical point. In this instance, we have to check the concavity on either side[51] of the critical point.

In doing so, you might see that if n is even, then there is a minimum at $t = 0$ since the concavity is always positive. If n is odd then we get a negative concavity to the left of $t = 0$ and a positive concavity to the right of $t = 0$. This concavity change means that for odd n, $t = 0$ is an inflection point.

2.10 CONSTRAINED OPTIMIZATION

Our ultimate goal and why we've sprinted through this treatment of derivatives is ultimately to generalize many dimensions and really develop a full[52] picture of differential calculus.

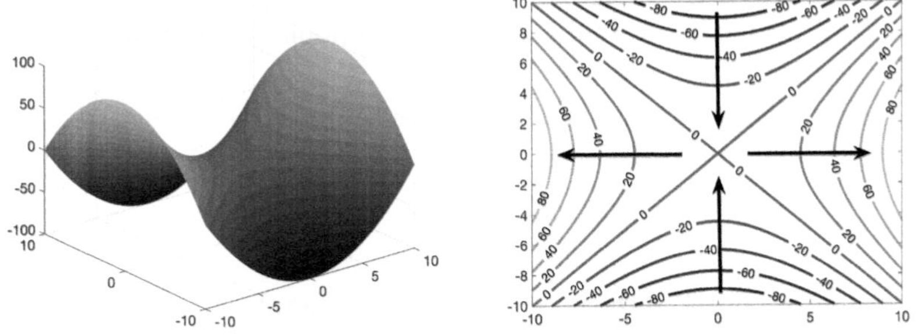

Figure 2.16: This surface has a point at the origin that looks like a maximum in some directions, and a minimum in other directions. The notion of critical points gets muddied.

IN THREE-DIMENSIONS, THE IDEA OF AN OPTIMAL VALUE TAKES ON SOME NEW COMPLEXITIES. For instance, let's look at the function plotted in figure 2.16. In some directions it's increasing, in other directions it's decreasing. If we look in the x-direction, then the origin looks like a minimum. If we look in the y-direction, the origin looks like a maximum.

[50]Unless an inflection point is what we are looking for

[51]A little bit to the left and a little bit to the right

[52]But ultimately shallow

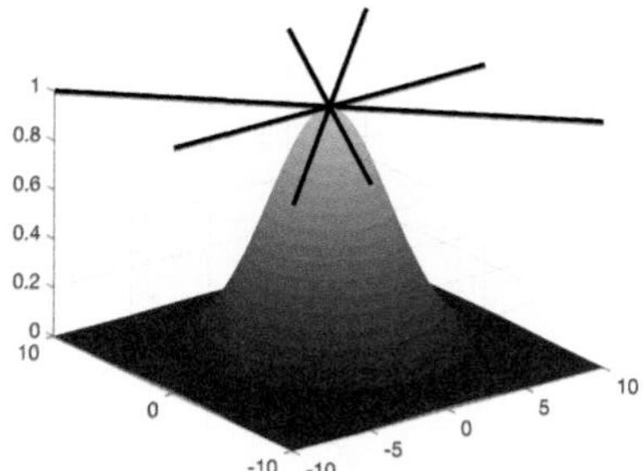

Figure 2.17: The notion of a tangent line gets a little ambiguous in higher dimensions. A single point has infinitely many tangent lines.

In fact, even in cases where there is a clear maximum, the notion of a tangent line becomes a little bit ambiguous. In figure 2.17, we see there are many possible lines that all satisfy the definition of the tangent line. We drew these at the peak for simplicity, but every point on the surface falls victim to the same ambiguous notion of a tangent line.

Since the idea of a derivative gets more complicated in higher dimensions, the idea of optimization also gets more complicated. For instance, given the function

$$A(l, w) = lw$$

which you should recognize as the formula for the area of a rectangle, is there an optimal length and width?

In general, the answer is no, which we can see from the surface in figure 2.18. But what if we limit ourselves to a perimeter of length P? In this case, we have a second function

$$p(l, w) = 2w + 2l = P$$

where P is just a number. We can rearrange this to get

$$w = \frac{P - 2l}{2}$$

Now, w is a function of l; i.e. $w = w(l)$.

The contour plot, figure 2.19, shows the contours of $A(w, l)$, along with the perimeter line, $w(l)$. If we set the perimeter, then we are looking for the maximum value of area *along the perimeter function.* The maximum area, if it exists, is given by the contour that is tangent to the constraint equation $w(l)$. In figure 2.19, the constraint equation is given in grey,

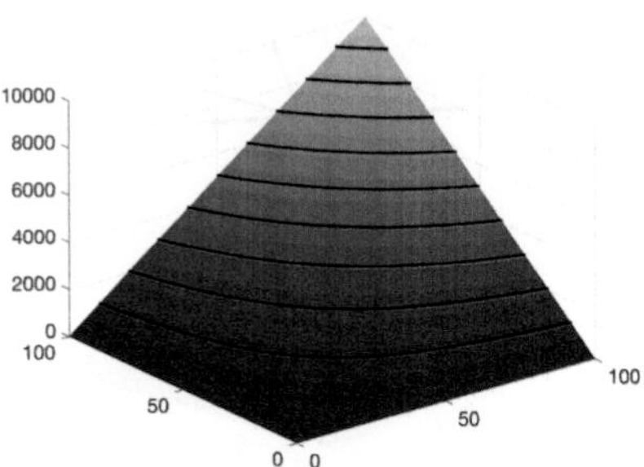

Figure 2.18: A surface of the area equation for a rectangle. There is clearly no maximum area; if we continue to increase width, w or length l, our area will continue to increase.

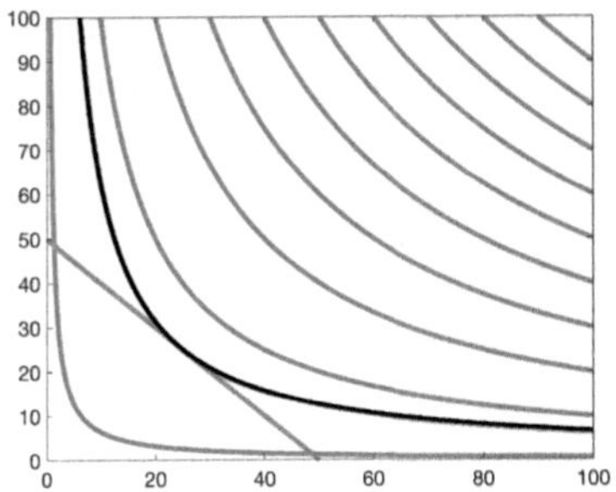

Figure 2.19: The constraint equation, $w(l)$, is given in grey, the contours of $A(w, l)$ in blue, and the contour that is tangent to the constraint equation is given by the black contour. This black contour is the maximum area given the perimeter constraint. The point (w^*, l^*) where the contour and the constraint equation meet are the optimal length and width that will maximize area for a given perimeter.

the contours in blue, and the contour that is tangent to the constraint equation is given by the black contour.

WE CAN DEAL WITH THIS ALGEBRAICALLY INSTEAD, and in some cases this is simpler than trying to determine things graphically. If we replace the variable w in $A(l, w)$ with the function $w(l)$ from our constraint equation, we get

$$A(l, w(l)) = l\left(\frac{P - 2l}{2}\right) = A(l)$$

The function we wish to maximize is now once again in one-dimension. We know how to find the maximum in one-dimension, so we can carry on as we would normally: we take a derivative, set the derivative equal to zero and solve for our input variable.

$$\frac{\mathrm{d}A}{\mathrm{d}l} = \frac{P}{2} - 2l = 0 \implies l = \frac{P}{4}$$

To find the optimal width, we then plug to optimal length into our constraint equation

$$w\left(\frac{P}{4}\right) = \frac{P}{2} - \frac{P}{4} = \frac{P}{4}$$

So we maximize area when $l = w = {}^{P}\!/_{4}$.

IN THE REAL WORLD, THIS HAS A PRETTY SIGNIFICANT INTERPRETATION: a square has the largest area of any rectangle for a given perimeter. Figure 2.20 shows several rectangles of the same perimeter. The square will always have the largest area.

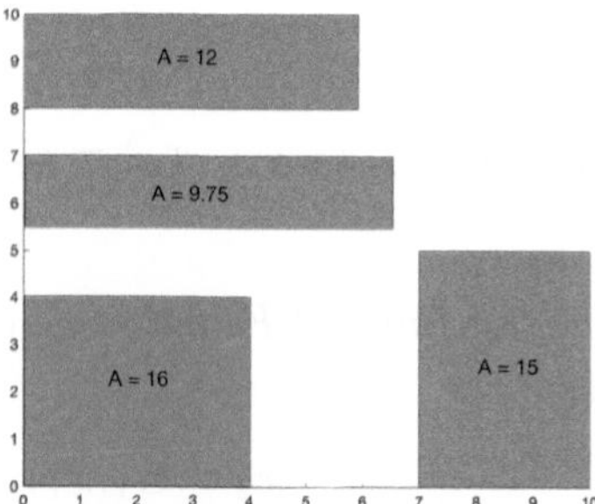

Figure 2.20: All of these rectangles have perimeter 16*, but the square has the largest area.*

This means that if we want to maximize, say, a grazing area for cows using the least amount of fencing we should build a square yard. Using the fencing to make anything other than a square will result in a smaller grazing area.

We'll do another example shortly, because this constrained optimization stuff always seems to be a sticking point, but I want to briefly have a high-level discussion about this.

AS WE'VE ALREADY DISCUSSED AND HOPEFULLY AGREED UPON, taking derivatives in higher dimensions needs some refining of our notion of a derivative (as we saw in figure 2.17 and because of this, our notion of optimization suffers (like in figure 2.19). We can be saved by the fact that sometimes[53] we don't just want to maximize or minimize some function on its own, but we have to find the best possible function value that satisfies a separate condition.

Constrained optimization is exactly that: it is optimizing some function while keeping in mind, and adhering to certain constraints. If we have one n-dimensional function to optimize, and $n-1$ constraints, then we can use the above technique to find optimal inputs to maximize or minimize our function.

To put this in mathematical language, if we would like to optimize

$$f(x_1, x_2, x_3, \cdots, x_n)$$

subject to the constraints

$$\begin{aligned} g_1(x_1, x_2, x_3, \cdots, x_n) &= c_1 \\ g_2(x_1, x_2, x_3, \cdots, x_n) &= c_2 \\ g_3(x_1, x_2, x_3, \cdots, x_n) &= c_3 \\ &\vdots \\ g_{n-1}(x_1, x_2, x_3, \cdots, x_n) &= c_{n-1} \end{aligned}$$

Using a thing called the *Implicit Function Theorem*[54], we can rewrite the set of conditions as

$$\begin{aligned} x_2 &= G_1(x_1) \\ x_3 &= G_2(x_1) \\ &\vdots \\ x_n &= G_{n-1}(x_1) \end{aligned}$$

and plugging these all back into the function we would like to optimize gives us

$$\begin{aligned} f(x_1, x_2, x_3, \cdots, x_n) &= f(x_1, G_1(x_1), G_2(x_1), \cdots, G_{n-1}(x_1)) \\ &= f(x_1) \end{aligned}$$

[53]Often times, really, if we're talking about the real world

[54]Don't worry about this; it's just a thing you can throw back to in case you take more math courses.

Since we have a function of one variable, we can find an optimal value of x_1 that will maximize/minimize f by setting

$$\frac{\mathrm{d}f}{\mathrm{d}x_1} = 0$$

Then, we may use our G functions to find the optimal values for x_2 to x_n. This ensures that we satisfy all constraints and maximize/minimize f simultaneously.

A soup can

A can of Campbell's condensed chicken noodle soup holds $318mL$ of soup. Assuming we need to use a cylindrical can, what should the dimensions of the can be to minimize the material used?

In this case, we want to minimize the surface area of a can, subject to a volume constraint. We should always start by identifying and writing out the function we wish to minimize or maximize:

$$SA(r,h) = 2\pi r^2 + 2\pi rh$$

Next, we should identify our constraint, and confirm it is a function of the same variables.

$$V(r,h) = \pi r^2 h = 318$$

We can solve our constraint equation for one of our variables. In this case, it so happens to be easier to solve for h, so we'll do that:

$$h(r) = \frac{318}{\pi r^2}$$

Plugging this into our equation for surface area, we get

$$SA(r,h(r)) = 2\pi r^2 + \frac{636}{r} = SA(r)$$

which we can then minimize!

$$\begin{aligned} SA'(r) = 4\pi r - \frac{636}{r^2} &= 0 \\ \frac{4\pi r^3 - 636}{r^2} &= 0 \end{aligned}$$

$$r = \left(\frac{636}{4\pi}\right)^{1/3} \approx 3.69\text{cm}$$

An actual can of Campbell's condensed chicken noodle soup has a radius of $3.41cm$. Why isn't it perfectly optimized?

We need to remember there are many more things that go into manufacturing than just optimizing material, perhaps we are limited by our machinery or we also need to account for thickness of material or some property of the liquid inside that further constrains our optimization[55]. The fact that we got *this* close suggests that minimizing material was indeed top priority.

IT'S IMPORTANT TO REMEMBER that most of these very simple constrained optimization problems usually won't[56] feel very "worldly." We are trying to reduce every problem to a one-dimensional problem, which only works in the absolute simplest of cases. The important thing here is not the precise applications but the ideas that can be used to solve aspects of bigger, more complex problems. We'll get closer to this in good time, but I felt like this was important to point out.

Just to hammer it home, if we wanted to package soup in a box instead of a cylindrical can, and we had volume as a constraint, and wanted to minimize surface area just the same, we already have a two-dimensional problem:

$$V(h, l, w) = lwh = 284$$

$$SA(h, l, w) = 2lw + 2hw + 2hl$$

We would need a second constraint, because we have *three* input variables.

2.11 ELASTICITY

A derivative tells us the rate of change of a function per unit change in the input, but sometimes we would like to know the *percent change* of a function per *percent change* in the input. We do this through a function called **elasticity**.

[55] For condensed soups, the amount of water that needs to be added is exactly one can; this isn't a coincidence, but rather by design.

[56] And shouldn't

Definition 2.11.1. Given a function $f(x)$, the **elasticity** of $f(x)$ is given by

$$E = \left| \frac{x}{f(x)} \frac{\mathrm{d}f}{\mathrm{d}x} \right|$$

and tells us the percent change in $f(x)$ that we expect to occur if we change x by 1%.

Often times, we use elasticity not necessarily to measure relative changes in our variables, but how relative changes in our *parameters* may affect our mathematical models.

For instance, take the function

$$P(t) = \frac{K}{1 + (K-1)\mathrm{e}^{-rKt}}$$

which models the growth of a population in an environment with limited resources. The parameter K is the number of individuals the environment can carry, and r is the natural growth rate of the population if given sufficient resources. Usually K and r are measured or estimated from measurements and are thus prone to error. Elasticity can help us determine which parameters in a model we should measure most carefully.

For our example, we can determine the elasticity of parameter K if we *recontextualize* our function as a function of K instead of t

$$P(K) = \frac{K}{1 + (K-1)\mathrm{e}^{-rKt}}$$

then elasticity is

$$E_K = \frac{K}{P(K)} \frac{\mathrm{d}P}{\mathrm{d}K}$$

The equation itself will not tell us much, but we can sketch a sample graph of the elasticity and determine what we might expect to happen. The elasticity is shown in figure 2.21. We see that for small times t, a 1% error in the value of K is negligible, producing $< 1\%$ error in the output of the function. There is a maximum, however, where things are at their worst, and as t runs for a very long time, we see that our error persists in proportion to the error in K. Therefore, if K is in error of $p\%$ then the results of the function are off by the same percentage.

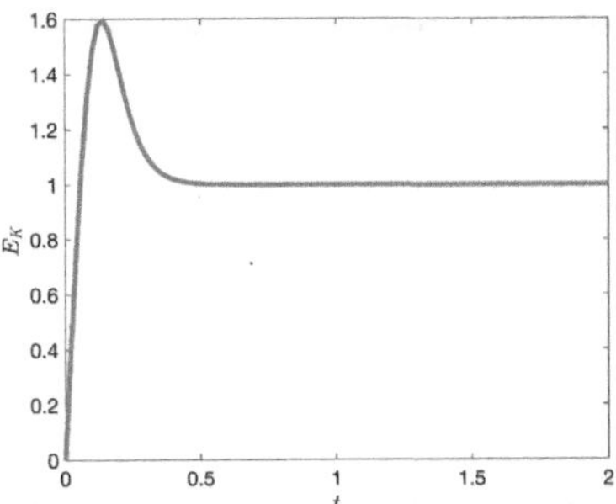

Figure 2.21: The elasticity of K in our population model. We see that as we run our model over time, a 1% error in the value of K persists as a 1% error in the value returned by the function.

We can do the same sort of analysis for the growth rate r by recontextualizing our function into a function of r.

$$P(r) = \frac{K}{1 + (K - 1)\mathrm{e}^{-rKt}}$$

then elasticity is

$$E_r = \frac{r}{P(r)} \frac{\mathrm{d}P}{\mathrm{d}r}$$

Again, we can sketch an example curve by plugging in values for r and K to see what the elasticity tells us about our model. We see in figure 2.22 that while a 1% error in the value of r will produce errors of roughly 1% for intermediate values of t, if we are interested in the long-term behaviour

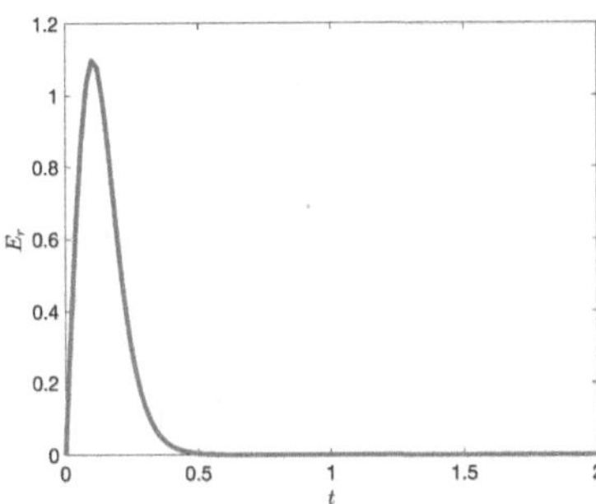

Figure 2.22: The elasticity of r in our population model. We see that as we run our model over time, a 1% error in the value of r is washed out.

of our population, then the growth rate in effectively unimportant. Any error in r will be washed out over the long term.

We can conclude from this that if we needed to allocate limited resources into measuring K or r – in this particular case – our efforts would be better spent making sure K is as accurate as possible.

2.12 PARTIAL DERIVATIVES

Finally, we can *start* to discuss what it means to take a derivative in two dimensions. Let's consider the contour plot of a function $f(x, y)$ shown in figure 2.19. When we were doing constrained optimization, we saw that if we overlaid another curve (our constraint equation), $y = y(x)$, on top of the contour diagram, we could reduce our problem to a one-dimensional problem. In other words, a constraint equation coupled with our two dimensional function, $f(x, y)$, gave us $f(x, y) = f(x, y(x)) = f(x)$. So maybe, we should ask if there is some way to generalize this?

What if we were to make an arbitrary, simple "constraint" equation? The simplest one possible?[57] Let's use the constraint equation

$$y(x) = C$$

What does this do for us? Well, it allows us to make the following substitution

$$f(x, y) = f(x, C) = f(x)$$

since C is a constant and not variable. Along this line, I could, with enough contours plot $x, f(x)$ and get the function in 2.23. Then, taking a derivative is extremely easy.

What did we actually do here? By setting the "constraint"[58] to a horizontal line, I can estimate the rate of change in the x-direction.

Similarly, I could impose a slightly different "constraint":

$$x = x(y) = K$$

[57]This is kind of a situation where my knowledge of other things is motivating this, and so it might seem like it's coming out of nowhere. In some sense, that's exactly what's happening given your limited, highly curated exposure so far. But bear with me. I hope I'll be able to give you a big picture eventually.

[58]This word has now twice appeared in quotations because it's a self-imposed constraint like my keto diet, or my commitment to keeping this PG.

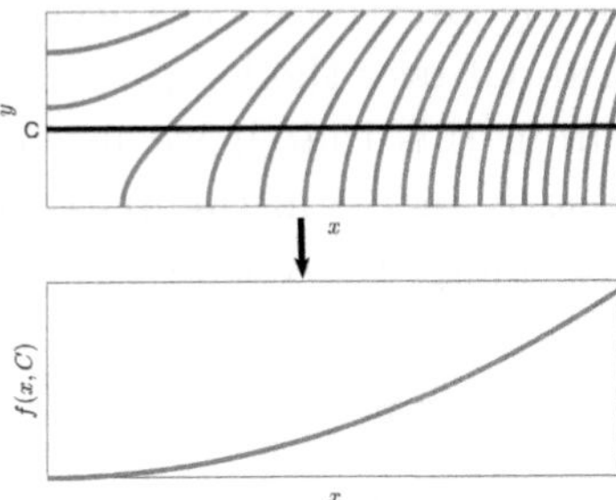

Figure 2.23: By fixing $y = C$, we are able to treat the function as a function of one variable, for which we can easily calculate a derivative.

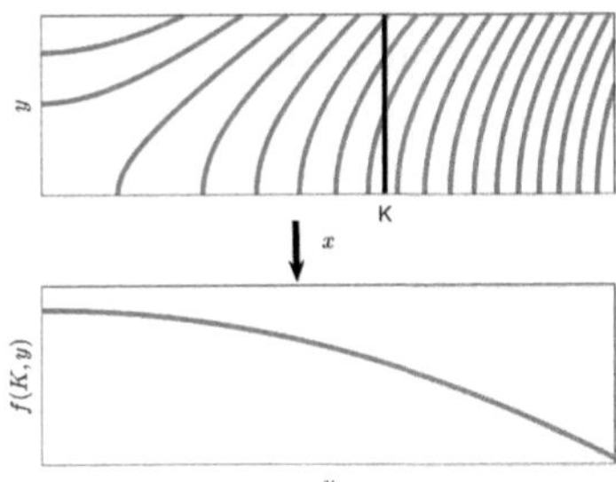

Figure 2.24: By fixing $x = K$, we are able to treat the function as a function of one variable, for which we can easily calculate a derivative.

which would allow me to estimate my rate of change in the y-direction. When plugging this into $f(x, y)$ I get

$$f(x, y) = f(K, y) = f(y)$$

The hurdle here, I think, is answering the question *is there any significance in calling these constants C or K?*

The answer is no, and so why call them anything at all? Why not just keep in our heads that if we would like to know the rate of change in the x-direction, that is the same as pretending y is a constant and not a variable. Similarly, if we would like to know the rate of change in the y-direction, we can just pretend for a moment that x is constant without actually making any substitutions. After all, what is the difference between y and C or x and K conceptually? They are just strokes on a page used to represent a direction.

In order to formalize this notion of "taking the derivative in a direction" we use a similar but different notation when taking derivatives of

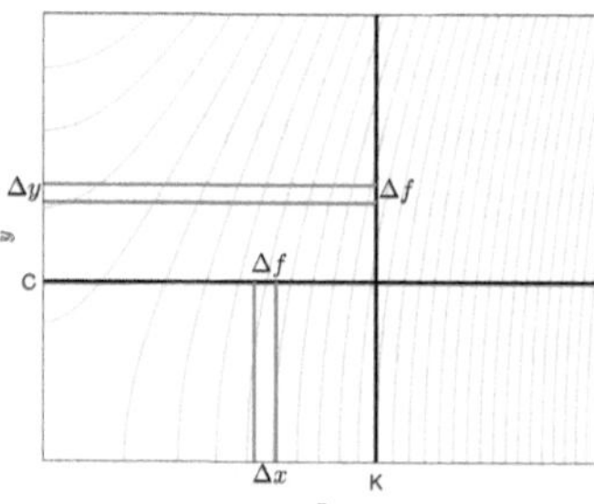

Figure 2.25: A summary of how we find derivatives in the x and y directions respectively using a contour plot. We fix one of the inputs and then measure the change in output with respect to the input that we didn't fix. If our contour lines are close enough together, we get the derivative in that direction.

multivariate functions. These derivatives in particular variables are called **partial derivatives**.

Definition 2.12.1. The **partial derivative in x** of a two-dimensional function, $f(x, y)$, is written as

$$\frac{\partial f}{\partial x}$$

and interpreted as *the rate of change of $f(x, y)$ in the x-direction if y is held constant.* Similarly, the **partial derivative in y** of a two-dimensional function, $f(x, y)$, is written as

$$\frac{\partial f}{\partial y}$$

and interpreted as *the rate of change of $f(x, y)$ in the y-direction if x is held constant.*

While the symbol is slightly different, that has more to do with the *context* rather than the mechanics involved. Taking a partial derivative is the same process and follows the same rules are taking a normal derivative.

As an example, let's consider the function

$$f(x, y) = x^2 + y^3 + xy^2$$

If we would like to know the rate of change in the x-direction, we will treat y as if it's constant and take a derivative normally.

$$\frac{\partial f}{\partial x} = 2x + 0 + y$$

The first term comes about because the derivative of x^2 is $2x$ as we expect, the derivative of y^3 *with respect to* x is 0 in the second term because we assume y is a constant. In the third term, we have xy^2. We again assume y^2 is constant and take the derivative like so:

$$\frac{\partial}{\partial x}(xy^2) = y^2 \frac{\partial}{\partial x}(x) = y^2 \cdot 1 = y^2$$

If we wish to know the rate of change in the y-direction, we do the same thing, but this time hold x constant while we treat y as a variable.

$$\frac{\partial f}{\partial y} = 0 + 3y^2 + 2xy$$

Once again, x^2 from the first term is considered constant when taking a partial derivative with respect to y and so its derivative is 0. The second term is made up exclusively of y and so we take the derivative normally. In the third term, x is treated as a constant multiplying a function of y and so it will also multiply the derivative.

PRACTICE PROBLEMS

Limits

1. What is the limit of

$$f(x) = \frac{x^2 - 4}{x - 2}$$

as $x \to 2$.

2. Does the limit

$$\lim_{x \to 0} \frac{x^2 + 1}{x^3}$$

exist? If so, what is it? If not, why not?

3. Does the limit

$$\lim_{x \to \infty} \frac{x^2 + 1}{x^3}$$

exist? If so, what is it? If not, why not?

Basic Derivatives

1. The distance travelled by a car can be modelled by the function

$$x(t) = \sqrt{t}$$

what function then models the car's speed at time t?

2. As $t \to \infty$, what speed does the car approach?

3. Can you explain why the limit as $t \to \infty$ of distance travelled and speed seem to be contradictory?[59]

4. A population of rabbits grows as

$$R(t) = A\mathrm{e}^{bt}$$

where t is measured in years. How many new rabbits are born in year 3?

5. What is the function that describes how many new rabbits are born per year?

[59] It's not; this is just more weirdness that happens at ∞.

6. What is the function that describes how the birth rate changes?

7. Consider the function

$$D(x) = D_0 x^{-2}$$

   This function models the density of a population, people per square kilometre, as a function of distance, x, from a coast. How quickly does the density decrease with distance from the coast?

8. At what distance is the density changing by less than 1 $^{\text{person}}/_{\text{km}^2}$?

Product Rule & Chain Rule

1. Each of the below functions is of the form $f(g(x)$. Determine what $f(x)$ and $g(x)$ are.

   (a) $e^{\sqrt{x}}$
   (b) $\sin(x^2)$
   (c) $3e^{x^2/3}$
   (d) $\ln\left(4/x^2\right)$

2. Combine the following functions into compound functions $f(g(x))$ or $f(g(h(x)))$.

   (a) $f(x) = 2x$, $g(x) = 4x$
   (b) $f(x) = x^2$, $g(x) = x/4$, $h(x) = \ln(x)$
   (c) $f(x) = sin(x)$, $g(x) = x^2$
   (d) $f(x) = \ln(x)$, $g(x) = e^x$, $h(x) = e^x$

3. Take the derivatives of all the functions in questions 1 and 2.

4. It's obviously not a great idea to establish a settlement directly on the coast. Most populations are slightly off the coast to protect against flooding, tidal erosion and other problems. A better function to model population density as an output and distance from the coast as an input would be

$$D(x) = D_0 x^2 e^{-kx}.$$

   By plugging in different values of x, find when the population density is growing fastest and when it is decreasing fastest.

5. Most population growth is limited by resources. We can model a population size over time as

$$P(t) = \frac{1000}{1 + e^{-1.1(t-10)}}$$

Find the derivative to give the rate of population growth

6. Can you rewrite the derivative as a function of P instead of t?

7. How fast is the population growing at $t = 0$?

8. Plot the function

$$S(t) = t^{1.3} \sin\left(2\pi t/12\right) + t^2 + 10$$

If S is sales and t is in dollars, why might the function oscillate like it does?

9. Take the derivative of $S(t)$. Which term might account for the company growth? Which for the seasonal variations?

10. How many more sales should we expect in year 6 from year 5, on average without seasonal variations?

11. Write out an equation that you *could* solve to find optimal p_0^*. Simplify your expression by finding a common denominator.

12. Kleiber's law[24] states that the metabolic rate, μ, of an animal is roughly equal to its mass to the power 0.75. In other words,

$$\mu(M) = kM^{3/4}$$

We also know that the optimal flight speed, v, of a bird is dependent on mass[2]

$$v(M) = nM^{3/25}$$

Write a function $v(\mu)$ that relates speed to metabolic rate.

13. What metabolic rate minimizes flight speed?

14. Is there a maximum? Why might this not be realistic?

Optimization

1. One family of functions used to model drug concentration is

$$C(t) = t^n \exp(-t^m)$$

At what time t is drug concentration at a maximum?

2. A population in a resource-limited environment can be modelled by

$$P(t) = \frac{Kr - \mu}{(Kr - \mu - r)e^{-(rK-\mu)t} + r}$$

where $P(t)$ is the population size, r is the reproduction rate of the species (i.e. how many babies are made per year), μ is the death rate per year and K is the capacity of the environment.

   (a) Calculate the growth rate of the population, $\dfrac{dP}{dt}$

   (b) When is the **growth rate** the largest?

3. Let's assume we have $n = 3$ subspecies in a population. Then the Shannon Index of diversity is given by

$$D(p_0, p_1, p_2) = -(p_0 \ln(p_0) + p_1 \ln(p_1) + p_2 \ln(p_2))$$

And more over, let's assume that p_2 is a mutation arising from the offspring of $p1$ so that

$$p_2 - \alpha p_1 = 0$$

   (a) Turn $D(p_0, p_1, p_2)$ into a function of one variable: $D(p_0)$

   (b) What values of p_0, p_1 and p_2 maximize or minimize diversity?

CHAPTER 3

Linear Algebra

As we've seen in the previous chapter, a function can take many inputs and produce an output. A natural question might be to ask if we can take multiple inputs and generate multiple outputs. Well, yes, we can by defining multiple functions. Is there a way to work on groups of functions together? If they are related to one another, is there hidden information within the relationships? If the functions are linear, then we have an entire branch of mathematics dedicated to dealing with many linear functions all at once. This branch is called **linear algebra**, and it is arguably the most important branch of mathematics in the modern world.

Linear Algebra is the study of linear relationships. When we want to learn something about multiple outputs that are all related to the same inputs, but in different ways, we use linear algebra[1]. Linear algebra is extremely powerful and finds its way into almost every branch of science. Here, we will cover the basics but I can whole-heartedly recommend a course in linear algebra for any budding scientist.

3.1 VECTORS

The standard definition of a vector that you'll find in most physics courses is

[1]So long as (as the name suggests) the relationships are linear

DOI: 10.1201/9781003265405-3

Definition 3.1.1. A **vector** is a quantity with both magnitude and direction. It is either written as

$$\vec{v}$$

or

$$\mathbf{v}$$

This means that quantities like "40 km north" and "2 floors down" are vector quantities.

IN MATH, WE LIKE TO QUANTIFY THINGS, and directions like "north," "up," "down," and "60° to the left" are not very quantitative. The directions as given are a little bit vague. We can ask things like *up from where?* or 60° *to the left of what?* It helps to have a frame of reference, something we may all agree upon and something we are all familiar with. Turns out this was codified long ago by Descartes. We call it the Cartesian coordinate system[2].

Using Cartesian coordinates, we can codify vectors as a collection of numbers where the order of the numbers determines how many steps you take in a direction *starting from the origin.* In this sense, we have an agreed upon a starting point, the point $(0, 0)$ in $2D$ and $(0, 0, 0)$ in $3D$ and have completely quantified the magnitude and direction.

Definition 3.1.2. A **vector** is a collection of quantities that, when plotted in Cartesian space, result in an arrow starting from the origin and point to (x, y, z) in three-dimensions or (x, y) in two dimensions. We represent the vector with the notation

$$\mathbf{v} = \vec{v} = [x, y, z]^T = \begin{bmatrix} x \\ y \\ z \end{bmatrix}$$

---

[2]Why is it not the Descartesian coordinate system? Did the Baroque cultural movement favour alliteration? It probably had more to do with Europe's love of Latin and Descartes Latinized name being Cartesius.

In other words, we take x steps in the x-direction, y steps in the y direction and, if need be, z steps in the z-direction. The resulting straight line path from the origin to where we end up is the vector $[x, y, z,]^T$.

The ^{T} means **transpose**. When a vector is written horizontally, it is called a **row vector**; when written vertically, it is called a **column vector**. The operation ^{T} changes a row vector into a column vector and vice versa. This is going to be important when we start doing things with vectors.

OF COURSE, THERE IS NO REASON TO LIMIT OURSELVES TO DIMENSIONS WE CAN SEE. We can write vectors in 4, 5, 6 dimensions, or any dimension we choose. A vector in n-dimensions is usually written as

$$\vec{v} = [x_1, x_2, x_3, \cdots, x_{n-1}, x_n]^T$$

We take x_1 steps in the x_1-direction, x_2 steps in the x_2-direction, and so on until we get to x_n steps in the x_n-direction. We can't really visualize what the x_n direction looks like, or what it means to take a step in that direction but we need not have a physical, spatial interpretation of a vector in order to use it.

WHILE THE STANDARD INTERPRETATION OF A VECTOR IS AN ARROW from the origin to a particular point in space, vectors are not anchored to a starting point. Vectors can be moved around the Cartesian plane. The starting point is called the **tail** of a vector, and the end of the vector (the arrow head) is called the **tip**.

Two vectors are considered **parallel** if they point in the same direction. This means that the vectors are **scalar multiples** of each other[3].

Vector Addition and Scalar Multiplication

So we have vectors, great. They can describe directions in n-dimensional space. In order for them to truly be useful, we need to be able to *do* things with vectors. Since vectors are just a collection of numbers, maybe it is best if we try to define some of the operations we have for numbers, but for vectors.

---

[3]The components of one vector appear in the same ratio as components of the parallel vector.

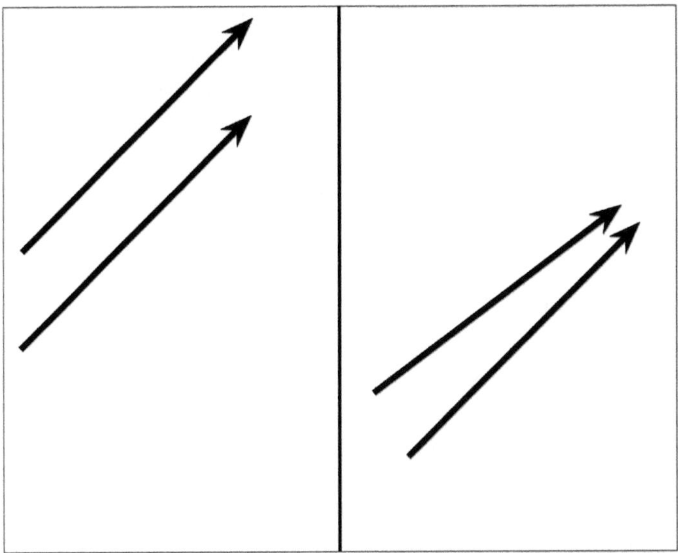

Figure 3.1: The vectors on the left are parallel, the vectors on the right are not.

WE'LL START WITH ADDITION BECAUSE IT'S EASIEST. Vector addition works **component-wise**: if we have two column vectors, we add things straight across. Say we have vectors

$$\vec{u} = \begin{bmatrix} x \\ y \\ z \end{bmatrix} \qquad \vec{v} = \begin{bmatrix} a \\ b \\ c \end{bmatrix}$$

then

$$\vec{u} + \vec{v} = \begin{bmatrix} x + a \\ y + b \\ z + c \end{bmatrix}$$

Graphically, we add two vectors when we want the resultant straight path of two[4] vectors that are connected *tip-to-tail* as seen in figure 3.4. The addition of vectors are always given as if the tail of the resultant vector starts at the origin.

[4]Or more

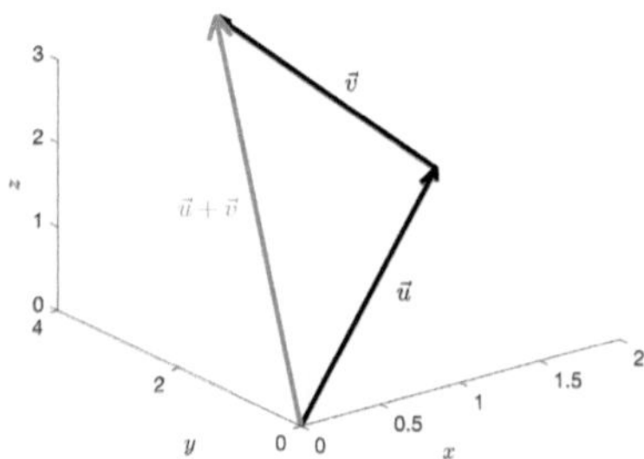

Figure 3.2: The addition of two vectors gives the resultant straight line path between the origin and the final position.

This is the natural definition of addition of vectors because it acts as we would expect in the real world. Adding two vectors, $[a, b]^T$ and $[c, d]^T$, is the same as asking *if I travel a units in the x-direction and b units in the y-direction* **and then** *travel c units in the x-direction and d units in the y-direction, what is the equivalent straight-line travel?* Defining vector addition in this way is not magic or arbitrary, it is the way it is because it gives the result that we expect intuitively.

Likewise, any vector that is floating around in space is the *difference* between two vectors connected *tail-to-tail*, and starting at the origin as illustrated in figure 3.3.

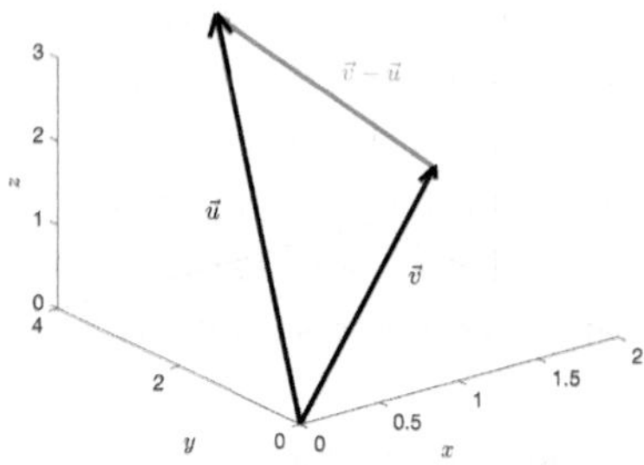

Figure 3.3: Any vector can be written as the difference between two vectors that both start at the origin.

The difference is again given component-wise.

$$\vec{u} - \vec{v} = \begin{bmatrix} x - a \\ y - b \\ z - c \end{bmatrix}$$

With differences, unlike with sums, the order matters: $\vec{u} - \vec{v}$ points in the opposite direction of $\vec{v} - \vec{u}$.

WE CAN ALSO MULTIPLY VECTORS BY A NUMBER. If we do this, then multiplication is also component-wise.

$$c\vec{v} = c \begin{bmatrix} x \\ y \\ z \end{bmatrix} = \begin{bmatrix} cx \\ cy \\ cz \end{bmatrix}$$

Geometrically[5] multiplication by a scalar[6] stretches or compresses a vector *but does not change its direction.*

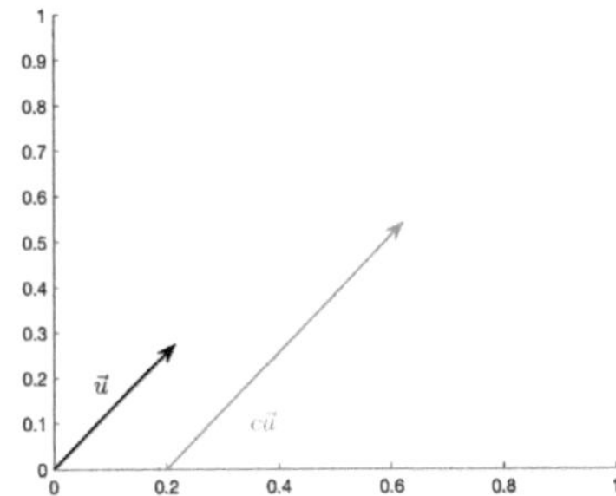

Figure 3.4: Multiplication by a scalar value (i.e. a number) changes the length but not the direction of a vector.

Again, it's worth repeating that this definition of multiplication is the way it is because it acts as we would expect[7]. In the same way that $t + t + t = 3t$, $\vec{s} + \vec{s} + \vec{s} = 3\vec{s}$. If we add a vector to itself[8], do we expect the direction of travel to change? Or do we expect to just go twice as far in the same direction? If you can answer these questions, then scalar multiplication of vectors becomes very intuitive.

Vector Multiplication – The Dot Product or Scalar Product

I'D VERY MUCH LIKE TO CONTINUE WITH THE THEME of *we define this operation on vectors a certain way because it gives the result we*

[5]i.e. Graphically

[6]i.e. Number

[7]You may not *know* you expect things to act this way, but that's probably because you haven't thought about it enough to decide what to expect

[8]Remember a vector must be parallel to itself.

expect intuitively, but that breaks down when it comes to multiplying two vectors. It breaks down because what is now known as the dot product came from the study of *quaternions*. In fact, the dot product is defined as such because it *does* work the way we intuitively expect it to work but in order to see it we need to have some intuition about complex numbers[9]. I don't expect you to have intuition about complex numbers, so we will have to make do with the fact that this definition comes from (seemingly) nowhere.

SO, WITH THAT OUT OF THE WAY, how do we multiply two vectors together in a way that makes sense? It turns out there are multiple ways to do this, but there is exactly *one* that works in every dimension from two to n: the dot (or scalar) product.

It's named as such because the dot product uses the $\cdot$ symbol for multiplication and returns a single scalar value as the product of two vectors.

Definition 3.1.3. The **dot product** between vectors $\vec{v} = [x_1, y_1, z_1]^T$ and $\vec{u} = [x_2, y_2, z_2]^T$ is defined as

$$\vec{v} \cdot \vec{u} = x_1 x_2 + y_1 y_2 + z_1 z_2$$

It is the sum of the pairwise products of the components.

What does this tell us? This gives us an idea of "how much" one vector "supports" another vector. That's a lot of quotation marks, and that's because none of that is very mathematical at all.

TAKING A FAIRLY SIMPLISTIC VIEW OF SOLAR PANELS might help us gain some intuition here. Without getting into too much detail, a solar panel is basically a piece of glass that takes in light and turns it into electrical power[10]. The more direct light it receives, the more power it can produce. The figure 3.5 shows a snapshot of the sun at a certain time of day, and a solar panel on the surface of the earth. "Direct light" is

[9] In two dimensions at least. We need intuition about quaternions if we want to see this in dimensions greater than two.

[10] You may even say that it is a physical function that takes a concentration of sunlight as an input and gives back some quantity of electrical power as output.

light that hits the panel perpendicular to its surface, but the sun puts out rays radially, the most direct rays from the sun may not directly in line with the solar panel. In this case, how much direct light is the solar panel receiving?

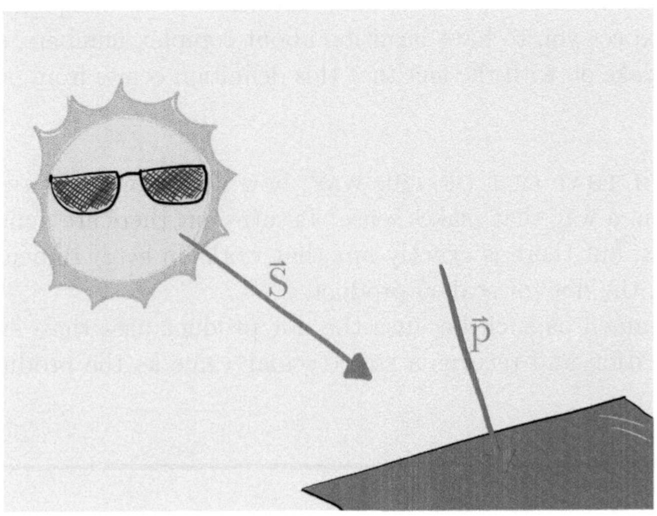

Figure 3.5: Solar panels work best when they receive direct sunlight, but that is not always the case.

If $\vec{S}$ is the amount of solar energy hitting the surface of the Earth directly[11], and $\vec{p}$ is the direction in which sunlight would hit our solar panel directly[12], then the amount of direct solar energy hitting the solar panel is given by $\vec{S} \cdot \vec{p}$[13].

If $\vec{S} = [S_x, S_y, S_z]^T$ and $\vec{p} = [p_x, p_y, p_z]^T$ then the amount of direct solar energy hitting the solar panel is

$$\vec{S} \cdot \vec{p} = S_x p_x + S_y p_y + S_z p_z$$

If in our very simple, synthesized example if we assume that a solar panel can *only* use direct solar rays then $\vec{S} \cdot \vec{v}$ gives exactly effective solar energy[14].

[11]Or, better the average solar energy hitting the Earth's surface directly for a given time of day

[12]We assume that the sun is sufficiently far enough away from the earth that we can treat the light as a very wide beam, as if the sun were a very powerful sky flashlight.

[13]This assumes that since our vector $\vec{p}$ isn't "real," it is of unit length. We'll get to why eventually

[14]i.e. the energy that can be absorbed; if it's not absorbed by the panel, then we don't care about it.

Another way to look at the dot product is a measure of how close two vectors are to the same direction. If the number is positive, then the two vectors are pointing generally in the same direction. If the dot product is negative, the two vectors are pointing in generally opposite direction, as shown in figure 3.6

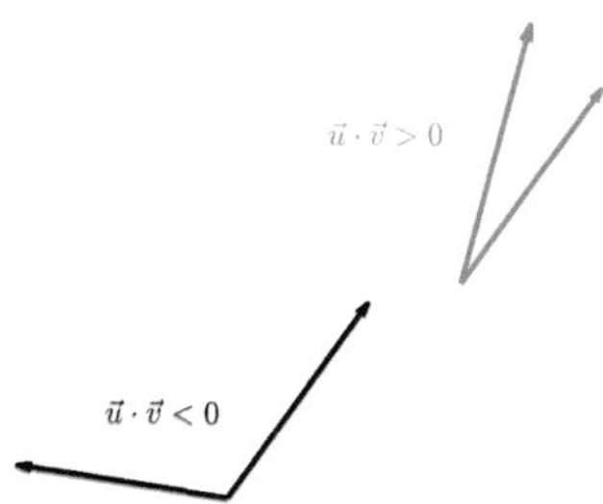

Figure 3.6: If two vectors are generally pointing in the same direction, their dot product is positive, if they are generally pointing in opposite directions, the dot product is negative. The dot product gets larger as the two vectors get closer to parallel and smaller (more negative) as the two vectors get closer to anti-parallel.

For vectors that are of fixed length, but able to rotate the dot product will get larger as rotation brings two vectors closer to parallel, and the dot product will get more negative as rotation brings the two vectors closer to anti-parallel (pointing in exact opposite directions). *If the length of two vectors is fixed*, the dot product achieves a maximum when $\vec{u}$ and $\vec{v}$ are parallel and a minimum when $\vec{u}$ and $\vec{v}$ are anti-parallel.

We have an idea of what a positive dot product means, and of what a negative dot product means, what of a zero dot product?

Remark. If

$$\vec{v} \cdot \vec{u} = 0$$

then $\vec{v}$ and $\vec{u}$ are **perpendicular**, or **orthogonal**

We can use the dot product to determine if a vector is perpendicular to another vector! For our purposes, we can treat the word orthogonal as the math-y synonym to perpendicular.

One way we can use this to check if two vectors are perpendicular to each other. For instance, the vectors $\vec{a} = [5, 6]^T$ and the vector

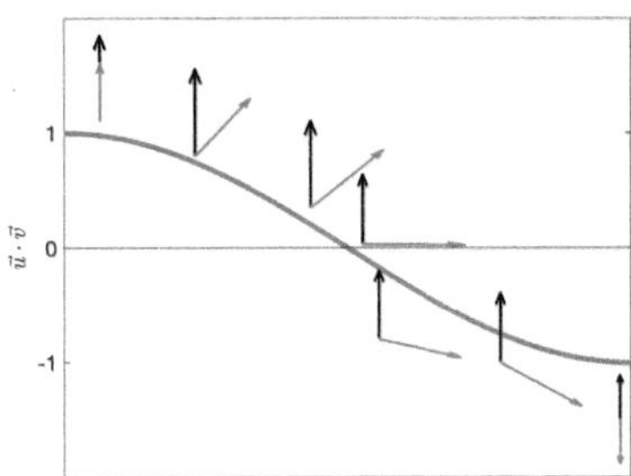

Figure 3.7: If two vectors are generally pointing in the same direction, their dot product is positive, if they are generally pointing in opposite directions, the dot product is negative. The dot product gets larger as the two vectors get closer to parallel and smaller (more negative) as the two vectors get closer to anti-parallel.

$\vec{b} = [12, -10]$ are perpendicular because

$$\vec{a} \cdot \vec{b} = 5 \times 12 + 6 \times (-10) = 60 - 60 = 0$$

Notice that the vector $\vec{O} = [0, 0]$ is orthogonal to *everything.*

We also use the dot product to *make* two vectors perpendicular to each other. For instance, we can find the vector $\vec{b}$ which is perpendicular to $\vec{a} = [k, n]$.

$$\vec{a} \cdot \vec{b} = [k, n] \cdot [b_1, b_2] = kb_1 + nb_2 = 0$$

Here we have one equation, $kb_1 + nb_2 = 0$ but two unknowns: b_1 and b_2. This means we are free to choose either b_1 or b_2[15]. Let's choose $b_1 = 1$ for simplicity, then $b_2 = -k/n$. So the vector $\vec{b} = [1, -k/n]$[16] is perpendicular to $\vec{a}$.

Why might this be useful? Windmills work best when the face of the blade is oriented perpendicular to the direction of the wind. Since wind does not typically blow up or down, we can express the direction of the average wind speed in a particular region as a vector

$$\vec{w} = [w_1, w_2] = [N, E]$$

[15] This is because technically there are infinitely many parallel vectors which are all perpendicular to $\vec{a}$.

[16] Or any multiple of it

where the first component represents north/south wind, and the second component is east/west[17]. We want to orient our windmill then such that the blade, $\vec{b}$, is perpendicular to the average direction of the wind.

Like we saw above, we will have some freedom in how we choose to orient our windmill. Using the dot product, we get

$$\vec{w} \cdot \vec{b} = [N, E] \cdot [b_1, b_2] = Nb_1 + Eb_2 = 0.$$

Solving for b_2 gives

$$b_2 = \frac{-Nb_1}{E}.$$

If we choose b_1 so that we can get rid of the denominator, $b_1 = E$, then we get $b_2 = -N$. Obviously the wind doesn't always blow in one direction, and this is why you will often see windmills pointing in slightly different directions in a wind farm; but if the wind farm is erected intelligently, the positioning of the windmills will not be accidental.

Notice in both these examples that b_2/b_1 is the **negative reciprocal** of k/n or w_2/w_1. In $2D$, this is always true: two vectors are perpendicular if their slopes, i.e. $rise/run$, i.e. the ratio of their components, are the negative reciprocals of each other. We can show this generally if

$$\vec{a} = [a_1, a_2]\, \vec{b} = [b_1, b_2]$$

and we force them to be perpendicular

$$\vec{a} \cdot \vec{b} = a_1 b_1 + a_2 b_2 = 0$$

we can rearrange to get

$$\frac{a_1}{a_2} = -\frac{b_2}{b_1}$$

In three-dimension (3D), things get more complicated. If we want the vector $\vec{c} = [c_1, c_2, c_3]$ to be perpendicular to $\vec{a} = [A, B, C]$, we have one equation and *three* unknowns!

$$\vec{a} \cdot \vec{c} = Ac_1 + Bc_2 + Cc_3 = 0$$

Which means we can pick *two* of the components of $\vec{c}$, say c_1 and c_2 and solve for c_3

$$c_3 = -\frac{Ac_1 + Bc_2}{C}$$

[17]North and east are positive values; south and west are negative values.

and this works for *any* values of c_1 and c_2. In this case, not all perpendicular vectors to $\vec{a}$ are parallel to each other as in the $2D$ case. This is because, there is a whole plane of perpendicular vectors. The possibilities are infinite.

Magnitude of a vector

How long is the vector $[x, y, z]^T$?

In two dimensions, this is a pretty easy question to answer if we know a little bit about right-angle triangles. Since x-direction and y-direction are **perpendicular**[18,19] to each other, the length of a vector is just the hypotenuse of the triangle formed by moving x units, then y units[20]. So

---

Definition 3.1.4. The **magnitude** of a vector

$$\mathbf{v_2} = \vec{v_2} = [x, y]^T$$

is given by

$$||\mathbf{v_2}|| = \sqrt{x^2 + y^2}$$

Similarly, since z is perpendicular to both x and y, the magnitude of a 3D vector is given by

$$||\mathbf{v_3}|| = \sqrt{x^2 + y^2 + z^2}$$

Figure 3.8 shows how the magnitudes of the vectors are given by the corresponding right angle triangles.

KNOWING THE MAGNITUDE OF A VECTOR ALSO PROVIDES US with another definition of the dot product. For two vectors $\vec{u}$ and $\vec{v}$,

$$\vec{u} \cdot \vec{v} = ||\vec{u}||||\vec{v}|| \cos(\theta)$$

[18] Or **orthogonal**
[19] Perpendicular
[20] Or vice versa

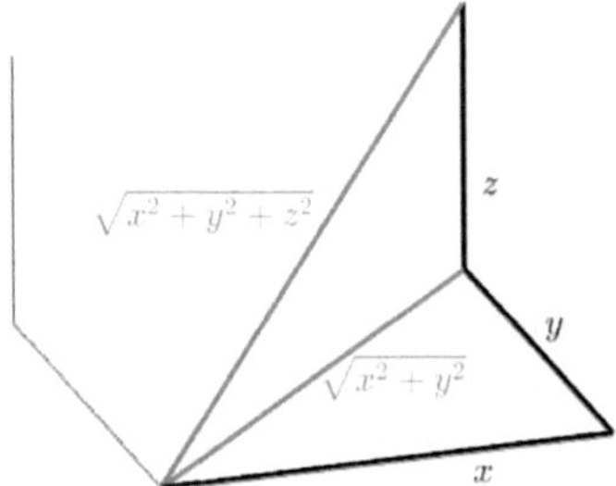

Figure 3.8: The darker blue line is the vector $\mathbf{v_2}$*, and the lighter blue line is the vector* $\mathbf{v_3}$*. We can see with the dark black lines the corresponding right angle triangles that give rise to the magnitude formulas for each.*

where θ is the angle between the vectors when they are put tail-to-tail. This formulation is rarely used to find the dot product itself, but often used when we need to know the angle between two vectors.

This formulation of the dot product can be extremely useful when rearranged as

$$\cos(\theta) = \frac{\vec{u} \cdot \vec{v}}{||\vec{u}||||\vec{v}||}$$

This formulation is known as **cosine similarity** and can be used as a crude way to detect possible plagiarism in, say, student assignments.

For instance, let's say we have the following three pieces of text:

A

> O how I wish that ship the Argo had never sailed off to the land of Colchis,

B

> Tush! never tell me; I take it much unkindly That thou, Iago, who hast had my purse

C

> Loud the rowers' cry Who through the storm-swept paths of Mio Bay Ride to the rising sea.

Each row of the vector represents a unique word that appears in one of the three above quotes, we put a into that row the number of times the word appears in that passage. We will do this as a table first, as it would be easier to see how this works: I have left out the names that appear in each quote since, because these were written in three different time periods in three different parts of the world, we would not expect these names to be common[21].

Translating this table into vectors, we get

$$\vec{v}_A = [1,1,1,1,1,1,2,1,1,1,1,1,1,1,0,0,0,0,0,0,0,0,0,0,0,0,0,0,0,0,0,0,0,0,0]$$
$$\vec{v}_B = [0,0,1,0,1,0,0,1,1,0,0,0,0,0,1,1,1,1,1,1,1,1,1,1,1,1,0,0,0,0,0,0,0,0,0]$$
$$\vec{v}_C = [0,0,0,0,0,0,3,0,0,0,0,1,0,1,0,0,0,0,0,0,0,0,1,0,0,0,1,1,1,1,1,1,1,1,1]$$

What is the cosine similarity between $\vec{v}_A$ and $\vec{v}_B$, $\vec{v}_A$ and $\vec{v}_C$, and $\vec{v}_B$ and $\vec{v}_C$,?

$$\cos(\theta_{AB}) = \frac{\vec{v}_A \cdot \vec{v}_B}{||\vec{v}_A||||\vec{v}_B||} \tag{3.1}$$
$$= \frac{4}{\sqrt{17}\sqrt{16}} = \frac{1}{\sqrt{17}} \tag{3.2}$$
$$= 0.243 \tag{3.3}$$

$$\cos(\theta_{AC}) = \frac{\vec{v}_A \cdot \vec{v}_C}{||\vec{v}_A||||\vec{v}_C||} \tag{3.4}$$
$$= \frac{8}{\sqrt{17}\sqrt{21}} \tag{3.5}$$
$$= 0.423 \tag{3.6}$$

$$\cos(\theta_{BC}) = \frac{\vec{v}_B \cdot \vec{v}_C}{||\vec{v}_B||||\vec{v}_C||} \tag{3.7}$$
$$= \frac{1}{\sqrt{16}\sqrt{21}} \tag{3.8}$$
$$= 0.055 \tag{3.9}$$

What does this tell us? It tells us that passage A and C are most similar at 43% similarity, and B and C are least similar at about 5% similarity. Why might this be? We should preface answering this question

[21]It would be quite curious to see the name Othello in an ancient Greek tragedy.

Word	$\vec{v}_A$	$\vec{v}_B$	$\vec{v}_C$
O	1	0	0
how	1	0	0
I	1	1	0
wish	1	0	0
that	1	1	0
ship	1	0	0
the	2	0	3
had	1	1	0
never	1	1	0
sailed	1	0	0
off	1	0	0
to	1	0	1
land	1	0	0
of	1	0	1
Tush	0	1	0
tell	0	1	0
me	0	1	0
take	0	1	0
it	0	1	0
much	0	1	0
unkindly	0	1	0
thou	0	1	0
who	0	1	1
hast	0	1	0
my	0	1	0
purse	0	1	0
Loud	0	0	1
rowers	0	0	1
cry	0	0	1
through	0	0	1
storm-swept	0	0	1
paths	0	0	1
Ride	0	0	1
rising	0	0	1
sea	0	0	1

with the fact that we have taken not nearly enough of a sample in *any* case to draw any real conclusions. I have taken the first 2–3 lines of three different plays, and they weren't even randomly chosen[22]. But some insights might be gleaned nonetheless[23]. Passage A and passage C are both translated into English; A from Ancient Greek, and C from Japanese. They are both modern translations and thus will share more common language than not.

Passages A and B both happen to be from Western Europe, A being Greek and B being Shakespeare. We know that the Greeks has a large influence on Western European culture and may be why these two passages appear more similar than B and C. Of course, all we can say for certain is that this only holds for the first few words of these particular plays. In the second half of this book, we will see how to extend these notions to much larger sets of data so that we can speak more generally. This is merely an example of a non-obvious use of the dot product and vectors which, at scale, can have some really cool implications.

The clever reader may have already noticed that

$$||\vec{v}||^2 = \vec{v} \cdot \vec{v}$$

AN APPLICATION OF VECTORS

An *antigen* is a part of a virus that can bind to an antibody or antigen receptor on a T-cell – a cell of the immune system. The *antigenic distance* is a measure of how different the antigens on two viral strains are. We won't get into how this is measured, because this is not a course on immunology, but we can talk about *why* it is measured.

When we plot in two dimensions of antigenic distance, we often see patterns about the evolution of a viral strain emerge. In the left panel of figure 3.9, adapted from [25], we see the evolution of Influenza A H3N2 over time.

If we look at the centres[24], we can draw vectors from one subtype to the next, and track the evolution of Influenza A, as in the right panel of figure 3.9. With an image like this, it is fairly clear to see the pattern in antigenic evolution of the virus. We may even be able, just from looking

[22] They just happen to be ones I know of, and like.

[23] But I need to reiterate that much, much more work would need to be done to say anything for certain

[24] The centre of a "blob" is a little bit vague. This is again a situation where we are going to estimate by looking. There are rigorous mathematical ways of doing things like finding the centre of a set of points or even measuring the distance between sets of points, but we haven't been very rigorous up to this point, so why bother starting now?

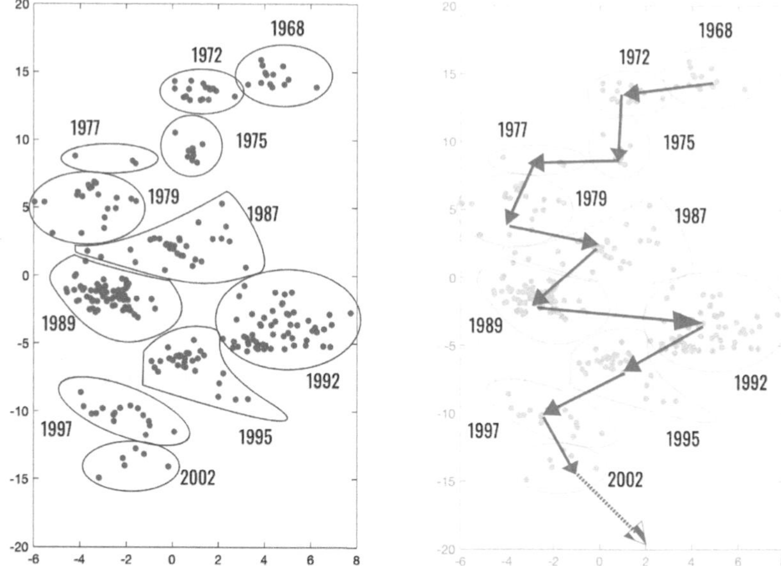

Figure 3.9: An antigenic cartograph for Influenza A H3N2. Reproduced using data from Smith et al.

at the image, predict where the next strain of virus may land on the antigenic plot.

3.2 MATRICES

A collection of numbers is a vector. What is a collection of vectors? This is called a **matrix** and there are a few ways to think about a matrix and what it does. One way, as we just said, is to think of it as a collection of row vectors stacked on top of one another

$$\begin{matrix} \vec{v}_1 \\ \vec{v}_2 \\ \vec{v}_3 \end{matrix} \begin{bmatrix} a & b & c \\ l & m & n \\ x & y & z \end{bmatrix}$$

or as a collection of column vectors placed beside each other.

$$\begin{matrix} \vec{w}_1 & \vec{w}_2 & \vec{w}_3 \end{matrix}$$
$$\begin{bmatrix} a & b & c \\ l & m & n \\ x & y & z \end{bmatrix}$$

Matrices work in a similar way to vectors and many of the operations carry over in some form or another. Typically, when we use variables to indicate a matrix we use capital letters[25].

The **size of a matrix** is given by the number of rows it has and the number of columns it has, as *rows* × *columns*. The number of rows always comes first, and the number of columns always comes second like figure 3.10.

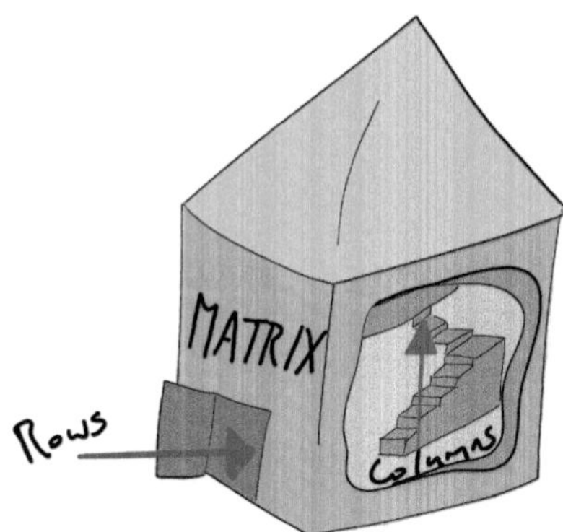

Figure 3.10: An easy way to remember how to write the size of a matrix is to think of a house: you need to go in the house (rows), before you can go up the stairs (columns). So the size of the matrix is rows then columns! This also holds true for the elements of a matrix. The value in the position $(4, 3)$ *is in the fourth row and third column, or* $a_{5,2}$ *is the fifth row, second column.*

Adding matrices

Adding a matrix, M, to another matrix, N is the same as adding two vectors. We add each component to the corresponding component in the other matrix.

[25]So, a number would be x, a vector would be $\vec{x}$, and a matrix would be X. This isn't a rule so much as a convention.

If we have

$$M = \begin{bmatrix} a & b \\ c & d \end{bmatrix}$$

and

$$N = \begin{bmatrix} w & x \\ y & z \end{bmatrix}$$

then

$$M + N = \begin{bmatrix} a+w & b+x \\ c+y & d+z \end{bmatrix}.$$

and

$$M - N = \begin{bmatrix} a-w & b-x \\ c-y & d-z \end{bmatrix}.$$

Notice that we need to have a matching component in *both* matrices in order for addition to work. In other words, **we can only add two matrices together when we have we same number of rows and the same number of columns**. Otherwise, addition is incompatible[26].

For instance, let's look at the matrices

$$A = \begin{bmatrix} a & b & c \\ d & f & g \end{bmatrix}$$

and

$$B = \begin{bmatrix} 6 & 2 \\ 5 & 9 \\ 22 & \sigma \end{bmatrix}$$

we can ask the question *what is* $A + B$*?* But this isn't a question we can answer. If we try, this is what happens:

$$\begin{aligned} A + B &= \begin{bmatrix} a & b & c \\ d & f & g \end{bmatrix} + \begin{bmatrix} 6 & 2 \\ 5 & 9 \\ 22 & \sigma \end{bmatrix} \\ &= \begin{bmatrix} a+6 & b+2 & c+?? \\ d+5 & f+9 & g+?? \\ ?!+22 & ?!+\sigma & ?!+?? \end{bmatrix} \end{aligned}$$

[26] An allegory for scalar numbers is trying to add things with different units. What is 8 kilograms plus 0.25 metres? Nothing but trouble.

everything is red doesn't exist! Much like we have to be careful when manipulating numbers, we have to be careful when manipulating matrices. We have to make sure that the questions we are asking make sense[27].

3.3 MULTIPLICATION: NUMBERS AND MATRICES

This is another operation that is identical to the case for vectors. If we have a number, a, and multiply it by a matrix, M, we just have to multiply each of the components by the number (in this case a).

$$M = \begin{bmatrix} x & y \\ m & n \end{bmatrix}$$

$$aM = \begin{bmatrix} ax & ay \\ am & an \end{bmatrix}$$

Simple! I think that's about all we need to say about that[28].

3.4 MULTIPLICATION: MATRIX AND VECTORS

The multiplication of a matrix with a vector is a lot like the dot product but, as with matrix addition, we need to be sure that what we are doing makes sense.

In order for the multiplication to make sense for the multiplication of $M\vec{v}$, the number of columns of M must **match** the number of rows of $\vec{v}$. Some examples,

$$\begin{bmatrix} 2 & 4 \\ 6 & 1 \end{bmatrix} \begin{bmatrix} 12 \\ 42 \end{bmatrix} \quad ✓$$

$$\begin{bmatrix} 2 & 4 & k \\ 6 & 1 & 9 \end{bmatrix} \begin{bmatrix} 12 \\ 42 \\ a \end{bmatrix} \quad ✓$$

$$\begin{bmatrix} 2 & 4 \\ 6 & 1 \\ 9 & 2 \\ a & b \end{bmatrix} \begin{bmatrix} 12 & 3 & 5 & 99 \end{bmatrix} \quad ✗$$

[27]The old adage that rings true, especially in applied math, is *garbage in* and *garbage out*. If you are asking the wrong questions, you will get the wrong answers, even if all your machinery is working properly.

[28]For now...

In the first case we have a 2×2 matrix multiplying a 2×1 vector, so everything will work out[29]. In the second case, we have a 2×3 matrix multiplying a 3×1 vector so again everything is fine. In the third case we see a 4×2 matrix multiplying a 1×4 vector. This doesn't work out because the number of columns of the matrix, 2, doesn't match the number of rows of the vector, 1.

We can also multiply $\vec{v}M$, and in general this is **not** the same as $M\vec{v}$. In the case of $\vec{v}M$, the number of **columns** of the vector, $\vec{v}$ must match the **rows** of the matrix, M. Some examples,

$$\begin{bmatrix} 3 \\ h \\ 534 \\ 2 \end{bmatrix} \begin{bmatrix} 4 & 3 & a \\ 123 & 987 & 2x \\ 9 & 4 & x+y \\ 5 & 3 & 1 \end{bmatrix} \quad ✗$$

$$\begin{bmatrix} 3 & h & 534 & 2 \end{bmatrix} \begin{bmatrix} 4 & 3 & a \\ 123 & 987 & 2x \\ 9 & 4 & x+y \\ 5 & 3 & 1 \end{bmatrix} \quad ✓$$

$$\begin{bmatrix} 3 & h & 534 \end{bmatrix} \begin{bmatrix} 4 & 3 & a \\ 123 & 987 & 2x \\ 9 & 4 & x+y \end{bmatrix} \quad ✓$$

In the first case our vector is of size 4×1 and our matrix is size 4×3. Since $4 \neq 1$, this doesn't work. In the second case where we flip the vector[30] the size becomes 1×4 and now the number of columns of $\vec{v}$ matches the number of rows of M and so this multiplication is valid. The third case is also valid because $\vec{v}$ has 3 columns and M has 3 rows.

In general, as long as the *inner* dimensions of the multiplication match, then the multiplication is valid.

$$\begin{bmatrix} 3 & 9 & a \\ x & 67 & \sigma \\ 2 & 0 & 1 \end{bmatrix} \begin{bmatrix} x \\ y \\ z \end{bmatrix} \qquad \begin{bmatrix} 2 & 1 \end{bmatrix} \begin{bmatrix} a & b & c & d & f \\ g & h & k & l & m \\ 3 & 6 & \alpha & 98 & 11 \\ 4 & 7 & x & y & z \end{bmatrix}$$

Sizes: 3×3 3×1 ✓ Sizes: 1×2 4×5 ✗

$$\begin{bmatrix} 4 & 2 \\ 1 & 9 \\ 5 & 2 \end{bmatrix} \begin{bmatrix} x \\ y \\ z \end{bmatrix} \qquad \begin{bmatrix} 2 & 1 \end{bmatrix} \begin{bmatrix} a & b & c \\ l & m & n \end{bmatrix}$$

Sizes: 3×2 3×1 ✗ Sizes: 1×2 2×3 ✓

[29]Keep in mind we don't yet know *how* it will work out, just that it will.

[30]i.e. take the *transpose*

So how do we actually *do* this multiplication? It's basically a series of dot products!

When doing the multiplication $M\vec{v}$, we take the dot product of each row of the matrix with the vector and that becomes an entry in the resultant vector.

$$\begin{matrix}\vec{r}_1\\ \vec{r}_2\\ \vec{r}_3\end{matrix}\begin{bmatrix} a & b & c \\ m & n & l \\ p & q & r \end{bmatrix}\vec{v} = \begin{bmatrix} \vec{r}_1 \cdot \vec{v} \\ \vec{r}_2 \cdot \vec{v} \\ \vec{r}_3 \cdot \vec{v} \end{bmatrix}$$

$$\begin{bmatrix} a & b & c \\ m & n & l \\ p & q & r \end{bmatrix}\begin{bmatrix} x \\ y \\ z \end{bmatrix} = \begin{bmatrix} ax + by + cz \\ mx + ny + lz \\ px + qy + rz \end{bmatrix}$$

This works for any size matrix, not just 3×3.

$$\begin{matrix}\vec{r}_1\\ \vec{r}_2\\ \\ \vec{r}_m\end{matrix}\begin{bmatrix} a_{1,1} & a_{1,2} & \cdots & a_{1,n} \\ a_{2,1} & a_{2,2} & \cdots & a_{2,n} \\ \vdots & \vdots & \ddots & \vdots \\ a_{m,1} & a_{m,2} & \cdots & a_{m,n} \end{bmatrix}\vec{v} = \begin{bmatrix} \vec{r}_1 \cdot \vec{v} \\ \vec{r}_2 \cdot \vec{v} \\ \vdots \\ \vec{r}_m \cdot \vec{v} \end{bmatrix}$$

What does this *do?* It's a way to turn vectors into other vectors[31]. A matrix can rotate and scale a vector at the same time like in figure 3.11 or it can flatten a vector from 3D into 2D (as well as scale and rotate it) like in figure 3.12.

For instance,

$$\begin{aligned} \begin{bmatrix} 3 & 2 & 1 \\ 7 & 8 & 9 \end{bmatrix}\begin{bmatrix} 10 \\ 20 \\ 30 \end{bmatrix} &= \begin{bmatrix} 30 + 40 + 30 \\ 70 + 160 + 180 \end{bmatrix} \\ &= \begin{bmatrix} 100 \\ 410 \end{bmatrix} \end{aligned}$$

[31] You may ask, can't addition of vectors do this? Yes, but with addition we're stuck in a particular dimension; with matrices we aren't!

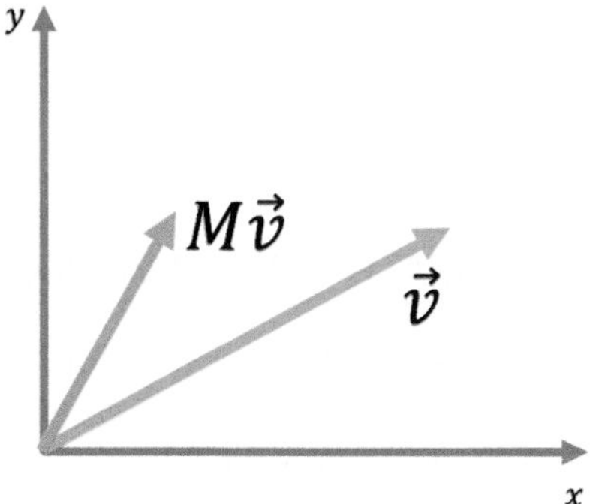

Figure 3.11: Multiplying a vector by a matrix rotates and scales the vector to give us a new vector.

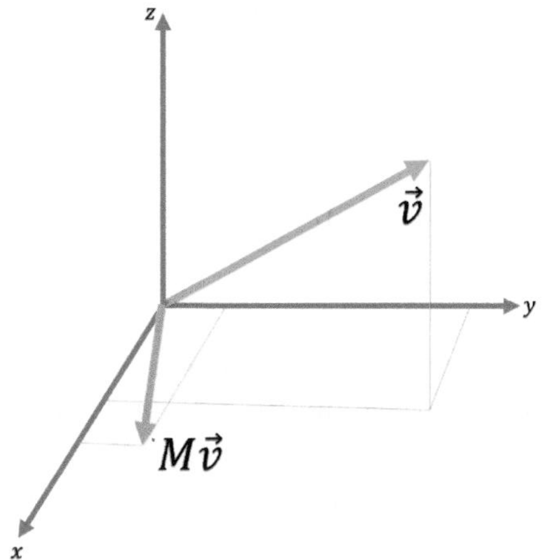

Figure 3.12: Along with rotating and scaling a vector, a matrix can change the dimension of a vector.

brings a 3D vector into the plane, where as

$$\begin{aligned} \begin{bmatrix} 3 & 2 \\ 8 & 9 \\ 1 & 5 \end{bmatrix} \begin{bmatrix} 10 \\ 20 \end{bmatrix} &= \begin{bmatrix} 30+40 \\ 80+180 \\ 10+100 \end{bmatrix} \\ &= \begin{bmatrix} 70 \\ 260 \\ 110 \end{bmatrix} \end{aligned}$$

brings a vector from 2D into 3D space.

The following problem shows a rotation and a scaling:

$$\begin{bmatrix} 0 & -2 \\ 2 & 0 \end{bmatrix} \begin{bmatrix} 1 \\ 3 \end{bmatrix} = \begin{bmatrix} -6 \\ 2 \end{bmatrix}$$

3.5 MULTIPLICATION: MATRIX AND MATRIX

Multiplying two matrices is a natural extension of matrix and vector multiplication. We treat the matrix on the *right* of our multiplication as a collection of vectors[32]. Let's assume we have two matrices, A and B. We can write B as a collection of column vectors

$$B = \begin{bmatrix} \vec{v}_1 & \vec{v}_2 & \cdots & \vec{v}_{n-1} & \vec{v}_n \end{bmatrix}$$

We know that the product $A\vec{v}_i$ is itself a vector. These vectors make up the columns of the resulting matrix.

$$\begin{aligned} AB &= A \begin{bmatrix} \vec{v}_1 & \vec{v}_2 & \cdots & \vec{v}_{n-1} & \vec{v}_n \end{bmatrix} \\ &= \begin{bmatrix} A\vec{v}_1 & A\vec{v}_2 & \cdots & A\vec{v}_{n-1} & A\vec{v}_n \end{bmatrix} \\ &= C \end{aligned}$$

As we matrix and vector multiplication, we need to make sure that the *sizes* of our matrices are compatible. In order to compute the product AB, the number of columns of A must match the number of rows of B. If A is of size $n \times m$ and B is of size $m \times p$, then the product AB has size $n \times p$.

Remark. Matrix multiplication is **not** generally commutative. This means that **the order matters** when multiplying. For two matrices A and B

$$AB \neq BA$$

3.6 LESLIE MATRICES

Matrix models are very popular in population ecology for determining the age distribution of individuals over time in a population. It's an example worth the time because it has a really nice interpretation!

[32]Exactly the interpretation we used above

Let's say we have a human population that we can split up into the following age groups:

h_0 : `humans aged 0–12`
h_1 : `humans aged 13–19`
h_2 : `humans aged 20–29`
h_3 : `humans aged 30–39`
h_4 : `humans aged 40–64`
h_5 : `humans aged 65+`

which we put into a vector

$$\vec{h} = \begin{bmatrix} h_0 \\ h_1 \\ h_2 \\ h_3 \\ h_4 \\ h_5 \end{bmatrix}$$

We can build a matrix,

$$L = \begin{bmatrix} l_{1,1} & l_{1,2} & l_{1,3} & l_{1,4} & l_{1,5} & l_{1,6} \\ l_{2,1} & l_{2,2} & l_{2,3} & l_{2,4} & l_{2,5} & l_{2,6} \\ l_{3,1} & l_{3,2} & l_{3,3} & l_{3,4} & l_{3,5} & l_{3,6} \\ l_{4,1} & l_{4,2} & l_{4,3} & l_{4,4} & l_{4,5} & l_{4,6} \\ l_{5,1} & l_{5,2} & l_{5,3} & l_{5,4} & l_{5,5} & l_{5,6} \\ l_{6,1} & l_{6,2} & l_{6,3} & l_{6,4} & l_{6,5} & l_{6,6} \end{bmatrix}$$

When we multiply this matrix by $\vec{h}$, we get

$$L\vec{h} = \begin{bmatrix} l_{1,1} & l_{1,2} & l_{1,3} & l_{1,4} & l_{1,5} & l_{1,6} \\ l_{2,1} & l_{2,2} & l_{2,3} & l_{2,4} & l_{2,5} & l_{2,6} \\ l_{3,1} & l_{3,2} & l_{3,3} & l_{3,4} & l_{3,5} & l_{3,6} \\ l_{4,1} & l_{4,2} & l_{4,3} & l_{4,4} & l_{4,5} & l_{4,6} \\ l_{5,1} & l_{5,2} & l_{5,3} & l_{5,4} & l_{5,5} & l_{5,6} \\ l_{6,1} & l_{6,2} & l_{6,3} & l_{6,4} & l_{6,5} & l_{6,6} \end{bmatrix} \begin{bmatrix} h_0 \\ h_1 \\ h_2 \\ h_3 \\ h_4 \\ h_5 \end{bmatrix} = \begin{bmatrix} n_0 \\ n_1 \\ n_2 \\ n_3 \\ n_4 \\ n_5 \end{bmatrix}$$

We have five equations here that we can analyse. The first is

$$n_0 = l_{1,1}h_0 + l1,2h_1 + l_{1,3}h_2 + l_{1,4}h_3l_{1,5}h_4 + l_{1,6}h_5$$

If we treat n_0 as *the number of individuals aged* 0–12 *in the next generation*, we can see that the number of humans aged 0–12 we expect next generation depends on how many of each age class we have currently. We can fill in the matrix L accordingly:

- Do we expect humans aged 0–12 to produce new humans aged 0–12?
  - For so many, many reasons we do not, but we expect some individuals in this age class to remain in this age class in the next generation. So we should set $l_{1,1} = m_0$
- Similarly, we don't expect individuals 65+ to be producing children, so $l1, 6 = 0$ as well.
- Do we expect humans aged 15–64 to produce offspring?
  - Yes, but at varying rates. $l_{1,2} = r_1$, $l_{1,3} = r_2$, $l_{1,4} = r_3$, $l_{1,5} = r_4$

So the first row of our matrix is now set. To get n_1 we multiply the second row of the matrix L with our vector $\vec{h}$ using the dot product to get

$$n_1 = l_{2,1}h_0 + l_{2,2}h_1 + l_{2,3}h_2 + l_{2,4}h_3 + l_{2,5}h_4 + l_{2,6}h_5$$

From here we can see that the number of humans aged $13-19$ in the next generation has contributions from each of the current age classes. The element $l_{2,1}$ is the number of individuals who were aged $0-12$ originally who are now aged $13-19$, some may have died in the interim years and some may not have aged out of the previous age class, so it won't be everyone (i.e. the value of $l_{2,1} \neq 1$), so we'll say $l_{2,1} = s_0$.

Just like with those aged $0-12$, we expect some people to still be in this age class one generation later[33]. For this reason we set $l_{2,2} = m_1$. Do we expect there to be a contribution from the 20 to 29 year olds to the 13–19 year olds? Well, people can't be born as a teenager, and time marches ever forward so the h_2's cannot contribute to the next generation of h_1's. Therefore, $l_{2,3} = 0$. The same logic holds for all the other age classes so we get $l_{2,4} = l_{2,5} = l_{2,6} = 0$.

For the age class $20-29$ we don't expect *any* individuals from 0 to 12 to jump over 13–19 and go directly to 20+ in one generation[34]. So, we should set $l3, 1 = 0$.

[33] Depending on how long we define one generation to be

[34] Again, depending on what we define as a generation, but usually we define it as less than the span of an age class.

We can make these same arguments for all the rows in our matrix to get the far more tractable

$$L = \begin{bmatrix} m_0 & r_1 & r_2 & r_3 & r_4 & 0 \\ s_0 & m_1 & 0 & 0 & 0 & 0 \\ 0 & s_1 & m_2 & 0 & 0 & 0 \\ 0 & 0 & s_2 & m_3 & 0 & 0 \\ 0 & 0 & 0 & s_3 & m_4 & 0 \\ 0 & 0 & 0 & 0 & s_4 & m_5 \end{bmatrix}$$

We now have an interpretable matrix. If we write some headings to the left and top of the matrix to make things clearer

$$\left[\begin{array}{c|cccccc} & h_0 & h_1 & h_2 & h_3 & h_4 & h_5 \\ \hline n_0 & m_0 & r_1 & r_2 & r_3 & r_4 & 0 \\ n_1 & s_0 & m_1 & 0 & 0 & 0 & 0 \\ n_2 & 0 & s_1 & m_2 & 0 & 0 & 0 \\ n_3 & 0 & 0 & s_2 & m_3 & 0 & 0 \\ n_4 & 0 & 0 & 0 & s_3 & m_4 & 0 \\ n_5 & 0 & 0 & 0 & 0 & s_4 & m_5 \end{array}\right]$$

If we read down and left, we see that each entry in the matrix is the contribution of each current age class to each age class of the next generation. For instance, age class 0–12 contributes m_0 percent of itself to the next generation of 0–12 year olds, and those 65+ contribute nothing to next generation's 0–12 year olds.

If we want to know the population of different age classes k time steps from now, we can just multiply the vector $\vec{h}$ by L k times. Exponentiation of matrices is just *repeated multiplication*, just like it is for numbers. Let's call our vector of populations after k time steps $\vec{p}$ then

$$\vec{p} = \overbrace{LLLL \cdots L}^{k} \vec{h} = L^k \vec{h}$$

3.7 THE DETERMINANT

A Leslie matrix lets us look forward in time very easily. But what if we knew $\vec{p}$ and wanted to find out what our population had to look like k time steps *in the past* in order to currently see the population distributed according to $\vec{p}$ currently.

What we know is that

$$\vec{p} = L^k \vec{h} = R\vec{h}$$

where $L^k = R$ is still a 6×6 matrix. We have to solve this for $\vec{h}$. If these were all numbers, all we would do is divide both sides by R and call it a day, but we are not dealing with numbers we are dealing with matrices and there is no such thing as matrix division.

That doesn't mean all hope is lost, many matrices have what are called a **matrix inverse**.

Definition 3.7.1. The inverse of an $n \times n$ matrix, A, is a matrix B, such that

$$AB = \begin{bmatrix} 1 & 0 & 0 & \cdots & 0 & 0 & 0 \\ 0 & 1 & 0 & \cdots & 0 & 0 & 0 \\ 0 & 0 & 1 & \cdots & 0 & 0 & 0 \\ & \vdots & & \ddots & & \vdots & \\ 0 & 0 & 0 & \cdots & 1 & 0 & 0 \\ 0 & 0 & 0 & \cdots & 0 & 1 & 0 \\ 0 & 0 & 0 & \cdots & 0 & 0 & 1 \end{bmatrix} = I_n$$

The inverse of matrix A is denoted as A^{-1}.

The matrix on the right of the definition is called the **identity matrix**. It is an $n \times n$ matrix which has entries of 1 along the main diagonal, and 0s everywhere else. It has the special property that for any matrix M

$$IM = MI = M$$

and for any column vector $\vec{v}$

$$\vec{v}^T I = \vec{v}^T I\vec{v} \qquad\qquad = \vec{v}$$

It is the matrix multiplication equivalent of the number 1.

Using this definition, once we have matrix R, we can solve our Leslie Matrix problem backwards by invoking the inverse:

$$\begin{aligned} \vec{p} &= R\vec{h} \\ R^{-1}\vec{p} &= R^{-1}R\vec{h} \\ R^{-1}\vec{p} &= I\vec{h} \\ \vec{h} &= R^{-1}\vec{p} \end{aligned}$$

How we find the inverse of a matrix is the subject of a bona fide book on Linear Algebra. Here, we will only discuss under what conditions does the inverse exist[35].

There is a special operation that exists for matrices and it comes up in some of the most important applications for matrices. It is called the determinant.

For $2D$ matrices we get the determinant by taking the product of the diagonal and off-diagonal and subtracting them.

Definition 3.7.2. For the matrix

$$A = \begin{bmatrix} a & b \\ c & d \end{bmatrix}$$

The determinant is given by

$$det(A) = \begin{vmatrix} a & b \\ c & d \end{vmatrix} = ad - bc$$

The determinant has one particularly special property for our purposes, and that is

Remark. For an $n \times n$ matrix, A, if

$$det(A) = 0$$

then A^{-1} does not exist.

Notice that we've given the condition for existence of an inverse for $n \times n$ matrices, but only calculated the determinant for a 2×2 matrix.

Finding the determinant in higher dimensions is algorithmic, but tedious. For a 3×3 matrix,

$$B = \begin{bmatrix} a & b & c \\ l & m & n \\ x & y & z \end{bmatrix}$$

we pick a row or column[36]. Starting from the first element in that row/column, we cross out the row *and* column containing that element.

[35] Besides, in the modern world computers can compute inverses of matrices.

[36] Generally, it helps to pick the row or column with the most zeros in it.

This leaves us with four un-crossed-out numbers which look like a 2×2 matrix. We take the determinant of that resulting 2×2 matrix and multiply it by the first element of our chosen row or column.

We then move on to the second element in our chosen row or column and do the same thing: we cross the row *and* column containing that element, and take the determinant of the resulting 2×2 matrix[37]. We multiply this determinant by the second element in our chosen row or column. As a quick housekeeping exercise, we now have two pieces:

$$a(mz - ny)$$
$$l(bz - cy)$$

Finally, we do the same thing for the third element in our chosen row or column and get

$$x(bn - mc)$$

We now have three terms which we need to add together in the right way. That right way depends on which row or column we chose. Whether we add or subtract an element depends on what I'm going to call the sign matrix. Starting from the top-left[38] we put a + sign in that position. We then alternate between + and − until we have our 3×3 matrix:

$$Sg_3 = \begin{bmatrix} + & - & + \\ - & + & - \\ + & - & + \end{bmatrix}$$

Since we used the first column, we will add the first term, subtract the second term, and add the third term. This will give us our determinant

$$det(B) = a(mz - ny) - l(bz - cy) + x(bn - mc)$$

Knowing now how to compute the determinant for a 3×3 matrix, we can compute the determinant of a 4×4 matrix by first picking a row or column, and using the same procedure as above to reduce the determinant of a 4×4 matrix to the determinants of 4 3×3 matrices and then add them up according to a 4×4 sign matrix

$$Sg_4 = \begin{bmatrix} + & - & + & - \\ - & + & - & + \\ + & - & + & - \\ - & + & - & + \end{bmatrix}$$

[37]You may have to squint to push the numbers together into a 2×2 matrix this time.

[38]The first element in the matrix

In fact, we can iteratively build up this process[39] to get an algorithm for an $n \times n$ matrix. So long as we know how to compute the determinant of a 2×2 matrix, we theoretically can compute the determinant of *any* square matrix.

The computation of the determinant of an $n \times n$ matrix reduces to the computation of determinants of n $(n-1) \times (n-1)$ matrices, which is just the determinants of $n-1$ $(n-2) \times (n-2)$ matrices, which continues all the way down to 3 2×2 matrices. Again, this is always doable and fairly brainless, but tedious.

IT'S HARD TO FIND EXAMPLES OF REAL-WORLD APPLICATIONS OF THE DETERMINANT. That's because it's a **tool** that we will use to analyse models. It's like discussing the real-world applications of one piece of some industrial machine. What is the real-world application of a screw? On its own, not much it just holds things together, but it's a small tool that can go a long way when we are building things.

In order to use our determinant tool effectively, we need to be well versed in its computation. The examples here won't be dressed up in a story like most of the others throughout this book for these reasons.

Let's consider the matrix

$$A = \begin{bmatrix} 2 & 3 \\ 6 & 10 \end{bmatrix}$$

The determinant is given by

$$9 \cdot 2 - 6 \cdot 10 = -42$$

We can also ask questions like, *for what values of x is the matrix*

$$B = \begin{bmatrix} x & 0 & 5 \\ 7 & 2 & 3 \\ y & 6 & 10 \end{bmatrix}$$

not invertible?

Since this is a 3×3 matrix, we have a choice on which row we pick to build our determinant. The wise choice is always the row or column with a zero in it. In this case, it is the first row. We start with x, strike-through the corresponding row and column and take the determinant of the smaller matrix.

[39]Or pare down the process, depending on which way you look at it

this gives us

$$d_1 = -42x$$

Moving to the next element of the first row, we see a 0. We can ignore this completely since 0 times anything will just give us zero[40].

We move on to the third row, strike-through the corresponding row or column and take the determinant of the resulting matrix.

$$\begin{aligned} d_2 &= 5 \cdot (7 \cdot 6 - 2y) \\ &= 300 - 10y \end{aligned}$$

To get the determinant, we add our pieces together according to the appropriate sign matrix. We end up with

$$det(B) = 300 - 42x - 10y$$

We know that matrix B cannot be inverted if $det(B) = 0$. By setting the determinant to zero and solving for x we get

$$x = 5 - \frac{5}{21}y$$

Therefore, B is not invertible along the line $x = 5 - (5/21)\, y$.

3.8 EIGENVALUES & EIGENVECTORS

Consider the Leslie matrix

$$L = \begin{bmatrix} 0 & 1 & 1 & 0 \\ 1 & 0 & 0 & 0 \\ 0 & 1 & 0 & 0 \\ 0 & 0 & 1 & 0 \end{bmatrix}$$

Imagine we have a population distribution

$$\vec{p}_0 = \begin{bmatrix} 2.3247 \\ 1.7549 \\ 1.3247 \\ 1 \end{bmatrix}$$

[40] And then 0 plus anything doesn't change anything.

What does our population look like after one time step?

$$\vec{p}_1 = L\vec{p}_0 = \begin{bmatrix} 0 & 1 & 1 & 0 \\ 1 & 0 & 0 & 0 \\ 0 & 1 & 0 & 0 \\ 0 & 0 & 1 & 0 \end{bmatrix} \begin{bmatrix} 2.3247 \\ 1.7549 \\ 1.3247 \\ 1 \end{bmatrix} = \begin{bmatrix} 3.07953009 \\ 2.32471603 \\ 1.75483009 \\ 1.3247 \end{bmatrix}$$

But if we look carefully, this is just

$$\vec{p}_1 = \begin{bmatrix} 3.07953009 \\ 2.32471603 \\ 1.75483009 \\ 1.3247 \end{bmatrix} = 1.3247 \begin{bmatrix} 2.3247 \\ 1.7549 \\ 1.3247 \\ 1 \end{bmatrix} = 1.3247\vec{p}_0$$

We can see that our first age class, a_1, makes up $\approx 36\%$ of our total population in p_0 and in p_1. The only thing that has changes is that we have *more* individuals. The *distribution* of individuals among the age classes doesn't change. This is true if we look at p_2 as well

$$\vec{p}_2 = L\vec{p}_1 = L \cdot 1.3247\vec{p}_0 = 1.3247L\vec{p}_0 = 1.3247\vec{p}_1 = 1.3247^2\vec{p}_0$$

In fact, if we try to compute the population at time step n, we just get that

$$\vec{p}_n = 1.3247^n\vec{p}_0$$

THERE IS OBVIOUSLY SOMETHING VERY SPECIAL ABOUT THIS PARTICULAR VECTOR AND THIS PARTICULAR MATRIX. No matter how many time steps we look at, the distribution of individuals among the age classes doesn't change. Our age class a_1 always makes up $\approx 36\%$ of our population. We can imagine why this kind of information[41] might be useful. If we are planning a neighbourhood, it would be nice to know if the proportion of children will stay the same. It makes it very easy to plan schools, parks, etc. If this were an animal population, it helps to know that we should expect the same proportion of offspring each time step. If we scale our resources by the same amount, we should be able to keep up with the population growth.

These special vectors are useful in other applications as well. Let's say we have a lens as in figure 3.13. Light from a source at point K[42] hits the circular lens at a certain trajectory and bends due to the curvature

[41]What population distribution is unchanging in time?

[42]For, say, *K*rab

in the glass. How much it curves depends on where we hit the lens. We can do this with a matrix.

$$A = \begin{bmatrix} 1 & 0 \\ -P & 1 \end{bmatrix}$$

The beam travelling along

$$\vec{t}_{old} = [1, 1]]^T$$

will change direction and travel along

$$\vec{t}_{new} = A\vec{t}_{old} = [1, 1 - P]$$

as shown in figure 3.13.

Figure 3.13: A circular lens changes the direction of a beam of light. If we represent the beam as a vector, then we can use a matrix to describe the lens and its effect on light.

For a general beam that starts at point K and travels in the direction

$$\vec{t}_{old} = [y, x]^T$$

will travel in the direction

$$\vec{t}_{new} = [y, x - Py]^T$$

The light beam that comes directly at the lens isn't affected at all. If

$$\vec{t}_{old} = [0, 1]^T$$

Then

$$\vec{t}_{new} = R\vec{T}_{old} = [0, 1]^T = \vec{t}_{old}$$

we move along the exact same direction!

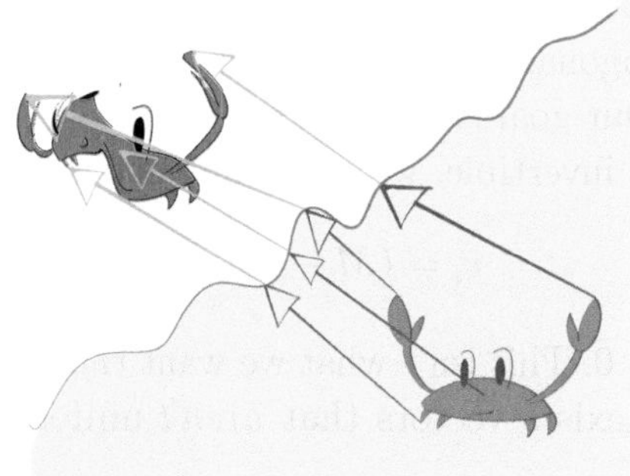

Figure 3.14: The surface of water can be viewed as infinitely many tiny lenses that all reflect light from each point of our crab slightly differently. This is why looked at something that is underwater often looks distorted. The more turbulent the surface of the water, the more distorted the image will look.

These special vectors are called **eigenvectors**[43]. They are special vectors that only change their **magnitude** when multiplied by the associated matrix, but not their trajectory. The magnitude changes a known amount called an **eigenvalue**. To formalize this

---

Definition 3.8.1. For an $n \times n$ matrix, M, there is a special set of vectors, $\vec{v}_i$, such that

$$M\vec{v}_i = \lambda_i \vec{v}_i \tag{3.10}$$

The vector $\vec{v}_i$ is known as an **eigenvector** of matrix M and λ_i is the associated **eigenvalue**.

So far, finding an eigenvector and eigenvalue seems like magic[44], but there is a systematic way of finding them. If we rearrange equation (3.10), we can write it as

$$\left(M - \lambda_i I_n\right)\vec{v}_i = \vec{0}$$

[43]Eigen is German for *own*. This is a matrix's "own" vector in that it turns into itself.

[44]Because we have just guessed, conveniently, and found them.

The matrix $M - \lambda_i I_n$ is the same as M, but we subtract λ_i from every element on the diagonal[45]. The symbol $\vec{0}$ represents a vector where every element is zero. Our goal is to solve for $\vec{v}_i$.

If $M - \lambda_i I_n$ is invertible, we can easily show that

$$\vec{v}_i = (M - \lambda_i I_n)^{-1} \vec{0}$$

which tells us $\vec{v}_i = \vec{0}$. This isn't what we want though, because we've seen above that there exists vectors that *aren't* uniformly zero that satisfy this property.

If $\vec{v}_i$ is not a vector of zeroes, that must mean that $M - \lambda_i I_n$ cannot be invertible which means $\det(M - \lambda_i I_n) = 0$.

Remark. The eigenvalues, λ_i of a matrix M are solutions to

$$det\,(M - lambda_i I_n) = 0$$

Knowing the eigenvalues, we can then find the eigenvectors by using the definition, i.e. equation 3.10, to solve for the components of $\vec{v}_i$.

Let's consider a population divided into two age groups: children and adults; the corresponding Leslie matrix might look like

$$L = \begin{bmatrix} 0.8 & 1 \\ 0.2 & 0.6 \end{bmatrix}$$

We can interpret this to mean that in each time step, 80% of children are still considered children and 20% move on to adulthood, each adult produces on average one child per time step, and 60% of adults survive to the next time step. The eigenvectors of L will tell us possible conserved population distributions, and the eigenvalues will tell us the overall growth rate of a population in that distribution.

We start by writing

$$L - \lambda_i I_2 = \begin{bmatrix} 0.8 & 1 \\ 0.2 & 0.6 \end{bmatrix} - \begin{bmatrix} \lambda_i & 0 \\ 0 & \lambda_i \end{bmatrix} = \begin{bmatrix} 0.8 - \lambda_i & 1 \\ 0.2 & 0.6 - \lambda_i \end{bmatrix}$$

then we take the determinant

$$\begin{aligned} det\,(M - \lambda_i I_2) &= (0.8 - \lambda)(0.6 - \lambda_i) - 0.2 \\ &= \lambda_i^2 - 1.4\lambda_i + 0.28 \\ &= 0 \end{aligned}$$

[45]Recall that I_n is the $n \times n$ identity matrix.

This is just a quadratic equation. We can solve for λ_i using the quadratic formula. We will get **two** possible λ's; both are eigenvalues of matrix L.

$$\lambda_1 = \frac{1.4 + \sqrt{0.84}}{2} \approx 1.1582575695$$
$$\lambda_2 = \frac{1.4 - \sqrt{0.84}}{2} \approx 0.2417424305$$

From here, we can find the eigenvectors associated with these eigenvalues. We will get two different eigenvectors. Let's start with $\vec{v}_1$. From equation (3.10), we can write

$$L\vec{v}_1 = \lambda_1\vec{v}_1$$
$$\begin{bmatrix} 0.8 & 1 \\ 0.2 & 0.6 \end{bmatrix} \begin{bmatrix} a \\ b \end{bmatrix} = \begin{bmatrix} \lambda_1 a \\ \lambda_1 b \end{bmatrix}$$
$$\begin{bmatrix} 0.8a + b \\ 0.2a + 0.6b \end{bmatrix} = \begin{bmatrix} \lambda_1 a \\ \lambda_1 b \end{bmatrix}$$

This gives us two equations with two unknowns. Solving the first for b, we get

$$b = (\lambda_1 - 0.8)a$$

and we put this into the second equation

$$0.2a + 0.6(\lambda_1 - 0.8)a = lambda_1(\lambda_1 - 0.8)a$$

Collecting like terms gives us $0a = 0$, which is always true. This isn't a problem, it just means we get to pick either a or b to be whatever we want. If we choose $a = 1$, then $b = \lambda_1 - 0.8 \approx 0.3582575695$ and

$$\vec{v}_1 = \begin{bmatrix} \lambda_1 - 0.8 \\ 1 \end{bmatrix} \approx \begin{bmatrix} 0.3582575695 \\ 1 \end{bmatrix}$$

This means that if our population is approximately 74% adults and 26% children, it will remain at those proportions and grow by approximately 1.16 people per time step.

We can do the same thing for λ_2. The process is identical, except we use λ_2 in equation (3.10).

$$L\vec{v}_2 = \lambda_1 \vec{v}_2$$

$$\begin{bmatrix} 0.8 & 1 \\ 0.2 & 0.6 \end{bmatrix} \begin{bmatrix} a \\ b \end{bmatrix} = \begin{bmatrix} \lambda_2 a \\ \lambda_2 b \end{bmatrix}$$

$$\begin{bmatrix} 0.8a + b \\ 0.2a + 0.6b \end{bmatrix} = \begin{bmatrix} \lambda_1 a \\ \lambda_1 b \end{bmatrix}$$

Again, when solving these two equations we'll get $0a = 0$, and $b = (\lambda_2 - 0.8)a$. We can pick $a = 1$ and then we get $b = \lambda_2 - 0.8 \approx -0.5583$. So our second eigenvector, the one associated with λ_2 is

$$\vec{v}_2 = \begin{bmatrix} 1 \\ \lambda_2 - 0.8 \end{bmatrix} \approx \begin{bmatrix} 1 \\ -0.5583 \end{bmatrix}$$

If our population is in these proportions, then it will grow at a rate of approximately 0.2417.

... Yes, the second eigenvalue above is nonsensical *in context*. You cannot have negative population numbers and so this eigenvector is irrelevant to our current problem[46].

There's no need to stop at one time step, we could look at many very easily. As long as our population is distributed in the same ratio as the eigenvector, we have

$$L^n \vec{v}_1 = \lambda_1^n \vec{v}_1$$

THIS IDEA OF EIGENVALUES AND EIGENVECTORS IS UBIQUITOUS IN MATH. They can help us analyse how things change while anchoring us to something conserved[47].

[46]Probably a good thing too because over many, many time steps this eigenvector will lead to the population dying since $\lambda_2^n \to 0$ as $n \to \infty$.

[47]In the case of matrices, eigenvalues conserve *vector direction*.

PRACTICE PROBLEMS

Vectors and Vector Addition

1. In terms of cardinal directions, we can simplify things by considering south as negative north and west as negative east. Then we can write any position as a vector $\mathbf{v} = [N, E]$.
   (a) How would you write 5 km south and 8 km east as a two dimensional vector?
   (b) How would you write 2.5 km north and -5 km west as a two-dimensional vector?
   (c) If you were to travel in a straight line, instead of component-wise, how far would you travel?
2. Does $\vec{v} + c$ make sense? Can we add numbers and vectors? Why not?
3. If I set an origin to my current position (where forward is in front of me and right is my right and up is above me) and move to a point 3 km to the right, 4 km forward and 500 m up then move 5 km left, 1 km $forward$ and 0 km up or down, how far from my original position am I?
4. Sackville, NB is at coordinates $\vec{S} = [-64.37, 45.89]^T$ given as $[latitude, longitude]^T$. Toronto is at $[\vec{T} = [-79.38, 43.65]^T$. What is the vector that connects Sackville to Toronto?
5. What is the length of the resultant vector in question 4? The straight line distance in more natural units is 1,211km. Can you use this information and some light trigonometry to determine how many kilometres are between each latitude line? Each longitude line?

The Dot Product

1. The notes above make a point of talking about things being a certain way because that's what we should expect. Thing about the dot product of three numbers $\vec{a} \cdot \vec{b} \cdot \vec{c}$. Given what you know about *two* vectors and their dot product, what do you think this might mean? Do you expect $\vec{a} \cdot \left(\vec{b} \cdot \vec{c}\right) = \left(\vec{a} \cdot \vec{b}\right) \cdot \vec{c}$? Does it?
2. Is $\vec{u} \cdot k\vec{u}$ the same as $k(\vec{v} \cdot \vec{u})$? Should we expect this to be true?

3. Say you are trying to hit an archery target that is 0.41m wide, can you come up with a relationship between how far away you are from the target, l and how much leniency you are allowed in your aim θ? See figure 3.15 for a schematic.

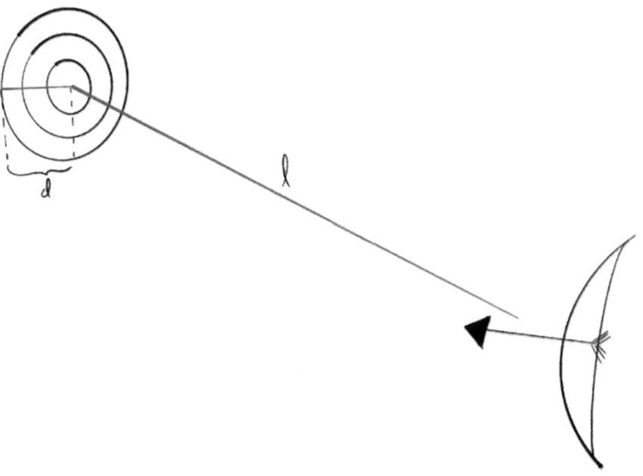

Figure 3.15: Figure for question 2

4. Given a vector $\vec{v} = [x, y]^T$, can you find a vector $\vec{v}_\perp$ that is perpendicular to $\vec{v}$ and has length $||\vec{v}_\perp|| = 1$?

5. The dot product formula can also be thought of as a function of the angle between two vectors θ.

$$\vec{a} \cdot \vec{b} = ||\vec{a}|| ||\vec{b}|| \cos(\theta) = f(\theta)$$

   What values of θ maximize/minimize the dot product? Which of these values correspond to maxima and which correspond to minima?

6. The length of a shadow of a building can be written as

$$S = ||\vec{s}|| \frac{\vec{s} \cdot \vec{g}}{\vec{s} \cdot \vec{b}}$$

   where $\vec{s}$ is the vector between the sun and the (flat) earth, $\vec{b}$ is the vector describing the building, and $\vec{g}$ is the vector of the ground.

Typically, the ground is given as a horizontal vector

$$\vec{g} = [1, 0]^T$$

and the building is given as

$$\vec{b} = [0, h]$$

where h is the height of the building. The sun moves through the sky throughout the day, starting at the horizon and ending at the horizon. If we ignore night-time, we can model the course of the sun through the sky with vector

$$\vec{s} = \left[\sin\left(\frac{\pi}{12}t\right)^2, \cos\left(\frac{\pi}{12}t\right)^2\right]^T$$

Using the definitions of S above and the vectors given, write a function $S(h, t)$.

7. A new by-law in our imaginary city where we are building these towers states that a house must have a minimum of 4 hours of sun in the morning (so $0 < t < 6$) how far away must we build from our building to ensure this?

8. If we build *two* buildings how far apart should they be to ensure that the space between them is never in complete shadow?

Matrices

1. Above there is an example of a matrix multiplication involving the matrix

$$M = \begin{bmatrix} 0 & 2 \\ -2 & 0 \end{bmatrix}$$

What vector do you get when you multiply M by the vector $\vec{v} = [x, y]^T$?

2. Call the answer to question 1 $\vec{w}$. What is $\vec{w}\dot{\vec{v}}$? What does this tell you about the rotation induced by the matrix M?

3. A Leslie matrix takes a vector of a population divided by age classes, $\vec{v}_t$, at a time t and gives you the population sizes for the age classes

at time $t+1$. Assume $\vec{v}_t = [B_t, T_t, C_t, YA_t, A_t, S_t]^T$ where B stands for babies, T for toddlers, C for children (school-aged and above), YA for young adults (university-aged to whatever is younger than me), A for adults, and S for seniors. Assume the following Leslie matrix:

$$L = \begin{bmatrix} 0 & 0 & 0 & 5000 & 10000 & 0 \\ 0.99 & 0 & 0 & 0 & 0 & 0 \\ 0 & 0.9 & 0 & 0 & 0 & 0 \\ 0 & 0 & 0.9 & 0 & 0 & 0 \\ 0 & 0 & 0 & 0.8 & 0 & 0 \\ 0 & 0 & 0 & 0 & 0.7 & 0 \end{bmatrix}$$

We know that $\vec{v}_{t+1} = \vec{v}_t = [B_{t+1}, T_{t+1}, C_{t+1}, YA_{t+1}, A_{t+1}, S_{t+1}]^T = L\vec{v}_t$. Write out v_{t+1} explicitly by carrying out the matrix multiplication.

4. Can you interpret the equations for B_{t+1}, T_{t+1}, C_{t+1}, YA_{t+1}, A_{t+1}, and S_{t+1}?

5. What is the angle between $\vec{v}_{t+1}$ and $\vec{v}_t$?

6. Which vector has a larger magnitude?

Eigenvalues and Eigenvectors

1. What are the eigenvalues of

$$M = \begin{bmatrix} 5 & 2 \\ 6 & 1 \end{bmatrix}$$

2. What are the eigenvalues of

$$L = \begin{matrix} 0 & 3 \\ 0.99 & 0.88 \end{matrix}$$

3. Find the eigenvectors of the above matrix.

4. If the above matrix is a Leslie matrix for Children and Adults in a population, what might the eigenvalues represent? What about the eigenvectors?

5. If you start from a vector $[C_0, A_0]$, and multiply by L^N, where N is very very large, what vector do you expect to end up at[48]?

[48] Start multiplying $L^m C_0$ for different values of m until you see a pattern.

CHAPTER 4

Derivatives in Multiple Dimensions

Combining linear algebra with calculus, we now have the necessary tools to describe how functions change in multiple dimensions.

We know that when we have a function of two variables, then we have two separate directional derivatives. In other words

$$\text{for a function } f(x, y) \text{ we have two derivatives: } \frac{\partial f}{\partial x} \text{ and } \frac{\partial f}{\partial y}$$

One is the rate of change in the x-direction, and the other is the rate of change in the y-direction.

Definition 4.0.1. We can group these two pieces of information in **vector of derivatives** as

$$\left[\frac{\partial f}{\partial x}, \frac{\partial f}{\partial y}\right]^T$$

This vector of partial derivatives is called the **gradient of** $\mathbf{f(x, y)}$, and it has the symbol

$$\nabla f = \left[\frac{\partial f}{\partial x}, \frac{\partial f}{\partial y}\right]^T$$

DOI: 10.1201/9781003265405-4

This doesn't just work in two dimensions either. If we have a function $g(x, y, z)$ with three partial derivatives:

$$\frac{\partial g}{\partial x} \quad \frac{\partial g}{\partial y} \quad \frac{\partial g}{\partial z}$$

then

$$\nabla g = \left[\frac{\partial g}{\partial x}, \frac{\partial g}{\partial y}, \frac{\partial g}{\partial z}\right]^T$$

If we can see the pattern here, we can write a definition in any dimension:

Definition 4.0.2. For a function of n-dimension, $f(x_1, x_2, x_3, \cdots, x_n)$ we define the gradient of f as

$$\nabla f = \left[\frac{\partial f}{\partial x_1}, \frac{\partial f}{\partial x_2}, \frac{\partial f}{\partial x_3}, \cdots, \frac{\partial f}{\partial x_n}\right]^T$$

The gradient of a function gives the **direction of steepest change.** This is often a useful quantity to know. For instance, consider the function

$$\begin{aligned} H(p_1, p_2, p_3) = &- p_1 \ln(p_1) - p_2 \ln(p_2) + p_3 \ln(p_3) \\ &- (1 - p_1 - p_2 - p_3) \ln(1 - p_1 - p_2 - p_3) \end{aligned}$$

This function is known as the **Shannon diversity index** for four species. If we know the proportions of which species 1, 2, and 3 appear in an ecosystem given by p_1, p_2, and p_3, respectively, then p_4 must be the rest; i.e. $p_4 = 1 - p_1 - p_2 - p_3$[1].

Generally, species diversity is a good a thing. If we are interested in a certain ecosystem, the gradient of the Shannon diversity index can tell us how we might increase diversity given proportions of different species currently present in the ecosystem.

The gradient is given by

$$\nabla H(p_1, p_2, p_3) = \begin{bmatrix} \frac{\partial H}{\partial p_1} \\ \frac{\partial H}{\partial p_2} \\ \frac{\partial H}{\partial p_3} \end{bmatrix} = \begin{bmatrix} -\ln(p_1) + \ln(1 - p_1 - p_2 - p_3) \\ -\ln(p_2) + \ln(1 - p_1 - p_2 - p_3) \\ -\ln(p_3) + \ln(1 - p_1 - p_2 - p_3) \end{bmatrix}$$

[1]But what if we know p_1, p_2, and p_4? Well, we can just re-number our species so p_4 is the unknown one!

Let's say we report that in our particular ecosystem of study, we see that of the 12 individuals present, five are triceratops, two are giraffes, four are cows, and the rest are puppies[2].

What strategy might we need to take to increase the diversity?

We can convert these counts to proportions by dividing by the total number of animals present:

$$\begin{aligned} p_1 &= \frac{5}{12} \\ p_2 &= \frac{2}{12} = \frac{1}{6} \\ p_3 &= \frac{4}{12} = \frac{1}{3} \\ p_4 &= 1 - p_1 - p_2 - p_3 = \frac{1}{12} \end{aligned}$$

Plugging these values into our gradient gives

$$\nabla H\left(\frac{5}{12}, \frac{1}{6}, \frac{1}{3}\right) = \begin{bmatrix} -\ln\left(^5/_{12}\right) + \ln\left(^1/_{12}\right) \\ -\ln\left(^1/_6\right) + \ln\left(^1/_{12}\right) \\ -\ln\left(^1/_3\right) + \ln\left(^1/_{12}\right) \end{bmatrix} \approx \begin{bmatrix} -1.61 \\ -0.69 \\ -1.39 \end{bmatrix}$$

[2]A strange ecosystem indeed.

This means that to increase diversity[3] we should move in the *negative* p_1-direction, the *negative* p_2-direction and the negative p_3-direction. Since we're dealing with real animals that can't be divided into pieces[4], we might think to take away one from each of species 1, 2, and 3, respectively, and replace them with puppies[5]. We would then get *new* proportions of

$$\begin{aligned}
{}_{new}p_1 &= \frac{4}{12} = \frac{1}{3} \\
{}_{new}p_2 &= \frac{1}{12} \\
{}_{new}p_3 &= \frac{3}{12} = \frac{1}{4} \\
{}_{new}p_4 &= 1 - {}_{new}p_1 - {}_{new}p_2 - {}_{new}p_3 = \frac{4}{12} = \frac{1}{3}
\end{aligned}$$

We can compute the Shannon Index for the old proportions and our new proportions

$$\begin{aligned}
H\left(5/12, 2/12, 4/12\right) &\approx 1.24 \\
H\left(4/12, 1/12, 3/12\right) &\approx 1.29
\end{aligned}$$

and see that indeed the new diversity index is greater than the old one! We could play the same game again to further increase the diversity, by plugging in ${}_{new}p_1$, ${}_{new}p_2$ and ${}_{new}p_3$ into our gradient vector.

$$\nabla H\left(\frac{4}{12}, \frac{1}{12}, \frac{3}{12}\right) = \begin{bmatrix} -\ln\left(4/12\right) + \ln\left(4/12\right) \\ -\ln\left(1/12\right) + \ln\left(4/12\right) \\ -\ln\left(3/12\right) + \ln\left(4/12\right) \end{bmatrix} \approx \begin{bmatrix} 0 \\ 1.39 \\ 0.29 \end{bmatrix}$$

This vector tells us that if we want to further increase diversity, we should leave the triceratops alone, and increase our giraffes and cows, then adjust puppies accordingly[6]. The will lead to new proportions that

---

[3] According to the Shannon Index, there are many measures of diversity.

[4] They can, I guess, but it's gory and sad

[5] It's important to keep the total number of individuals the same, or else we can't compare diversity.

[6] In case you're worried, we're not culling animals; we're just sending them on a nice holiday.

we will call P_i and are given as

$$P_1 = \frac{4}{12}$$
$$P_2 = \frac{2}{12}$$
$$P_3 = \frac{4}{12}$$
$$P_4 = 1 - P_1 - P_2 - P_3 = \frac{2}{12}$$

which yields a Shannon Index of

$$H(P_1, P_2, P_3) \approx 1.33$$

Higher still!

We can keep doing this to try and get a higher and higher diversity index by sending some animals on holiday, and bringing back others to take their place. Would it not but convenient if we knew exactly how many animals we would need and in what proportions we would need them in in order to maximize diversity? Could this not make our job much easier?

It can, and it will.

Critical Points

Critical points of functions of two variables, $f(x, y)$, have two values – an x value and a y value. Otherwise, the definition is the same

---

Definition 4.0.3. Given a function $f(x, y)$, the point (x^*, y^*) is a **critical point** of $f(x, y)$ if

$$\left.\frac{\partial f}{\partial x}\right|_{(x,y)=(x^*,y^*)} = 0$$
$$\left.\frac{\partial f}{\partial y}\right|_{(x,y)=(x^*,y^*)} = 0$$

or, equivalently,

$$\nabla f = \vec{0} = [0, 0]^T$$

Since we are looking for *two* values this time, it makes sense that we have *two* equations.

There are three types of critical points for functions of two variables: A critical point can either be:

- **A local maximum**
- **A local minimum**
- **A saddle point**

A local maximum/minimum is the same as in one dimension: it is a point that is higher/lower than all the points surrounding it[7]. A saddle point is a little different. A saddle point is a point that is a maximum in one direction and a minimum in another. All three are shown in figure 4.1.

In our previous example about diversity, we can use the gradient and the concept of critical points to find when the Shannon Index of Diversity is maximized[8].

As a reminder the Shannon Diversity Index for four species is

$$\begin{aligned} H(p_1, p_2, p_3) = &- p_1 \ln(p_1) - p_2 \ln(p_2) - p_3 \ln(p_3) \\ &- (1 - p_1 - p_2 - p_3) \ln(1 - p_1 - p_2 - p_3) \end{aligned}$$

The gradient, as we calculated above, is

$$\nabla H = \begin{bmatrix} -\ln(p_1) + \ln(1 - p_1 - p_2 - p_3) \\ -\ln(p_2) + \ln(1 - p_1 - p_2 - p_3) \\ -\ln(p_3) + \ln(1 - p_1 - p_2 - p_3) \end{bmatrix}$$

In order to find the critical point[9] we need to set $\nabla H = \vec{0}$. This gives us three equations that need to be solved simultaneously

$$\begin{aligned} -\ln(p_1) + \ln(1 - p_1 - p_2 - p_3) &= 0 \\ -\ln(p_2) + \ln(1 - p_1 - p_2 - p_3) &= 0 \\ -\ln(p_3) + \ln(1 - p_1 - p_2 - p_3) &= 0 \end{aligned}$$

From the third equation, we can say that

$$\begin{aligned} \ln(p_3) = \ln(1 - p_1 - p_2 - p_3) \\ p_3 = 1 - p_1 - p_2 - p_3 \\ p_3 = \frac{1 - p_1 - p_2}{2} \end{aligned}$$

[7]Remember, the derivative is a **local** operator; it only gives us information about a local region

[8]Or, possibly, minimized or saddle-ized

[9]Or points

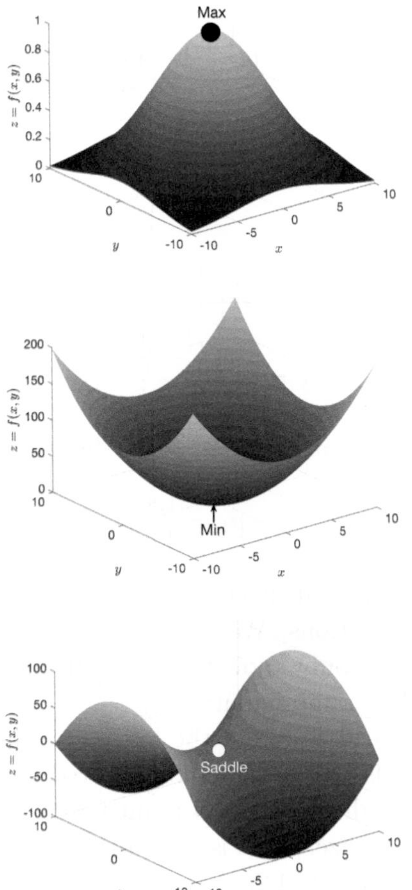

Figure 4.1: Some clear examples of a local maximum, local minimum, and saddle point for functions of two-dimensions. The functions are $f_{max}(x, y) = \mathrm{e}^{-(x^2 - +y^2}$, $f_{min}(x, y) = x^2 + y^2$ *and* $f_{saddle}(x, y) = x^2 - y^2$.

then plugging this into the second equation, we get

$$\begin{aligned}
\ln(p_2) &= \ln\left(1 - p_1 - p_2 - \frac{1 - p_1 - p_2}{2}\right) \\
p_2 &= \frac{1 - p_1 - p_2}{2} \\
p_2 &= \frac{1 - p_1}{3}
\end{aligned}$$

plugging the expressions for p_3 and p_2 into the first equation, we get everything in terms of p_1

$$\ln(p_1) = \ln\left(1 - p_1 - \frac{1-p_1}{3} - \frac{1-p_1-p_2}{2} - \frac{1-p_1-{}^{1-p_1}/_3}{2}\right)$$
$$p_1 = \frac{1}{4}$$

which then gives

$$p_2 = \frac{1-\frac{1}{4}}{3} = \frac{1}{4}$$
$$p_3 = \frac{1-\frac{1}{4}-\frac{1}{4}}{2} = \frac{1}{4}$$

Then using the face that $p_4 = 1 - p_1 - p_2 - p_3$ we see that

$$p_4 = \frac{1}{4}$$

What this tells us is that diversity[10] is maximized[11] when all species appear in equal proportions. We don't necessarily **know** that this is indeed a maximum, we only know that it is our only candidate. For all we know, this could be a minimum[12].

Checking values a little bit to the left and a little bit to the right in higher dimensions is a little more complicated than in one dimension because our idea of right or left of a point needs more clarification. In $2D$, which we can visualize, there is an entire circle of directions around our point as shown in figure 4.2.

The following sections will give us the tools we need to say definitively that this is a maximum[13].

Second Derivative

The second derivative gets a little more complicated when we are in two dimensions. Let's say we have a function $f(x, y)$ and take the partial derivative with respect to x. Then we have

$$\frac{\partial f}{\partial x}$$

[10] In the Shannon sense

[11] Or minimized, or saddle-ized

[12] Or a saddle

[13] Spoiler alert

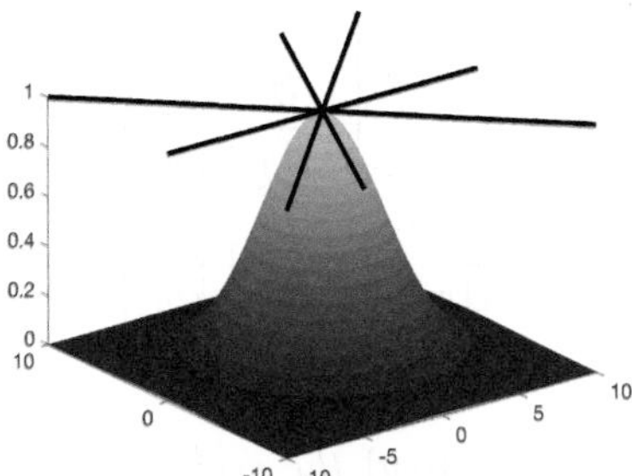

Figure 4.2: The concept of left and right of a point gets a little messy in higher dimensions. Left and right are completely dependent on your point of view and as a result there are infinitely many lefts and rights around a point.

We can also take a derivative with respect to y

$$\frac{\partial f}{\partial y}$$

These derivatives are, generally, still a function of x and y. In other words, they still exist in two dimensions.

In one dimension, to find the second derivative we would just take the derivative again, but in two dimensions we have to ask *which derivative do we take?* If we have a partial derivative with respect to x, then there are two options for the partial second derivative:

$$\frac{\partial^2 f}{\partial x^2} = \frac{\partial}{\partial x}\left(\frac{\partial f}{\partial x}\right)$$
$$\frac{\partial^2 f}{\partial y \partial x} = \frac{\partial}{\partial y}\left(\frac{\partial f}{\partial x}\right)$$

The latter is called the **mixed partial derivative**. We read the denominator from right to left: *we take a derivative with respect to x then with respect to y.*

Similarly, if we start with the partial derivative with respect to y,

$$\frac{\partial f}{\partial y}$$

we get two partial second derivatives:

$$\frac{\partial^2 f}{\partial y^2} = \frac{\partial}{\partial y}\left(\frac{\partial f}{\partial y}\right)$$
$$\frac{\partial^2 f}{\partial x \partial y} = \frac{\partial}{\partial x}\left(\frac{\partial f}{\partial y}\right)$$

And again for the latter, we read *the second derivative with respect to y then x.*

This may lead to a point of confusion when figuring out which differentiation comes first, and which comes second. Luckily, calculus isn't so cruel a romantic interloper that we need to worry about such small details. So as long as the derivatives

$$\frac{\partial^2 f}{\partial x \partial y} \text{ and } \frac{\partial^2 f}{\partial y \partial x}$$

are continuous, then

$$\frac{\partial^2 f}{\partial x \partial y} = \frac{\partial^2 f}{\partial y \partial x}$$

Therefore, there are only *three* second derivatives for a function of two variables, as opposed to four.

This derivative, $\frac{\partial^2 f}{\partial x \partial y}$ is called the **mixed partial derivative** of $f(x, y)$. The "mixed" implies that we are taking (at least) two derivatives. We can generalize the above fact to higher-order derivatives. Again, as long as everything is continuous, all mixed partial derivatives are equal. As an example,

$$\frac{\partial^3}{\partial x \partial y \partial x} = \frac{\partial^3}{\partial x^2 \partial y} = \frac{\partial^3}{\partial y \partial x^2}$$

THE INTERPRETATION OF THE SECOND DERIVATIVES is the same as in one-dimension 66.67%[14] of the time. The pure second derivatives:

$$\frac{\partial^2 f}{\partial x^2}$$
$$\frac{\partial^2 f}{\partial y^2}$$

[14] Rounded, obviously

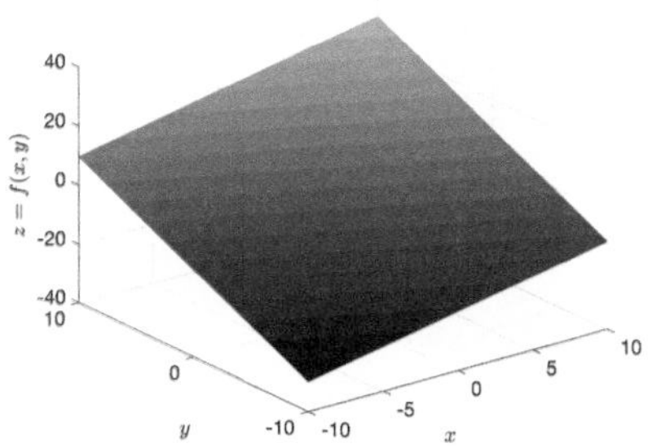

Figure 4.3: A flat surface $f(x, y) = x + 2y$

tell you about the *concavity* of the surface in the x and y directions respectively. The interpretation of the **mixed partial derivative** is a little less straight-forward[15]. One way to think of the mixed partial derivative is the amount of *twist* exhibited by a function.

For instance, think of the surface

$$flat(x, y) = x + 2y$$

This is just a flat surface like in figure 4.3. The first partial derivatives are

$$\frac{\partial flat}{\partial x} = 1$$
$$\frac{\partial flat}{\partial y} = 2$$

The pure second derivatives are

$$\frac{\partial^2 flat}{\partial x^2} = 0$$
$$\frac{\partial^2 flat}{\partial y^2} = 0$$

We expect this function to have zero concavity[16], and it does. Similarly,

$$\frac{\partial flat}{\partial y \partial x} = \frac{\partial flat}{\partial x \partial y} = 0$$

[15]You may notice a theme when we move from one dimension to multiple dimensions: things get less straight forward. This is not a coincidence. Generally, when we add dimensions, we gain possibilities. When we gain possibilities, we have to test more cases.

[16]Because it's flat

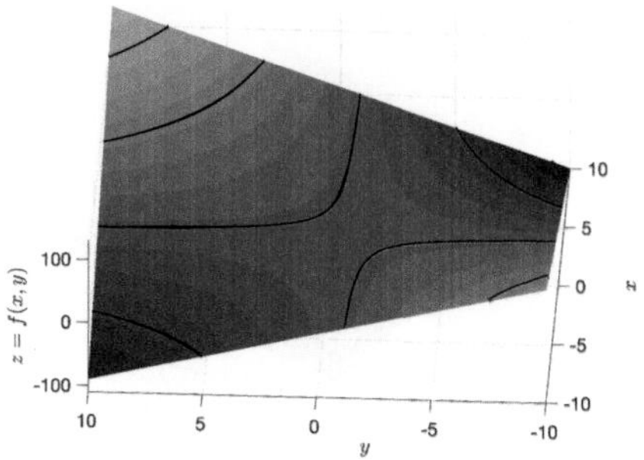

Figure 4.4: A flat surface $f(x, y) = x + 2y + xy$

The minimum term we could add to get some kind of mixed partial derivative is xy, so let's do that[17].

$$twist(x, y) = x + 2y + xy$$

The partial derivatives in this case are

$$\frac{\partial twist}{\partial x} = 1 + y$$
$$\frac{\partial twist}{\partial y} = 2 + x$$

and the two pure second partial derivatives are

$$\frac{\partial^2 twist}{\partial x^2} = 0$$
$$\frac{\partial^2 twist}{\partial y^2} = 0$$

Still zero concavity! Is the function completely flat though? No, because

$$\frac{\partial twist}{\partial x \partial y} = \frac{\partial twist}{\partial y \partial x} = 1$$

And we can see when this means when we look at figure 4.4. It looks like we have grabbed the edges of our flat plane, and twisted the ends in opposite directions. As above, this is one of those examples of the added

[17]Knowing to add this came from experience mostly; don't worry if you wouldn't know off the top of your head what to add; it comes with building up a mathematical intuition.

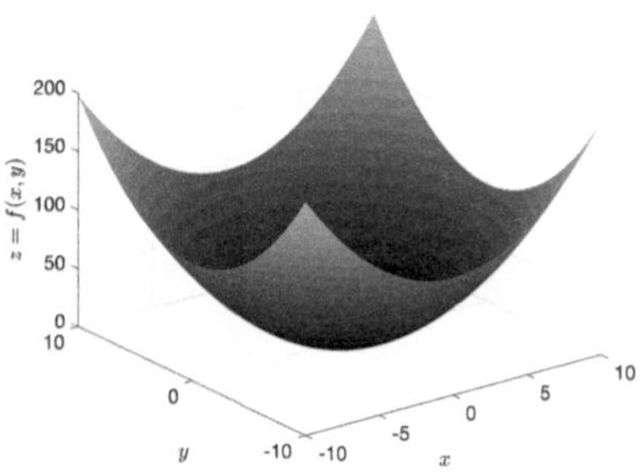

Figure 4.5: A flat surface $f(x, y) = x^2 + y^2$

dimensions complicating matters. Not only do we need to worry about bending a function[18], we also need to worry about *twisting*. It just so happens that in one dimension, there's no direction in which to twist.

Can we always see this twisting? Yes and no. Consider the function

$$c(x, y) = x^2 + y^2$$

The partial derivatives are

$$\begin{aligned} \frac{\partial c}{\partial x} &= 2x \\ \frac{\partial c}{\partial y} &= 2y \end{aligned}$$

and the three partial derivatives are

$$\begin{aligned} \frac{\partial^2 c}{\partial x^2} &= 2 \\ \frac{\partial^2 c}{\partial y^2} &= 2 \frac{\partial^2 c}{\partial x \partial y} = \frac{\partial^2 c}{\partial y \partial x} &= 0 \end{aligned}$$

In this case we expect the function to be concave up in both the x and y directions, and to not twist. We can see this in figure 4.5. If we again add our test term[19] we get the new function

$$c_{twist} = x^2 + y^2 + xy$$

[18]i.e. concavity

[19]Different, and less anxiety inducing, than a term test

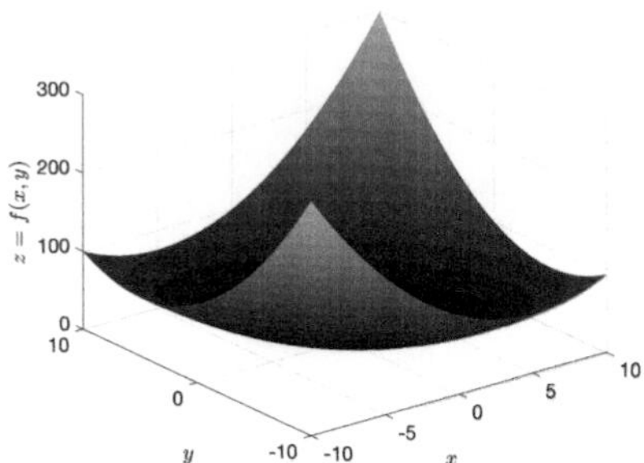

Figure 4.6: A flat surface $f(x, y) = x^2 + y^2 + xy$

which has partial derivatives

$$\frac{\partial c_{twist}}{\partial x} = 2x + y$$
$$\frac{\partial c_{twist}}{\partial x} = 2y + x$$

and second partial derivatives

$$\frac{\partial^2 c_{twist}}{\partial x^2} = 2$$
$$\frac{\partial^2 c_{twist}}{\partial y^2} = 2$$
$$\frac{\partial^2 c}{\partial x \partial y} = \frac{\partial^2 c_{twist}}{\partial y \partial x} = 1$$

If we plot this surface, as in figure 4.6 it doesn't look *that* much different than our surface without the twist.

What gives? Did I tell you something that only works sometimes[20]? If we look at our numbers, the curvature is literally two times the size of the twist, so it dominates in this example so we only *see* the curvature, even though curvature and twisting exist. In our first example, the twisting was extremely evident because there was no curvature. The twisting was *infinitely larger* than the curvature, and so it dominates the picture.

[20]I would never and I am appalled by the accusation.

When we move from one dimension to n dimensions, our first derivative became a vector of derivatives[21]. Our second derivatives become a matrix called the **Hessian matrix**.

Definition 4.0.4. For a function $f(x, y)$, the **Hessian matrix** of second derivatives is defined as the matrix of second partial derivatives

$$H_2 = \begin{bmatrix} \frac{\partial^2 f}{\partial x^2} & \frac{\partial^2 f}{\partial x \partial y} \\ \frac{\partial^2 f}{\partial y \partial x} & \frac{\partial^2 f}{\partial y^2} \end{bmatrix}$$

For an n-dimensional function, $f(x_1, x_2, \cdots, x_n)$ the Hessian matrix is defined as

$$H_n = \begin{bmatrix} \frac{\partial^2 f}{\partial x_1^2} & \frac{\partial^2 f}{\partial x_1 \partial x_2} & \cdots & \frac{\partial^2 f}{\partial x_1 \partial x_{n-1}} & \frac{\partial^2 f}{\partial x_1 \partial x_n} \\ \frac{\partial^2 f}{\partial x_2 \partial x_1} & \frac{\partial^2 f}{\partial x_2^2} & \cdots & \frac{\partial^2 f}{\partial x_2 \partial x_{n-1}} & \frac{\partial^2 f}{\partial x_2 \partial x_n} \\ \vdots & & \ddots & & \vdots \\ \frac{\partial^2 f}{\partial x_n \partial x_1} & \frac{\partial^2 f}{\partial x_n \partial x_2} & \cdots & \frac{\partial^2 f}{\partial x_n \partial x_{n-1}} & \frac{\partial^2 f}{\partial x_n^2} \end{bmatrix}$$

Second derivatives and critical points

THE NEXT THING WE MAY BE CONCERNED ABOUT IS HOW WE CAN USE THE SECOND DERIVATIVE TO GIVE US INFORMATION ABOUT A CRITICAL POINT. In one dimension this was fairly easy, we would look at the sign of the second derivative. This worked because there was only *one* second derivative. Now that we have three, we need a different rule.

Definition 4.0.5. The **second derivative test** for a function of two variables tells us when a critical point is a max/min/saddle or otherwise.
Let (x^*, y^*) be a critical point of $f(x, y)$. If we define

$$D(x, y) = \frac{\partial^2 f}{\partial x^2}\frac{\partial^2 f}{\partial y^2} - \left(\frac{\partial^2 f}{\partial x \partial y}\right)^2$$

[21] a.k.a. the gradient

then

- If $D(x^*, y^*) > 0$ and $\left.\dfrac{\partial^2 f}{\partial x^2}\right|_{x=x^*, y=y^*} > 0$, then the critical point is a minimum.
- If $D(x^*, y^*) > 0$ and $\left.\dfrac{\partial^2 f}{\partial x^2}\right|_{x=x^*, y=y^*} < 0$, then the critical point is a maximum.
- If $D(x^*, y^*) < 0$ then the critical point is a saddle.

The function used in the second derivative test doesn't come out of nowhere. It is the determinant of the Hessian matrix! Unfortunately, this particular test only works in two dimensions, but not all hope is lost in higher dimensions.

In higher dimensions, we have to look at the eigenvalues of the Hessian matrix. If **all** the eigenvalues are positive at the point (x^*, y^*), then the point is a minimum; if **all** the eigenvalues are negative, then the point (x^*, y^*) is a maximum; and if the eigenvalues are of mixed signs, then the point (x^*, y^*) is a saddle point[22].

WE REALIZE WE HAVEN'T REALLY DONE ANYTHING WITH THESE CONCEPTS YET but truth be told, we haven't really *done* anything. Other than the concept of a saddle point, and the second derivative test, there is nothing new here. We just made a quiche[23] of the things we've already studied. So, I've saved all the good applications for the next section.

4.1 APPLICATIONS

FIRST, WE SHOULD SEEK CLOSURE on the issue of the Shannon Diversity Index. For four species, we have a critical point when

$$p_1 = p_2 = p_3 = p_4 = \frac{1}{4}$$

and now we are in a position to verify that it is indeed a maximum.

[22]This is also true in two dimensions; the second derivative test is equivalent to checking the eigenvalues.

[23]Or equivalent combination food; a casserole, perhaps

The Hessian matrix of the Shannon Diversity Index is a 3×3 matrix,

$$H_3(p_1, p_2, p_3) = \begin{bmatrix} \frac{\partial^2 H}{\partial p_1^2} & \frac{\partial^2 H}{\partial p_1 \partial p_2} & \frac{\partial^2 H}{\partial p_1 \partial p_3} \\ \frac{\partial^2 H}{\partial p_2 \partial p_1} & \frac{\partial^2 H}{\partial p_2^2} & \frac{\partial^2 H}{\partial p_2 \partial p_3} \\ \frac{\partial^2 H}{\partial p_3 \partial p_1} & \frac{\partial^2 H}{\partial p_3 \partial p_2} & \frac{\partial^2 H}{\partial p_3^2} \end{bmatrix}$$

$$= \begin{bmatrix} \frac{1 - p_2 - p_3}{p_1(p_1 + p_2 + p_3 - 1)} & \frac{1}{p_1 + p_2 + p_3 - 1} & \frac{1}{p_1 + p_2 + p_3 - 1} \\ \frac{1}{p_1 + p_2 + p_3 - 1} & \frac{1 - p_1 - p_3}{p_1(p_1 + p_2 + p_3 - 1)} & \frac{1}{p_1 + p_2 + p_3 - 1} \\ \frac{1}{p_1 + p_2 + p_3 - 1} & \frac{1}{p_1 + p_2 + p_3 - 1} & \frac{1 - p_2 - p_3}{p_1(p_1 + p_2 + p_3 - 1)} \end{bmatrix}$$

$$H\left(\frac{1}{4}, \frac{1}{4}, \frac{1}{4}\right) = \begin{bmatrix} -8 & -4 & -4 \\ -4 & -8 & -4 \\ -4 & -4 & -8 \end{bmatrix}$$

The eigenvalues are then given by looking at the matrix

$$\hat{H}_3 = \begin{bmatrix} -8 - \lambda & -4 & -4 \\ -4 & -8 - lambda & -4 \\ -4 & -4 & -8 - lambda \end{bmatrix}$$

taking the determinant, and setting it equal to zero

$$\text{Det}\left(\hat{H}_3\right) = (\lambda + 4)^2(\lambda + 16)^2 = 0$$

The three eigenvalues are then

$$\lambda_1 = \lambda_2 = -4$$
$$\lambda_3 = -16.$$

Since they are all negative, our critical point $p_1 = p_2 = p_3 = p_4 = 1/4$ is indeed a maximum, just like we suspected all along[24].

That was a bit involved, using everything[25] we've learned so far. Let's do an example in two dimensions, as it might be easier to grasp.

Imagine we drop a match onto a very large metal plate and want to know how the heat from that match will dissipate through the plate[26]. We can model this with a function

$$H(t, x, y) = \frac{1}{4\pi t} e^{-x^2 + y^2/4t}$$

[24]Worthy of an Agatha Christie novel? No, but important all the same.

[25]Derivatives and eigenvalues, mostly

[26]There are a bunch of assumptions built-in to this model that we won't go into too much detail with: 1) the plate is infinitely large, 2) the match heats the plate at a single point, and 3) no heat is lost to the surrounding environment.

Obviously, if the match hits the plate at $(x, y) = (0, 0)$ at $t = 0$, then obviously that will be the hottest part of the plate at $t = 0$. We can ask, *does $(x, y) = (0, 0)$ remain the hottest part of the plate as the heat dissipates through the plate?*

In order to answer this, let's look for the maximum in x and y of $H(t, x, y)$. Taking a derivatives in x and y we get

$$\begin{aligned}\frac{\partial H}{\partial x} &= -\frac{2x}{16\pi t^2}\mathrm{e}^{-x^2+y^2/4t}\\ \frac{\partial H}{\partial y} &= -\frac{2y}{16\pi t^2}\mathrm{e}^{-x^2+y^2/4t}\end{aligned}$$

Which tells us $(x, y) = (0, 0)$ is a critical point[27]. In order to check if this is a maximum or minimum, we need the second derivatives

$$\begin{aligned}\frac{\partial^2 H}{\partial x^2} &= \frac{2(2x^2 - 4t)}{64\pi t^3}\mathrm{e}^{-x^2+y^2/4t}\\ \frac{\partial^2 H}{\partial y^2} &= \frac{2(2y^2 - 4t)}{64\pi t^3}\mathrm{e}^{-x^2+y^2/4t}\\ \frac{\partial^2 H}{\partial x \partial y} &= \frac{4xy}{64\pi t^3}\mathrm{e}^{-x^2+y^2/4t}\end{aligned}$$

We can use the second derivative test to determine if we have a maximum or minimum at $(x, y) = (0, 0)$. The second derivatives at $(x, y) = (0, 0)$ are

$$\begin{aligned}\frac{\partial^2 H}{\partial x^2}(0, 0) &= \frac{-8t)}{64\pi t^3}\\ \frac{\partial^2 H}{\partial y^2} &= \frac{-8t)}{64\pi t^3}\\ \frac{\partial^2 H}{\partial x \partial y} &= 0\end{aligned}$$

and the quantity D is then

$$D = \frac{t^2}{64\pi^2 t^6} > 0$$

Since this is positive and the x second derivative is negative, then $(0, 0)$ is a maximum. Since this works for all t, we can conclude that the hottest spot on the plate is always the spot that the match hit.

[27] Since exponentials are never truly 0 and we are assuming t is fixed

LET'S STEER AWAY FROM MATHEMATICAL MODELS FOR A MOMENT and address a question that I said would be left a mystery way back in Chapter 1. Say we have some data and we want to fit a function to it. In Chapter 1 we used our eyes and drew lines that "looked good" or used a couple of points and fit a line or exponential and said "it's probably fine." But, like we saw, there are many equally valid functions that "look good." We now have enough tools to take into account all of our data and find the *best* parameters to fit a model.

Curve Fitting and Least Squares Regression

Let's say we have data that comes in pairs: input data x and output data y. If we have n such data points we will write them as:

$$(x_1, y_1)$$
$$(x_2, y_2)$$
$$\vdots$$
$$(x_n, y_n)$$

How do we best fit a function to them?

Well, first we should probably decide on a general trend of the data. This is where our mathematical intuition, of which[28] we now have some, can be exceptionally handy. We should be able to look at some data and say *this looks exponential/linear/polynomial/Gaussian/etc.* Once we do that, we can confirm that we will *fit* the data with

$$\mathcal{Y} = f(\mathcal{X})$$

What's the deal with the fancy $\mathcal{X}$ and fancy $\mathcal{Y}$? Well, I want to differentiate our *model* input and output and our *observed* input and output. We use normal x and y for observed values, and fancy $\mathcal{X}$ and $\mathcal{Y}$ for our model values.

MODELS COME WITH PARAMETERS, AS WE HAVE SEEN THROUGHOUT THE COURSE. When we are fitting curves to data, these parameters become important. They are the values we are looking for and so we want to write them into our function notation, but also want to distinguish them from our input variable(s). Luckily, there is no shortage of punctuation

[28] Hopefully

out there and so we use a ; in our function notation to separate input variables from model parameters:

$$\begin{aligned} \text{outputs} &= f(\text{inputs}; \text{parameters}) \\ \mathcal{Y} &= f(\mathcal{X}; p_1, p_2, p_3, \cdots, p_m) \end{aligned}$$

We do this so communicate that the model function is indeed going to take $\mathcal{X}$ as an input, but that we are going to treat it as if it were a function of the parameters p_1, p_2, p_3, *etc.* for the purposes of fitting the model to our data. Generally, $n > m$. If not, then we will not have enough equations to specify all parameters.

With this information in hand, let's proceed. We now have a series of data points (x_i, y_i) where i goes from 1 to n, and a skeleton of a function $\mathcal{Y} = f(\mathcal{X}; p_1, p_2, p_3, \cdots, p_m)$ and we have to ask *what exactly are we trying to optimize?* If we naively try to optimize f, all we will get is parameters that give us the largest/smallest[29] $\mathcal{Y}$ for any $\mathcal{X}$. That won't necessarily give us the best fitting curve.

WHAT WE WANT TO DO IS get parameter values that allow the function to best match the observed data. In other words, in some sense, we want the *difference* between our y_i values and our $\mathcal{Y}_i = f(x_i; p_1, p_2, p_3, \cdots, p_m)$ to be as small as possible. Doing this is known as **least squares regression**. We try to minimize the total *residual*[30] between our curve and the data.

We define the total error as

$$S = \sum_{i=1}^{n} (y_i - f(x_i; p_1, p_2, p_3, \cdots, p_m))^2$$

All of the x_i's and y_i's are known. They are observations we have made. Which means S is *only* a function of our parameters

$$S = \sum_{i=1}^{n} (y_i - f(x_i; p_1, p_2, p_3, \cdots, p_m))^2 = S(p_1, p_2, p_3, \cdots, p_m)$$

and it is this function that we want to minimize. If we find a set of numbers p_1^*, p_2^*, p_3^* *etc.* that minimize S, then we can guarantee that our "error" is as small as possible for the function we chose. Since S is

[29] Or saddest

[30] For our purposes, it is absolutely okay to think of residual as another word for error between our model and our observations. In statistics, these two words are distinct, however.

just a function in m-dimensions, we merely have to solve the system of equations

$$\nabla S(p_1, p_2, \cdots, p_m) = \vec{0}$$

in order to find our parameter values.

We say that our curve $f(\mathcal{X}; p_1^*, p_2^*, \cdots, p_m^*)$ is the best fit in the **least squares** sense.

THE FACT THAT WE HAVE TO QUALIFY BEST FIT with *in the least squares sense* implies that this is not the only way to fit a curve to a set of data. It's not, but it's the easiest to digest and arguably one of the most widely used, especially for linear data.

As an example let's look at some data for the drug theophylline from a study done in the mid-90s [6]. Figure 4.7 shows the data. The graph

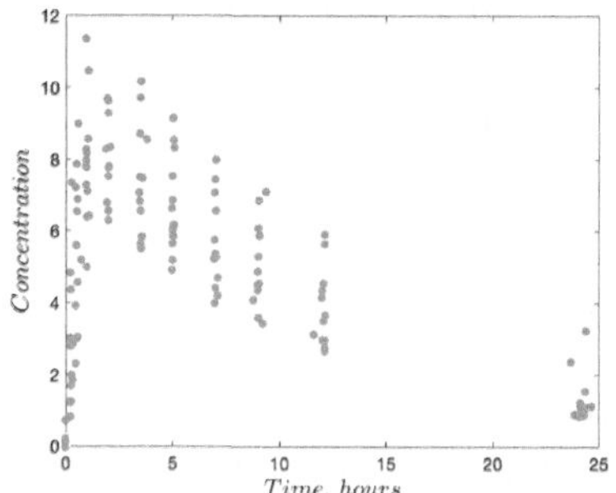

Figure 4.7: Data on the blood concentration of theophylline in 12 subjects.

can help us figure out what kind of function we want to use to model the data, but a table of values is far more useful when actually doing the calculations. There are 133 pairs of values in this data set. Let's only work with the first 24 since we are doing this example by hand. These data are shown below.

i	1	2	3	4	5	6	7	8	9	10	11	12
x	0	0.25	0.57	1.12	2.02	3.82	5.1	7.03	9.05	12.12	24.37	0
y	0.74	2.84	6.57	10.5	9.66	8.58	8.36	7.47	6.89	5.94	3.28	0
i	13	14	15	16	17	18	19	20	21	22	23	24
x	0.27	0.52	1	1.92	3.5	5.02	7.03	9	12	24.3	0	0.27
y	1.72	7.91	8.31	8.33	6.85	6.08	5.4	4.55	3.01	0.9	0	4.4

WE'VE SEEN ENOUGH PHARMACOKINETIC EXAMPLES that we should be able to look at figure 4.7 and guess that a decent function to model this would be

$$\mathcal{Y} = f(\mathcal{X}) = A\mathcal{X}\mathrm{e}^{-t^B}$$

This model has two parameters, A and B which we will want to fit to the data so we write

$$f(\mathcal{X}; A, B) = A\mathcal{X}\mathrm{e}^{-t^B}$$

Using the table of values and this skeletal function, we can define the function that we wish to minimize:

$$\begin{aligned} S &= \sum_{i=1}^{24}(y_i - f(x_i; A, B))^2 \\ &= \left(0 - A \cdot 0\mathrm{e}^{-0^B}\right)^2 \\ &\quad + \left(2.84 - A \cdot 0.25\mathrm{e}^{-0.25^B}\right)^2 \\ &\quad + \left(6.57 - A \cdot 0.57\mathrm{e}^{-0.57^B}\right)^2 \\ &\quad + \cdots + \left(4.4 - A \cdot 0.27\mathrm{e}^{-0.27^B}\right)^2 \\ &= S(A, B) \end{aligned}$$

Even 24 values were too much to type out to be honest, but I hope you get the idea. When we plug our data into this function, we are left with a function of our parameters only. Now we can minimize this function by first taking the gradient[31]

$$\nabla S = \begin{bmatrix} \sum_{i=1}^{24} 2(y_i - Ax_i\mathrm{e}^{-x_i^B})x_i\mathrm{e}^{-x_i^B} \\ \sum_{i=1}^{24} 2(y_i - Ax_i\mathrm{e}^{-x_i^B})Ax_i^{b+1}\ln x_i\mathrm{e}^{-x_i^B} \end{bmatrix} = \begin{bmatrix} 0 \\ 0 \end{bmatrix}$$

I know there are a lot of x's and y's and i's and letters and symbols floating around, but it's important to know that other than A and B, the other symbols are just placeholders for numbers. This is two equations[32] with two unknowns and in theory we can solve it for A^* and B^* which will minimize S. We would then *fit* our data with the model

$$f(\mathcal{X}) = A^*\mathcal{X}\mathrm{e}^{-\mathcal{X}^{B^*}}$$

[31]Yes, these derivatives aren't the easiest to read. Yes, you may have to work through the example yourself to fully understand it.

[32]Long ones, admittedly

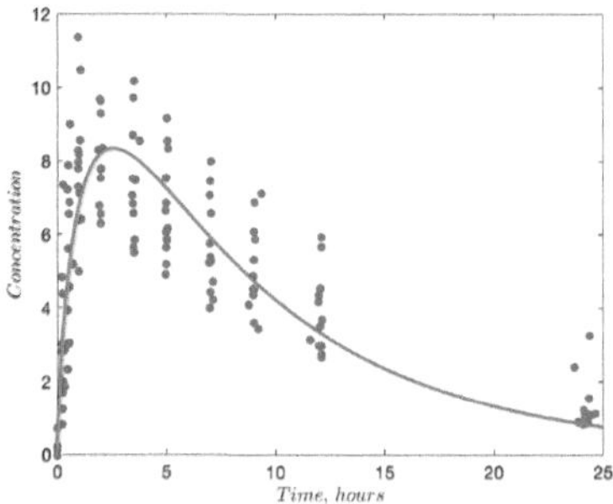

Figure 4.8: Data on the blood concentration of theophylline in 12 subjects and the best fit curve in the least squares sense.

when we do this for the *entire* dataset of 133 points, we get the following function

$$f(\mathcal{X}) = 18.187\mathrm{e}^{-\mathcal{X}^{0.57}}$$

which is plotted over the data in figure 4.8. What if we only used the 24 points above? In that case, we would get the function

$$f(\mathcal{X}) = 20.787\mathrm{e}^{-\mathcal{X}^{0.58}}$$

It's almost the same, but not quite because we're only using about 20% of the available data. This data is extremely well-behaved, and hence we get very similar parameter values even though we ignore 80% of our data. It's very similar though, so either the 24 points we've chosen were *very* representative of the data, or we're on the right track with the function that we picked as our skeleton[33].

YOU CAN VERIFY THESE VALUES BY HAND, BUT I DON'T SUGGEST IT. In reality, non-linear regression is *never* done by hand. In fact, I'm not even sure this particular example *can* be done by hand.

4.2 DISTRIBUTION FITTING, PROBABILITY, AND LIKELIHOOD

I HOPE BY NOW THAT WE HAVE AT LEAST COME TO TERMS WITH THE FACT THAT OUR MATHEMATICAL NOTATION IS CONTEXT-SENSITIVE.

[33]Or both

We can recontextualize everything we've learned so far to discuss probabilities. Let's say we have a function, $p(X;\theta)$ where X is some *event*[34] and p, the function, is the probability that event X will occur. We treat probability in-depth in Chapter 9.

AS A QUICK EXAMPLE, let's use the classic example of flipping a coin. There are two events that can occur on each flip: either the "heads" side is showing, or the "tails" side is showing. Assuming the coin is completely fair, the function $p(X)$ can be defined as

$$p(X) = \begin{cases} 0.5 & \text{if } X = \text{heads} \\ 0.5 & \text{if } X = \text{tails} \end{cases}$$

There is a 50% chance of coin landing heads and 50% chance the coin lands tails.

THAT WAS AN EXTREMELY SIMPLE EXAMPLE. So simple in fact that we quietly lost our *parameter* θ along the way. Usually p will be much more complicated than this.

Definition 4.2.1. A **probability distribution**, $p(X;\theta)$ is a special function which has the following properties:

- The *range* of $p(X; theta)$ is $[0, 1]$.
- The *total area* between the x-axis and the function is 1 when calculated on the function's *domain*[35].

We conceptualize probability distributions all the time, often without knowing it. For instance, most of us can probably agree that getting murdered by a serial killer is not advisable. Many of us probably[36] run through a series of criteria in our heads before getting into the Uber we just called to weigh our chances of being serial murdered. We may not have a function in mind when we're doing this, maybe not even a graph, but we do it none the less. It's 2 pm, the driver is a 20-year-old woman,

[34] Event here can be anything: An event might be picking a number, or an individual from a population and measuring their height/weight/any other property about them.

[35] The means that there is a 100% chance that *something* will happen.

[36] Hopefully

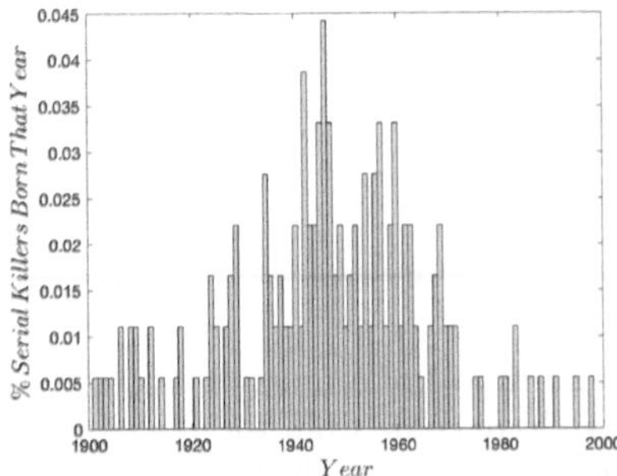

Figure 4.9: A histogram of the birth year of serial killers normalized to probability. This means that if we pick a serial killer born in the 20th century at random, the probability that they were born in a particular year is given by the value on the y-axis.

the car is an S-class Mercedes, it's clean and you are with a group of friends and there are no discrepancies between what the Uber app is telling you and the situation in front of you. Probability states that it's safe to get in the car. Compare this with waiting around at 2am, a rusted out 1968 VW Beetle, the driver looks to be a man in his 40s, you're alone, and license plate reads R4E338 instead of RAB33B, but it's close enough. Maybe you are less sure about getting in that car. The probability that you're in danger went up in your head based on the "events" or inputs in front of you. Let's look at some actual serial killer data now.

Figure 4.9 shows the probability that a serial killer was born in a given year. We are only looking over the 20th century and, obviously, only have data on those that have been caught and convicted. Just because there are gaps in the birth years in the 70s, 80s, and 90s doesn't mean that no serial killers were born in those years, just that they haven't been caught and convicted. Maybe we should model this data with a probability distribution, $p(X;\theta)$ and figure out which parameters best fit the data. Then we can fill in the probabilities for the missing years.

CONCEPTUALLY, WE FIT THIS THE SAME WAY AS BEFORE. We have to look at the probability data and pick a skeleton of a function that would fit. The difference in this case is that we need to make sure the function we pick satisfies the rules of a probability distribution. Then, we have to figure out a function to maximize or minimize in terms of the parameter value(s), θ. Then we optimize.

The biggest difference here is the function that we wish to maximize or minimize. Instead of looking at the least squares error, we are going to look at a function called the **log likelihood**, but first let's define likelihood.

Definition 4.2.2. The **likelihood** of a function is the probability that the event data, X_i for $1 \leq i \leq n$, we have come from the model function p with parameter choice θ. The **log likelihood** is given by the function

$$L(\theta) = \sum_{i=1}^{n} \ln(p(X_i;\theta))$$

If we maximize the log-likelihood, then we maximize the chances that the probability function we chose, $p(X,\theta)$ accurately represents the data. When we have one parameter we do this by taking the derivative and setting it to zero. When we have multiple parameters, we do this by setting the *gradient* of the log-likelihood to zero.

We haven't looked much at probability distributions, but I have used a few throughout without telling you. The most common is the **normal distribution**[37]. It looks like

$$p(X;\mu,\sigma) = \frac{1}{\sqrt{2\pi\sigma}} e^{-\dfrac{(X-\mu)^2}{2\sigma^2}}$$

The log-likelihood of a normal distribution is given by

$$\begin{aligned} L(\mu,\sigma) &= \sum_{i=1}^{n} \ln\left(\frac{1}{\sqrt{2\pi\sigma}} e^{-\dfrac{(X-\mu)^2}{\sigma^2}}\right) \\ &= \sum_{i=1}^{n}\left(\frac{1}{2}\ln(2\pi\sigma) - \frac{(X_i-\mu)^2}{2\sigma^2}\right) \end{aligned}$$

We have two unknowns: μ and σ, and we can get two equations to solve them by taking the gradient of L and setting it to zero

$$\nabla L = \vec{0}$$

[37] AKA a Gaussian distribution or a bell curve

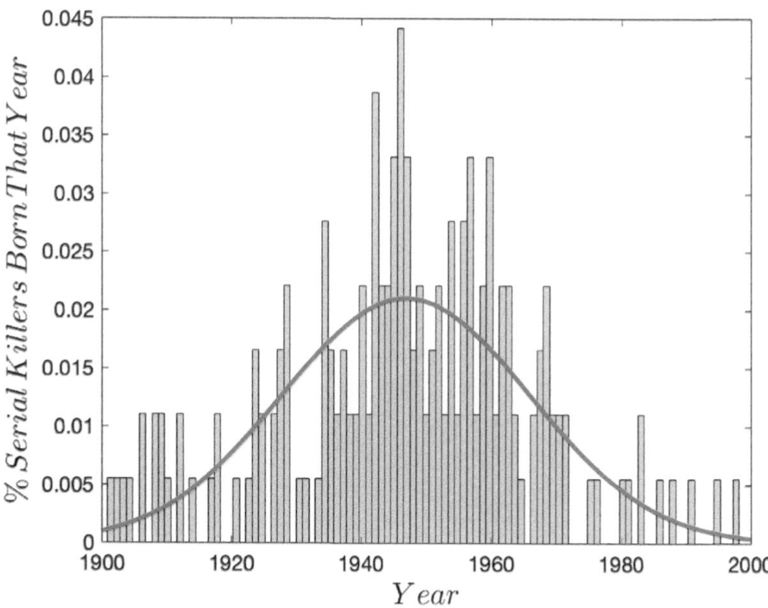

Figure 4.10: A histogram of the birth year of serial killers normalized to probability. This means that if we pick a serial killer born in the 20th century at random, the probability that they were born in a particular year is given by the value on the y-axis. The orange line is the fitted normal distribution.

For our serial killer data, we have 181 points. The biggest difference here is that our data does not come in pairs. We *only* have the years of birth, the probability comes from counting how many instances of a given year we have and dividing by 181. This is why there are no Y_i in our analysis. What remains true though is that we wouldn't usually do this by hand for a data set this big. Figure 4.10 shows the normal distribution fit for the data with maximum likelihood. The fitted normal looks like

$$p(X) = \frac{1}{\sqrt{2\pi 19}} e^{-\frac{(X-1946)^2}{19^2}}$$

You may look at the figure and say *that doesn't look like it fits AT ALL*. Maybe it doesn't, maybe the normal wasn't the right probability distribution to use, or perhaps the really tall spikes are more anomalous than they appear. We are drawn to them because they stand out, but if we remove them, we see that our normal fits well to the majority of the points[38].

A much deeper treatment of probability, distributions, and how to fit them is given in Chapters 9, 10, and 11.

[38]It is theorized that many serial killers were victims of lead poisoning. The graph presented here is often used as evidence of that. As cars and thus leaded gasoline became more popular, we see a correlated rise in the number of serial killers. Leaded gas was phased out and eventually banned starting in the 1970s. These days, lead is hardly used because of its toxicity.

PRACTICE PROBLEMS

1. Find the gradient of the following function

$$f(x, y) = \frac{3x^2}{y}$$

2. What are the second derivatives of

$$g(t, \tau) = \tau e^{t/\tau}$$

3. We saw that second mixed partial derivatives are symmetric: $\partial^2 f/\partial x \partial y = \partial^2 f/\partial y \partial x$. Use this fact to show that the mixed *third* derivatives are all equal. That is,

$$\frac{\partial^3 f}{\partial x \partial y \partial z} = \frac{\partial^3 f}{\partial y \partial x \partial z} = \cdots$$

4. For a function $f(x, y)$, the partial derivative with respect to x gives us rate of change in the x-direction. Similarly, the partial derivative with respect to y gives us the rate of change in the y-direction. The gradient, the vector of partial derivatives, gives us the direction of steepest change. What if we wanted to find the rate of change in an arbitrary direction, $\vec{u}$?

   In this case, we would use the **directional derivative**:

$$\nabla_{\vec{u}} f = \nabla f \cdot \frac{\vec{u}}{||\vec{u}||}$$

   For a function $f(x, y)$ and vector $\vec{u} = [u_1 u_2]^T$, expand the above formula for the directional derivative.

5. What is the directional derivative of $f(x, y) = \ln(x)e^y$ in the direction $\vec{u} = [a, 2a]^T$?

6. If the gradient of a function $f(x, y)$ is zero, can the directional derivative be non-zero in any direction? Why or why not?

7. It can be very difficult to think of examples of when the mixed partial derivatives are *not* equal. This is because most functions we think of are continuous, with continuous derivatives. One example where this fails is

$$f(x, y) = \frac{x^2 y - x y^2}{x + y}$$

   Using the limit definition show that

(a)

$$\left.\frac{\partial f}{\partial x}\right|_{(x,y)=(0,0)} = 0$$
$$\left.\frac{\partial f}{\partial y}\right|_{(x,y)=(0,0)} = 0$$

(b)

$$\left.\frac{\partial^2 f}{\partial x \partial y}\right|_{(x,y)=(0,0)} \neq \left.\frac{\partial^2 f}{\partial y \partial x}\right|_{(x,y)=(0,0)}$$

Hint The second partial derivative can be found through a limit with

$$\frac{\partial^2 f}{\partial y \partial x} = \frac{f_x(0,h) - f_x(0,0)}{h}$$

where f_x is the first derivative with respect to x.

8. Where is the critical point of

$$f(x,y) = y\ln(x)\mathrm{e}^{-y}$$

What type of critical point is it?

9. Where is the critical point of

$$f(x,y) = y\mathrm{e}^{-By} + Bx\ln(Ax)$$

What type of critical point is it?

10. There is a valley between two mountains and the height of the terrain can be modelled with the function

$$T(x,y) = \frac{x^2 + y^2}{2} + xy$$

What path should we take so ensure we remain at the same altitude for our entire hike? If we start at coordinates $(x,y) = (1,-1)$ and deviate from our path, do we fall down the hill or start climbing?

t	x	D
0	1	0.8
0	10	0.1
3	2	0.8
4	2	0.92
8	20	0.2

11. Later on our hike we encounter some terrain that looks like

$$\frac{x^2 + y^2}{2} + xy^2 + x^2y$$

which path allows us to stay on flat ground? Is this path the ridge of a mountain range? A valley? or something in between? How do you know?

12. The density of an urban centre usually changes radially from the core like

$$D(x, y, t) = 1 - \frac{1}{1 - exp(-r(x + y) + kt)}$$

where $D = 1$ is maximum density, x is distance from the core and t is the time in years. What do

$$\frac{\partial D}{\partial x} \text{ and } \frac{\partial D}{\partial t}$$

represent?

13. Calculate the two partial derivatives.

14. Urban sprawl is the idea that as population increases, the geographic area used by a city is increased. Does the model above capture this? Plot a few graphs of D vs. x for different values of t and explain how urban sprawl is captured by the graphs.

15. What are the two parameters of this model?

16. If we have the following table of data write out the equation for the error term.

17. Optimize r and k for the function D given the measurements given above.

18. The intensity of light at any point in a $3D$ room from four bulbs located at positions $(0,0,0)$, $(0,1,0)$, $(1,0.5,0)$, and $(0.5,0.5,1)$ is given by

$$\begin{aligned} I(x,y,z) =& \mathrm{e}^{-0.1(x^2+y^2+z^2)} \\ &+ \mathrm{e}^{-0.1(x^2+(y-1)^2+z^2)} \\ &+ \mathrm{e}^{-0.1((x-1)^2+(y-0.5)^2+z^2)} \\ &= +\mathrm{e}^{-0.1((x-0.5)^2+(y-0.5)^2+(z-1)^2)} \end{aligned}$$

What are the critical points of this function? Is the light intensity maximized or minimized at this point?

CHAPTER 5

Differential Equations

DETERMINING A SKELETON OF A FUNCTION WAS A *sine qua non*[1] FOR BOTH CURVE FITTING AND DISTRIBUTION FITTING. This is great as long as we know what functions to look for, but most of the time we probably don't. Usually we are asking questions that are particular enough that a line or an exponential are more than sufficient, but if we look at our pharmacological example, we needed a slightly more complicated function than just a line or an exponential in order to determine what kind of curve we should fit.

By developing theoretical models of physical systems, we are able to translate complicated physical systems into the language of mathematics. Historically, one of our greatest tools to do this has been **differential equations**.

In practice, it's very difficult to determine a function that captures all the important features of a natural phenomenon. Sometimes, such a function exists but we don't have the tools to write it down.

Take as an example the critical point of $f(x) = x^2 + \mathrm{e}^{-x}$. From the graph in figure 5.1 we can clearly see that the point exists, but the mathematical language that we know is insufficient for writing down this point. You can see this yourself if you take a derivative and try to solve for x.

For many phenomena, while it is either difficult or impossible to determine a function to describe the behaviour explicitly, it is often possible to describe its derivative.

[1]Latin for *absolutely essential*

DOI: 10.1201/9781003265405-5

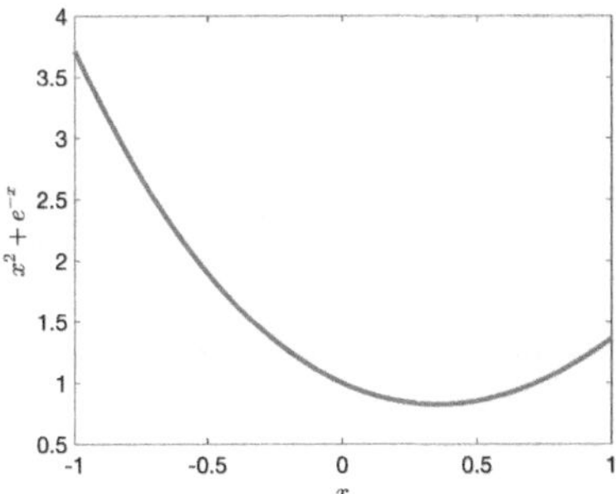

Figure 5.1: The critical point of this function, $f(x) = x^2 + \mathrm{e}^{-x}$ clearly exists – we can see it on the graph, but our mathematical language is insufficient for describing this point in any way other than $2x - \mathrm{e}^{-x} = 0$.

WHY IS A DERIVATIVE EASIER TO WRITE DOWN THAN A FUNCTION? We can think of this practically using a free-range drift[2] of pigs. We could, as the pigs' keepers, count how many pigs are in the pen at any given time. It takes time to count, and if the pigs are coming and going, we may never get an accurate count. It would be much easier to count a closed pen *once* and then watch the entrance and exit to add any new pigs that come in, and subtract any pigs that leave.

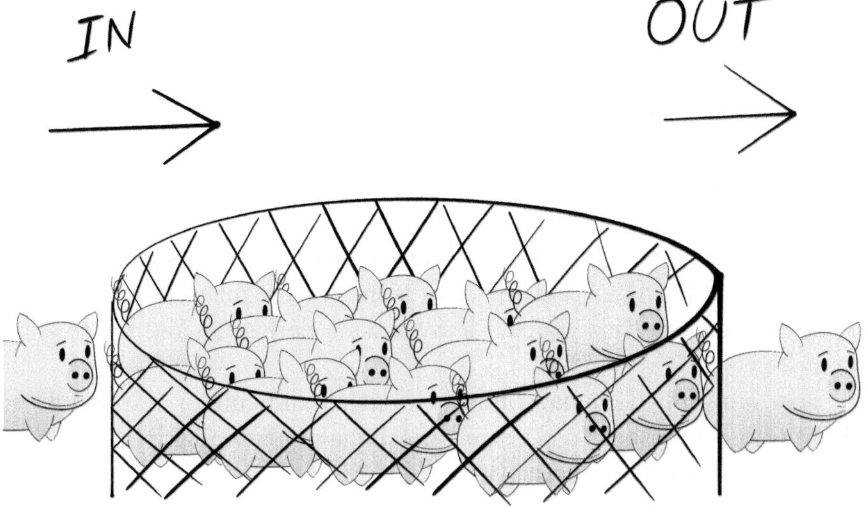

[2]Yes, it's the proper collective noun

FIRST ORDER DIFFERENTIAL EQUATIONS, ones that only involve the first derivative of a function, tend to be used for modelling because they simplify how we might think about physical processes. First-order differential equations can be interpreted as

$$\frac{\mathrm{d}f}{\mathrm{d}t} = \text{IN} - \text{OUT}$$

In order to model something with a function $f(t)$, we need only know how many new ones we get in one time step and how many we lose in one time step.

For instance, if we generalize our model above, we can say that in a population P, the per capita birth rate is B and the per capita death rate is D and thus the population follows

$$\frac{\mathrm{d}P}{\mathrm{d}t} = \underbrace{BP}_{\text{IN}} - \underbrace{DP}_{\text{OUT}}$$

As another completely different example, let's look at Newton's second law. It says that the force of an object is related to its mass, m, and its acceleration a. Since a is the derivative of speed, v we can write

$$F = m\frac{\mathrm{d}v}{\mathrm{d}t}$$

Now, if an object is falling through the sky and we know a thing or two about gravity, we can say that the force of gravity is acting to increase the object's speed with a constant force mg, meanwhile, air resistance[3] acts against the object to slow it down. The drag force is usually written as $\frac{1}{2}Kv^2$. We can write a differential equation for this system:

$$m\frac{\mathrm{d}v}{\mathrm{d}t} = \underbrace{mg}_{\text{IN}} - \underbrace{\frac{1}{2}Kv^2}_{\text{OUT}}$$

[3]a.k.a. the drag force

5.1 SOLVING BASIC DIFFERENTIAL EQUATIONS: WITH AN EXAMPLE

On average, humans are born at a rate of 0.0185 per year. This means, roughly that the human population is increasing by 1.85% each year[4]. Humans live, on average, to 73 years which gives us a rate of 0.014.

So, each year our human population *changes* by adding 1.85% of the current population, and subtracting 1.4% of the current population. In the language of functions,

$$\frac{\mathrm{d}P}{\mathrm{d}t} = 0.0185P(t) - 0.014P(t)$$

This has some biological intuition behind it and gives us a basis for describing our function.

THE EQUATION WE HAVE CREATED FOR $P(t)$ is a little bit different than equations that we're used to seeing. It is a relationship between the function, P and its own derivatives.

Usually, when we have equations with unknown quantities, those quantities are numbers. For example,

$$3x^2 + 2x = 5 - x$$

If asked, we could solve this equation for the variable x and we expect to get a number. A differential equation is not much different than the algebraic equation above. The difference being that our unknown quantity is a function, and we know some relationship between the unknown function and its derivative. When we solve a differential equation, we expect to get a function back, not a value.

THE SIMPLEST DIFFERENTIAL EQUATION IS OF THE FORM

$$\frac{\mathrm{d}f}{\mathrm{d}x} = kf(x)$$

How can we interpret this?

The differential equation is telling us that we are looking for a function that, when we differentiate it, does not change save for it is now multiplied

[4]A little more complicated: this number also means that the average age of reproduction is 27.

by a constant. We can do this by inspection. Let's think of functions that when we differentiate, get multiplied by a constant. Let's try

$$f(x) = x^2$$

Then

$$\frac{\mathrm{d}f}{\mathrm{d}x} = 2x = 2\sqrt{f(x)}$$

since $f(x) = x^2$. That doesn't work maybe it's because we need to be more general.

$$f(x) = x^k$$

This will insure that our derivative is multiplied by k at the very least.

$$\frac{\mathrm{d}f}{\mathrm{d}x} = kx^{k-1} = kf(x)^{(k-1)/k} \neq kf(x)$$

This clearly isn't working.

If your memory is better than mine, then you may have noticed that our differential equation looks awfully similar to a derivative rule that we long ago defined. There is a single function that when we take a derivative is equal to itself times a constant

$$f(x) = \mathrm{e}^{kx}$$

whose derivative is

$$\frac{\mathrm{d}f}{\mathrm{d}x} = k\mathrm{e}^{kx} = kf(x)$$

Therefore, the solution to the differential equation is $f(x) = \mathrm{e}^{kx}$.

Going back to our population example, we had

$$\frac{\mathrm{d}P}{\mathrm{d}t} = 0.0185P - 0.014P$$
$$\frac{\mathrm{d}P}{\mathrm{d}t} = (0.018 - 0.014)P$$
$$\frac{\mathrm{d}P}{\mathrm{d}t} = 0.004P$$

which has the same form as our differential equation above. We can infer that the solution is then

$$P(t) = \mathrm{e}^{0.004t}$$

This is one of the reasons why e^x is such an important function; one of its properties is that the function itself is related to its derivative. When we look at differential equations, we are usually looking for functions whose derivatives are related to the function itself. As a consequence, solutions to differential equations[5] usually involve e^x.

As an example, another differential equation related to population growth is

$$\frac{dP}{dt} = r\left(1 - \frac{P}{K}\right)P$$

This is the differential equation for **logistic growth**. It says that the population grows at rate r and is proportional to the current population size P, but there's a catch. The term $(1 - P/K)$ will become 0 when $P = K$, therefore when $P = K$, $\dfrac{dP}{dt} = 0$. In other words, the population stops growing[6]. This term means our population is limited by a resource in the environment. The environment can only support a population of K individuals. This number K is called the **carrying capacity** of the environment.

While we don't have the tools in hand to solve this equation, but it does exist and I can state it for you. The solution to the logistic equation is

$$P(t) = \frac{K}{1 - \left(\frac{K-P_0}{P_0}\right)e^{-rt}}$$

where P_0 is the population at time $t = 0$.

5.2 EQUILIBRIA AND STABILITY

Even though we don't have the tools to solve anything but the most basic differential equations, there are still things we can do. As mentioned above, there are many phenomena that we can describe using a differential equation but we don't have the language to write down the solution. This means that we need ways to draw conclusions from differential equations without actually knowing the solution.

[5] When we can find them, anyway

[6] Or changing at all for that matter

Equilibria

So far, our applications of derivatives have mostly[7] focused on when the derivative is zero. We use this to optimize processes or algorithms depending on the context. Well, we have derivatives in a differential equation... what if we set that to zero? What does this mean? Are we still optimizing something?

The types of differential equations we'll be discussing here are **first-order differential equations**. They involve some mix of our independent (input) variable, our function, and its first derivative. We will focus on equations when t is not *explicitly* stated in our equation; these are called **autonomous differential equations**. We can always write these types of differential equations as

$$\frac{\mathrm{d}f}{\mathrm{d}t} = F(f(t))$$

which says that the derivative of the function $f(t)$ is some function of $f(t)$ itself.

If we set the derivative equal to zero, that's the same as setting the multivariable function $F(f(t))$ to zero.

$$\frac{\mathrm{d}f}{\mathrm{d}t} = F(f(t)) = 0$$

We are going to look at this in a slightly different context than optimization. When

$$\frac{\mathrm{d}f}{\mathrm{d}t} = 0$$

This can (and does) also mean that $f(t)$ is not changing in t. Which means if we can solve

$$F(f(t)) = 0$$

for $f(t)$, then we can find a value of f which is independent of t. This represents a particular *output value* of $f(t)$ that doesn't change in time. We say any value f^* that satisfies $F(f^*) = 0$ is an **equilibrium** or **steady-state** of the differential equation.

---

[7]I mean, entirely

As an example, let's modify our above population model to account for a population-limiting resource. It's true that populations grow exponentially when there are strong population numbers and plentiful resources. What happens if a resource becomes scarce? Reproduction should decrease. When might a resource become scarce? Maybe when it's being shared by too many individuals. We modify our population model in the following way

$$\frac{\mathrm{d}P}{\mathrm{d}t} = B(P)P - DP$$

So our birth rate depends on the number of individuals. As our population numbers get larger, our resource gets spread thinner, and if everyone is getting less then reproduction becomes harder and slows down. The simplest function that captures this idea is a straight line.

$$B(P) = r(K - P)$$

where r is some base rate of reproduction and K is the number of individuals the resource can sustain (often called the **carrying capacity**).

Plugging this into the differential equation leads to

$$\frac{\mathrm{d}P}{\mathrm{d}t} = r(K - P)P - DP$$

We don't have the tools to solve this equation, but that doesn't mean we are without hope. We could ask[8] the question *is there a population size that is perfectly balanced?* Or, *at what population level is every death replaced by exactly one birth to that the total population size is unchanged?* To answer this, we find the equilibrium points.

$$r(K - P)P - DP = 0P(r(K - P) - D) \qquad = 0$$

Solving for P gives us two answers:

$$P_1 = 0$$
$$P_2 = \frac{rK - D}{r}$$

The first answer makes sense. If we have no individuals, we cannot create any new individuals. So a population of size 0 will remain a population of size 0. The second answer, P_2, gives us the population level where every

[8] And answer

one has just enough resources so that every death in the population is replaced with exactly one birth.

An equilibrium, or steady state, is as the name implies: if we are at this value, we will not stray from it unless the system changes. So long as r, K and D stay constant, if we are at P_2 we will stay at P_2 forever. What happens if we *don't* start at P_2. Can we say anything about the population without solving the differential equation[9]?

Stability

Let's work from our example.

$$\frac{\mathrm{d}P}{\mathrm{d}t} = r(K - P)P - DP$$

Let's plot this treating P as the input and $\frac{\mathrm{d}P}{\mathrm{d}t}$ as the output. We already know where the function crosses the x-axis (i.e. where $\frac{\mathrm{d}P}{\mathrm{d}t} = 0$ and we know it is a parabola that opens downward, so we can sketch it fairly easily). By picking different points on the P axis, we can identify whether or not $\frac{\mathrm{d}P}{\mathrm{d}t}$ is positive or negative by looking at the value of the curve at a particular point. If $\frac{\mathrm{d}P}{\mathrm{d}t} > 0$ it means that P is *increasing* in time. Which means that over time we would move to the right on the x-axis. Contrarily, if $\frac{\mathrm{d}P}{\mathrm{d}t} < 0$, for a given value of P, then we would move to the left on the x-axis. This is summarized in figure 5.2.

We see that if we pick different values of P and see how we move along the P-axis, we always tend towards P_2 and away from P_1. In this case, P_2 is **locally asymptotically stable** and P_1 is **unstable**.

Definition 5.2.1. An equilibrium point f^* of the differential equation

$$\frac{\mathrm{d}f}{\mathrm{d}t} = F(f)$$

is **stable** if we tend towards it for values of f *nearby* f^*.
An equilibrium point is **unstable** if it is not stable.

---

[9] Yes, yes we can.

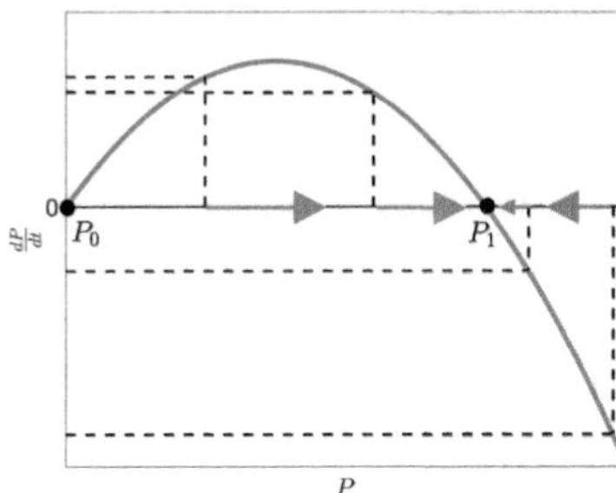

Figure 5.2: When plotting $\frac{\mathrm{d}P}{\mathrm{d}t}$ *vs. P, we can determine whether P(t) is increasing or decreasing as a function of time for a given value of P. We can use this information to move left or right on the P-axis. In this particular case, we will move towards the equilibrium* P_2*. When we get there, P is no longer changing in time, and thus we do not move away from this point.*

The *nearby* qualifier is an important one in the definition of stability. If we check points too far away, we may drift off in different directions. Take, for example, the following model.

CONTINUING WITH OUR POPULATION MODEL EXAMPLE, if we start with a population very, very close to zero but still positive, we will still tend towards P_2; this isn't realistic for a number of reasons[10]. If we have a very small population, the probability of a viable birth goes down and the population might decrease. This phenomena of low populations leading to diminished birth rate is called an **Allée effect**. We don't need to worry much about a differential equation that can capture this, we can work just with the graph of the situation in figure 5.3.

If we start with a P such that $P > P_2$, then we tend towards P_3 if $P < P_2$ we tend towards P_1. P_3 and P_1 are stable so long as we start *nearby* each respective equilibrium point. What nearby means changes depending on the problem at hand, but we can at least guarantee that small random changes[11] in a stable population won't lead to drastic changes in the future.

We don't always have to draw a graph to look at the local asymptotic stability of an equilibrium. If we look at the two examples that we have,

[10] If we think about people, it doesn't work on the basis that inbreeding is bad.

[11] The study of the effect of these random changes on differential equations is the field of stochastic differential equations.

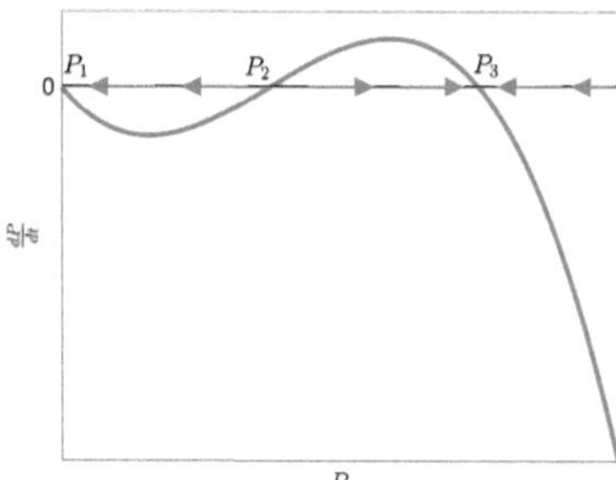

Figure 5.3: In the case of a strong Allée effect, we see that P_3 is stable only if we start with a population $P > P_2$. Otherwise, we tend towards P_1 (extinction). P_2 in this case is unstable because we will always move away from it either towards P_1 or P_3, depending on our starting population size.

we may notice a common feature of the unstable equilibria and the stable equilibria.

If we put these two graphs side by side, we may notice that our curve is always increasing at an unstable equilibrium, and decreasing at a stable equilibrium. In other words, an equilibrium is stable if $\frac{\mathrm{d}P}{\mathrm{d}t}$ is a decreasing function of P at the equilibrium and unstable if $\frac{\mathrm{d}P}{\mathrm{d}t}$ is an increasing function of P at the equilibrium. The derivative is our mathematical language for describing changes in a function (including increasing and decreasing).

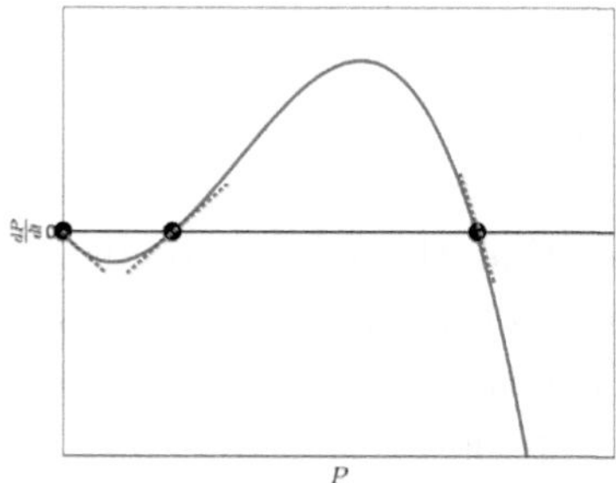

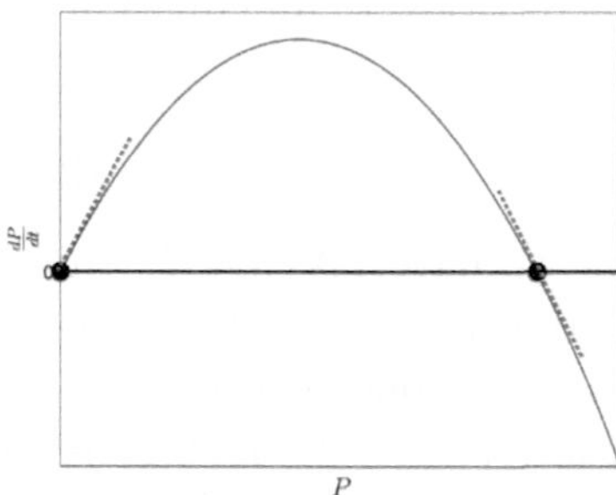

Figure 5.4: The way the derivative $\frac{\mathrm{d}P}{\mathrm{d}t}$ changes with respect to P can tell us whether an equilibrium is stable or unstable.

Remark. For a differential equation

$$\frac{\mathrm{d}P}{\mathrm{d}t} = F(P)$$

with equilibrium P^*, if

$$\left.\frac{\mathrm{d}F}{\mathrm{d}P}\right|_{P=P^*} < 0$$

then the equilibrium P^* is stable. If

$$\left.\frac{\mathrm{d}F}{\mathrm{d}P}\right|_{P=P^*} > 0$$

then the equilibrium P^* is unstable.

THE NOTION OF STABILITY BEING A NEARBY PHENOMENON is reinforced by this definition. Derivatives give local behaviour. If we try to use a derivative to estimate function values far away, we are likely going to be very wrong. The fact that derivatives play a role in determining stability should hint that stability, in this sense, is local. We cannot look too far away from an equilibrium point when talking about stability.

5.3 EQUILIBRIA AND LINEAR STABILITY IN HIGHER DIMENSIONS

In our population examples above, you may have noticed some unrealistic characteristics. In an equation like

$$\frac{\mathrm{d}P}{\mathrm{d}t} = bP - \mu P$$

where we have $IN-OUT$, this assumes that the *entire* population is contributing to the birth of new members, even those themselves born just a few seconds earlier. If that doesn't make sense, good. It shouldn't. Typically, at least for mammals[12] there is a period of growth and maturity before individuals are able to reproduce.

Our model clearly doesn't take this into account, but we saw earlier models that do: Leslie matrix models. We can combine the concepts behind the Leslie matrix models with the tools of differential equations

[12]And most birds and amphibians and reptiles

to create a **system of differential equations**. We can divide our population P into children, C, those unable to reproduce, and adults, A, those of reproductive age and write a differential equation for each

$$\frac{\mathrm{d}C}{\mathrm{d}t} = bA - mC$$
$$\frac{\mathrm{d}A}{\mathrm{d}t} = mC - \mu A$$

Now, children are birthed by adults at rate b, children mature into adults at rate m and adults die at rate μ. This is a slightly better model than the one we had before!

Moreover, we can write our two equations as a vector equation:

$$\begin{bmatrix} \frac{\mathrm{d}C}{\mathrm{d}t} \\ \frac{\mathrm{d}A}{\mathrm{d}t} \end{bmatrix} = \begin{bmatrix} bA - mC \\ mC - \mu A \end{bmatrix}$$

since, as we understand, a vector is just a way to collect information.

In much the same way that our Leslie Matrix was a multiplication of a matrix and a vector, the same is true here. Our vector is

$$\vec{P} = \begin{bmatrix} C \\ A \end{bmatrix}$$

and we can write our system of differential equations in a conveniently compact package

$$\frac{\mathrm{d}}{\mathrm{d}t}\vec{P} = M\vec{P}$$

where

$$M = \begin{bmatrix} -m & b \\ m & -\mu \end{bmatrix}$$

If we multiply the right-hand side, we will see we recover exactly the right-hard sides of our two differential equations.

The equilibria are found the same way, but now we have to find *two* things using *two* equations. In this case we would set

$$bA - mC = 0$$
$$mC - \mu A = 0$$

In this case there is one equilibrium and it is $C^* = A^* = 0$.

Stability is a little bit different in that it relies on the **eigenvalues** of the matrix M.

Definition 5.3.1. An **equilibrium** of a multi-dimensional linear system of differential equations of the form

$$\frac{\mathrm{d}}{\mathrm{d}t}\vec{X} = M\vec{X}$$

where M is made up solely of constants[13] is always $\vec{X} = \vec{0}$.
The **stability** is given by the eigenvalues of M.

- The equilibrium is **stable** if all eigenvalues have negative real part.
- The equilibrium is **unstable** if even one eigenvalue has positive real part.
- If not all eigenvalues have positive real parts, then the equilibrium is a saddle[14].

In our case, our matrix has two eigenvalues

$$\lambda_1 = -\frac{\mu + m}{2} + \frac{\sqrt{(m-\mu)^2 + 4bm}}{2}$$
$$\lambda_2 = -\frac{\mu + m}{2} - \frac{\sqrt{(m-\mu)^2 + 4bm}}{2}$$

Stability will depend on the values of m, b, and μ

5.4 THE JACOBIAN

The above method works for any linear system that can be written as the product of a matrix and a vector, but this isn't always possible. In fact, in *most* cases this can't be done directly. For instance, let's look at our logistic population model

$$\frac{\mathrm{d}P}{\mathrm{d}t} = r(K-P)P - \mu P$$

but split into children and adults.

$$\frac{\mathrm{d}C}{\mathrm{d}t} = r(K-(C+A))A - mC$$
$$\frac{\mathrm{d}A}{\mathrm{d}t} = mC - \mu A$$

[13] Numbers or parameters, not variables

[14] Which is still unstable, just with a special name

No matter what we do, or what tricks we try we cannot get a matrix of numbers multiplied by the vector $[C, A]^T$. Our matrix M will always have a C or A in it.

Of course, finding an equilibrium didn't depend on whether we could factor things, so we can at least do that just by setting the differential equations to 0.

$$r(K - (C + A))A - mC = 0$$
$$mC - \mu A = 0$$

In this case, as in the $1D$ case, we have two possible equilibria.

$$\begin{bmatrix} C_0 \\ A_0 \end{bmatrix}) = \begin{bmatrix} 0 \\ 0 \end{bmatrix}$$
$$\begin{bmatrix} C_1 \\ A_1 \end{bmatrix} = \begin{bmatrix} \frac{\mu}{r}\frac{Kr-\mu}{d+m} \\ \frac{m}{r}\frac{Kr-\mu}{d+m} \end{bmatrix}$$

In $1D$, we would check the derivative of the right-hand side evaluated at the desired equilibrium for stability, but what is the derivative of *two* equations?

When checking for maxima and minima in higher dimensions, we had to move beyond a single second derivative into a matrix of second derivatives. Here, we must do the same. We require what is called the **Jacobian** matrix of first derivatives.

Definition 5.4.1. For a system of differential equations

$$\frac{dx_1}{dt} = f_1(x_1, x_2, \cdots, x_n)$$
$$\frac{dx2}{dt} = f_2(x_1, x_2, \cdots, x_n)$$
$$\vdots$$
$$\frac{dx_n}{dt} = f_n(x_1, x_2, \cdots, x_n)$$

ith equilibrium $\vec{x}^* = [x_1^*, x_2^*, \cdots, x_n^*]^T$ such that

$$\begin{aligned} 0 &= f_1(x_1^*, x_2^*, \cdots, x_n^*) \\ 0 &= f_2(x_1^*, x_2^*, \cdots, x_n^*) \\ &\vdots \\ 0 &= f_n(x_1^*, x_2^*, \cdots, x_n^*) \end{aligned}$$

the stability of the equilibrium is given by the eigenvalues of the **Jacobian**, J, defined as

$$J = \left. \begin{bmatrix} \frac{\partial f_1}{\partial x_1} & \frac{\partial f_1}{\partial x_2} & \cdots & \frac{\partial f_1}{\partial x_n} \\ \frac{\partial f_2}{\partial x_1} & \frac{\partial f_2}{\partial x_2} & \cdots & \frac{\partial f_2}{\partial x_n} \\ \vdots & & \ddots & \vdots \\ \frac{\partial f_n}{\partial x_1} & \frac{\partial f_n}{\partial x_2} & \cdots & \frac{\partial f_n}{\partial x_n} \end{bmatrix} \right|_{\vec{x}=\vec{x}^*}$$

It is the matrix of all first partial derivatives of the right-hand sides, evaluated at the equilibrium values.
If all eigenvalues of the Jacobian matrix have negative real parts, then the equilibrium is stable. Otherwise it is unstable.

We can easily compute the Jacobian for our equilibrium

$$\begin{bmatrix} C_0 \\ A_0 \end{bmatrix}) = \begin{bmatrix} 0 \\ 0 \end{bmatrix}$$

using the fact that, from our system of equations

$$\begin{aligned} f_1(C, A) &= r(K - (C + A))A - mC \\ f_2(C, A) &= mC - \mu A \end{aligned}$$

All first partial derivatives here are given by

$$\begin{aligned} \frac{\partial f_1}{\partial C} &= -rA - m \\ \frac{\partial f_1}{\partial A} &= r(K - C - 2A) \\ \frac{\partial f_2}{\partial C} &= m \\ \frac{\partial F_2}{\partial A} &= -\mu \end{aligned}$$

which we can arrange into the Jacobian and substitute $C = C_0 = 0$ and $A = A_0 = 0$

$$J = \begin{bmatrix} \frac{\partial f_1}{\partial C} & \frac{\partial f_1}{\partial A} \\ \frac{\partial f_2}{\partial C} & \frac{\partial f_2}{\partial A} \end{bmatrix} \Bigg|_{(C,A)=(0,0)} = \begin{bmatrix} -m & rK \\ m & -\mu \end{bmatrix}$$

This produces eigenvalues

$$\lambda_1 = -\frac{m + \mu + \sqrt{(m-\mu)^2 + 4Kmr}}{2}$$
$$\lambda_2 = -\frac{m + \mu - \sqrt{(m-\mu)^2 + 4Kmr}}{2}$$

Assuming all the parameters of our model are positive, λ_1 has no choice but to be negative. This means that the stability of $C_0 = A_0 = 0$ depends on $\lambda_2 < 0$, or more particularly,

$$m + \mu - \sqrt{(m-\mu)^2 + 4Kmr} > 0$$

which can be simplified

$$\begin{aligned} m + \mu &> \sqrt{(m-\mu)^2 + 4Kmr} \\ (m+\mu)^2 &> (m-\mu)^2 + 4Kmr \\ (m+\mu)^2 - (m-\mu)^2 &> 4Kmr \\ 4m\mu &> 4Kmr \\ \frac{Kr}{\mu} &< 1 \end{aligned}$$

Therefore, the equilibrium $C_0 = A_0 = 0$ is only stable if $Kr/\mu < 1$.

Complex Eigenvalues in the Jacobian

You may have paused in the above definitions when you read that stability of an equilibrium depends on the sign of the *real part* of the eigenvalues. This caveat must be there because often our eigenvalues will be **complex**[15]. Luckily, the complex part doesn't change much. The biggest change is that instead of moving directly towards or away from an equilibrium, if our eigenvalues are complex we will spiral away[16]. Figure 5.5 shows the difference.

[15] i.e. they have the form $\lambda = a + ib$, where $i = \sqrt{-1}$.

[16] Or toward

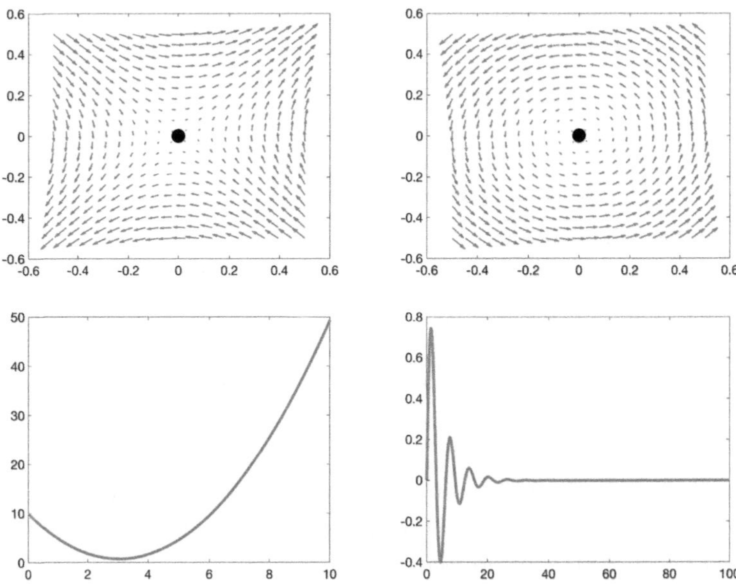

Figure 5.5: The top left panel shows the flow of a 2D system of ODEs when both eigenvalues are real. In this case, we see a saddle. The top right panel shows the rotational nature that occurs when eigenvalues are complex. The bottom row shows the solutions to one component of the ODE. You can see with the saddle, we get closes to the equilibrium at 0 *before shooting off to* ∞*. With the stable spiral, we oscillate until we eventually fall in to the equilibrium.*

PRACTICE PROBLEMS

1. Let's say we have a tank of water with a small hole in the bottom. Torricelli's law states that we lose water at a rate of $\sqrt{2gh}$ through the hole. If we also have a hose that is filling water at a constant rate C, write a differential equation for how the height of the water is changing over time (i.e. $\frac{dh}{dt}$).

2. At what height will the water stabilize at? Is this a stable or unstable equilibrium?

3. A population of honey bees grows like

$$\frac{dH}{dt} = L\left(\frac{H}{w+H}\right) - \alpha H$$

   What is the equilibrium value of honey bees?

4. If $L = 2000$, $w = 5000$, and $\alpha = 0.025$, what is the equilibrium value? Look up of the average size of a honey bee colony; how does this compare?

5. Is this equilibrium stable?

6. Reconsider the model

$$\frac{\mathrm{d}C}{\mathrm{d}t} = r(K - (C + A))A - mC$$
$$\frac{\mathrm{d}A}{\mathrm{d}t} = mC - \mu A$$

   and its equilibrium

$$\begin{bmatrix} C_1 \\ A_1 \end{bmatrix} = \begin{bmatrix} \frac{\mu}{r} \frac{Kr - \mu}{d + m} \\ \frac{m}{r} \frac{Kr - \mu}{d + m} \end{bmatrix}$$

   Compute the Jacobian for this equilibrium, the eigenvalues, and determine the conditions under which this equilibrium is stable.

7. A simple infection for which a person would get no immunity after recovery from infection[17] can be modelled by an SIS system, where a population is divided into Susceptible individuals and Infectious individuals. The equations for such a model are given by

$$\frac{\mathrm{d}S}{\mathrm{d}t} = B + rI - \beta SI - \mu S$$
$$\frac{\mathrm{d}I}{\mathrm{d}t} = \beta SI - (d + \mu)I - rI$$

   Find the equilibria of the model and check stability. Under what condition is the non-zero equilibrium stable?

8. In the SIS model, what might the non-zero equilibrium represent? Does it mean that some people are permanently sick?

9. Can the eigenvalues ever be complex? What would this mean in terms of the infectious population over time?

[17]Think strep throat, or chlamydia

CHAPTER 6

Integration

Most of what we've discussed so far[1] stems from the study of instantaneous change, or differentiation.

Back at the beginning of Chapter 2, I mentioned that there are two big ideas in calculus. The first is differentiation, or the study of change, and the second is **integration** or the study of accumulation. They are both complements of each other and opposites of each other. We use integration to answer questions like *if it snows for 3 hours, how deep will the snow be on the ground?* Or *if my phone uses 0.22 Watts per hour, how many hours will I get from a 5 Watt battery?*

The latter question is simply answered with a quick division, but that's only if we assume we use the exact same amount of energy every hour. What happens if our power consumption looks more like figure 6.1? Again, we would likely want to *model* our power consumption with a function and use some tools to determine accumulated change.

6.1 ACCUMULATED CHANGE

As mentioned above, the derivative tells us how much a function is expected to change at a given point. Another way to look at this is *how much does the function at one point contribute to the value of the function at later points?*

The complement, and opposite, of this idea is *how much does each value of a function contribute to the total?* This question is often just as important as the ones we've discussed thus far.

[1]With the obvious exception of vectors and matrices

DOI: 10.1201/9781003265405-6

Table 6.1: One week's worth of the time I spent playing video games.

Day	Hours Played
1	0.5
2	1.5
3	1.75
4	1.5
5	5
6	4.5
7	0

Table 6.2: One week's worth of the time I spent playing video games.

Day	Hours Played **per day**
1 - 3	1.25
4 - 5	3.25
6	4.5
7	0

When dealing with discrete events, accumulated change is simply the addition of those events. Let's look at the following table of data of my video game playing over a week in table 6.1.

To determine the accumulated time over one week, I would simply add these numbers together, for a total of 13.75 hours for the week.

Now what if the information in our table was presented in a slightly different way, as in 6.2. The biggest change here is that we are now measuring hours played per day, and some of the days are grouped. If I want to determine the weekly total now, by best option is to take the *rate* at which I played video games and multiply it by the number of days I played at that rate.

$$3 \cdot 1.25 + 2 \cdot 3.25 + 1 \cdot 4.5 + 1 \cdot 0 = 13.75$$

If we were to look at the plot of this in figure 6.1, we can see that the total isn't just the sum of hours, but the *areas* of each of the rectangles in our graph. In fact, if look at figure 6.2 which shows the data from table 6.1, we see that we did indeed add up the areas in this case as well, it just so happened that the width of our rectangles was 1.

Let's look at a different example using cell phone data rates. While you may be in a country where wireless data is unlimited, in Canada[2] it

[2]Where I am

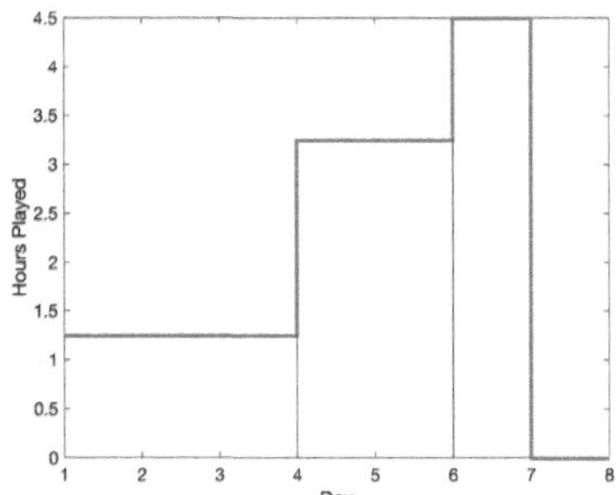

Figure 6.1: A visualilzation of Table 6.2

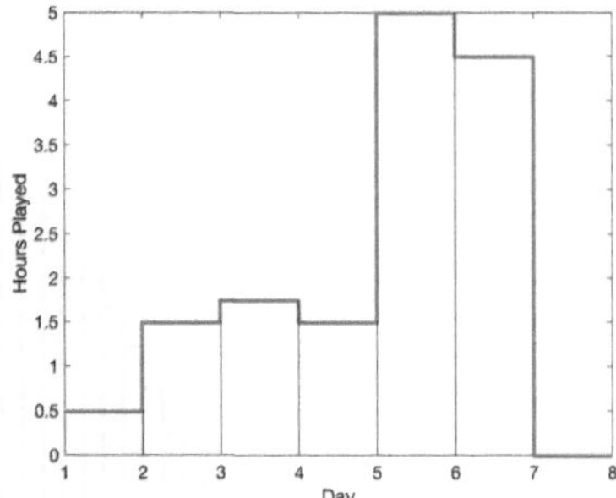

Figure 6.2: A visualilzation of Table 6.1

is most definitely not. Keeping track of how much data you are using is extremely important and becomes difficult when your phone is constantly connected to the Internet and almost constantly communicating.

The graph in figure 6.3 shows the speed at which downloading is happening, and how long that speed was measuring for using an app that tracks data usage. The app would track average data speed over a period of time, and report that speed. The speed is measured in Megabits per second, Mb/s, and time is measured in seconds. From this information, using the idea of accumulated change, and the area of rectangles, we can find out the total amount of data we used of the time period.

The more accurately, or more often we measure our data speed, the thinner the rectangles become, but at the same time the more rectangles we get, like in the panels of figure 6.4. The area of each rectangle is getting closer and closer to zero, but the number of rectangles is approaching infinity. This is conceptually opposite to the problem we faced with

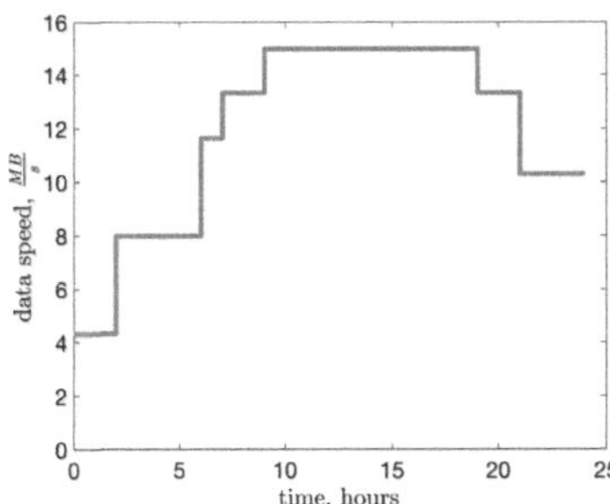

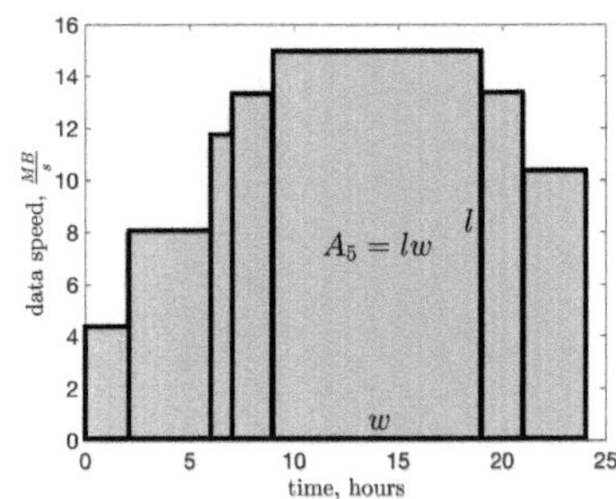

Figure 6.3: If we have discrete measurements, it is easy to measure total accumulated change. The total is nothing more than the area under the function, which is just the sum of the rectangular areas.

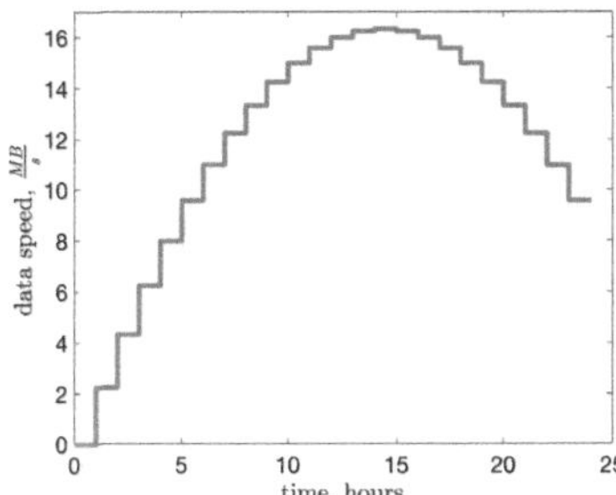

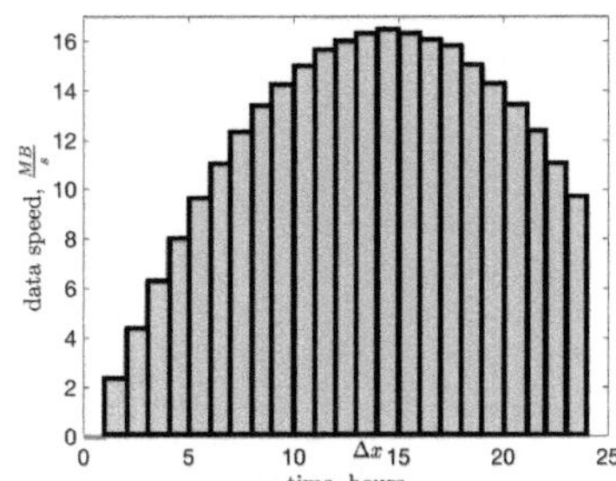

Figure 6.4: As our measurements get more precise and get closer to being continuous, we get smaller rectangles, but also many more rectangles.

derivatives. There we had a ratio that was approaching $0/0$ and saw that this approaches a number.

Here, we are adding up 0 infinitely many times; or looking at the product $0 \cdot \infty$. While generally, 0 times anything is zero, infinity behaves differently. In fact, in the continuum, when the rectangles have zero width, what we get is a smooth curve, like in Figure 6.5.

While there is zero area underneath any one point[3], there are *so many points* that clearly there is still an area underneath the curve and so we should be able to compute the area. That area is the limit of the area of rectangles as the width of the rectangles get smaller and smaller. That area is called the integral of a function $f(x)$ from a to b.

[3]Since there is no width

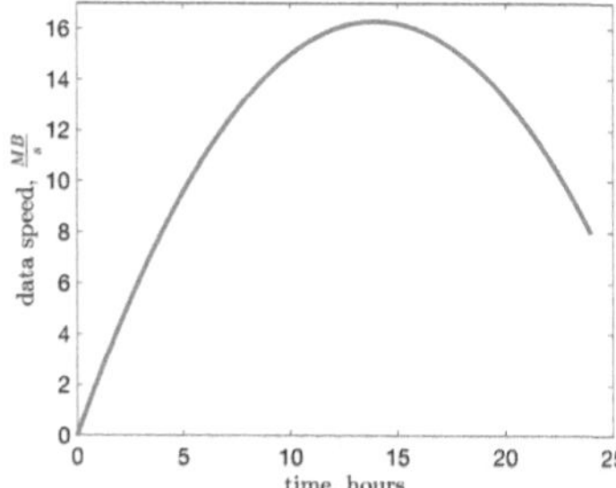

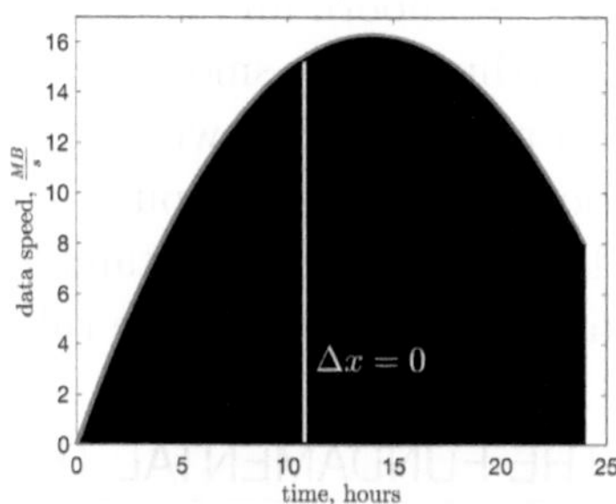

Figure 6.5: Eventually, with infinitely many measurements we would get a smooth curve. If we try to draw rectangles now, they would have zero area, like the white highlighted rectangle. This is offset by the fact that there are infinitely many rectangles. We can see that even though each individual rectangle has zero area, the sum of infinitely many of them clearly exists as the area below our curve. Limits, zero and infinity often work together to create things that are greater than the sum of their parts.

Definition 6.1.1. Let $f(x)$ be a function on an interval $[a, b]$. If we divide the interval up into n equal rectangles, then each rectangle will have width

$$\Delta x = \frac{b-a}{n}$$

We can definite the height of the kth rectangle as $f(a+(k-1)\Delta x)$. By adding up all n rectangles, we can approximate the area, A, under the curve. This sum is called a **left Riemann Sum**, R_L. Mathematically we write this as

$$R_L = \sum_{k=1}^{n} f(a + k\Delta x)\Delta x$$

The limit as n goes to infinity defines the area under the curve, or the integral of $f(x)$

$$A = \int_a^b f(x)\mathrm{d}x = \lim_{n\to\infty} \sum_{k=1}^{n} f(a + k\Delta x)\Delta x$$

The symbol $\int$ can be thought of as a fancy S for a fancy sum[4]. The symbol Σ is jagged and is used for discrete, wholly individual things; the

[4]At its heart, an integral is just a fancy sum.

symbol $\int$ is smooth and is used for the limit when there are fewer jags and everything looks smooth.

Typically, much like with derivatives, we do not want to use the limit definition in order to compute integrals. There are tools at our disposal so that we need only understand limits conceptually and computationally we can rely on more concrete rules.

6.2 THE FUNDAMENTAL THEOREM OF CALCULUS

The fundamental theorem of calculus is a statement that allows us to relate this concept of calculating areas to taking derivatives. As mentioned briefly in the last section, the mathematical expression for an area under a curve, $f(x)$ between two points a and b is given by

$$\int_a^b f(x)\mathrm{d}x$$

Let's say we specify a, but instead of having a parameter b as the right limit, we make this a variable, y, so that we can move it.

$$\int_a^y f(x)\mathrm{d}x$$

In doing so, we will obviously get different areas for different values of y. In fact, this is another function where y is the input and the area is the output. If we call this new function $F(y)$, we can say that

$$F(y) = \int_a^y f(x)\mathrm{d}x$$

The question we must ask ourselves is then *what is the relationship between $F(y)$ and $f(x)$?*

Theorem 6.2.1. *The* **fundamental theorem of calculus** *states that if*

$$F(y) = \int_a^y f(x)\mathrm{d}x$$

then

$$f(x) = \frac{\mathrm{d}F(x)}{\mathrm{d}x}$$

Therefore, if we can recognize that $f(x)$ is the derivative of something, then we can easily compute areas under curves.

As an example, let's say we want to compute the integral

$$\int_0^y \cos(x)\mathrm{d}x$$

If we look back at our table of derivative rules, we will see that $\cos(x)$ is the derivative of $\sin(x)$. Therefore, if

$$f(x) = \cos(x) = \frac{\mathrm{d}F}{\mathrm{d}x}$$

then

$$F(x) = \sin(x)$$

and

$$\int_a^y \cos(x)\mathrm{d}x = \sin(y)$$

In a sense, when computing integrals[5] we are working backwards from differentiation. We are given a derivative and are asked to find the function from which it came.

6.3 THE ANTI-DERIVATIVE

MOST MATHEMATICAL OPERATIONS CAN BE INVERTED. Meaning that if we do something to a mathematical object[6] there's usually a way to undo that something. We can add, and undo addition with subtraction. We can multiply and undo multiplication with division[7]. We can exponentiate, and then we can undo exponentiation with logarithms. Differentiation is no different. If we can take a derivative, we can undo the process by taking an anti-derivative[8].

Unfortunately, this process isn't quite as elegant as, say, division, or subtraction. It's not even as elegant as differentiation where we have a bunch of rules or, at worst, a limit problem to solve. The act of taking anti-derivatives requires intimate[9] knowledge of basic derivatives.

[5] Or areas or cumulative changes; they're all the same.

[6] A number, function, vector, matrix, *etc.*

[7] Usually

[8] Anti-derivative

[9] As in thorough, not wined and dined

For instance if

$$\frac{\mathrm{d}f}{\mathrm{d}x} = 3x^2 \tag{6.1}$$

what was $f(x)$? The only way to answer this question is to know that the derivative of x^3 is $3x^2$[10]. From here, we can conclude that the $f(x)$ must have been

$$f(x) = x^3 \tag{6.2}$$

This isn't always so easy, but sometimes it is. We learned in the last chapter that sometimes we can look at things and use our experience to determine an answer[11]. For example, if I give you the derivative

$$\frac{\mathrm{d}f}{\mathrm{d}x} = \mathrm{e}^{3x}$$

we recognize that e^x is its own derivative, and further to that fact we know that e^{3x} has a derivative $3\mathrm{e}^{3x}$. This is *almost* the derivative we are looking for, so logic dictates that $f(x) = \mathrm{e}^{3x}$ is *almost* the anti-derivative, the only problem is that the resulting derivative is three times too large. The obvious solution here is to undo that 3[12]. If we try

$$f(x) = \frac{\mathrm{e}^{3x}}{3}$$

we get a derivative of

$$\frac{\mathrm{d}f}{\mathrm{d}x} = \mathrm{e}^{3x}$$

which is exactly what we wanted! So the anti-derivative of e^{3x} is ${}^{\mathrm{e}^{3x}}\!/_3$. While we did this through a very good[13] guess, we could have arrived here in the same way by remembering that multiplying by a number *does not change derivatives*; so when we are three times too big, or even n times too big or small, we can just multiply or divide by the right number to fix our anti-derivative.

---

[10] Like we learned way, way back in Chapter 2

[11] We'll talk more about this in Part 2 of the book. This is true knowledge, and means you're well on your way to wisdom

[12] Tricimate? Or does that only work for 10s?

[13] Or maybe lucky, to some

Leveraging of the scalar multiplication rule of differentiation is more ubiquitous than you may first think. Consider the function

$$\frac{\mathrm{d}f}{\mathrm{d}t} = t^4$$

we know that $5t^4$ is the derivative of t^5. Again, it's almost what we want, but the resulting derivative is five-times too big. So, just divide by 5 since multiplying or dividing by a number does not affect the differentiation at all. With

$$f(t) = \frac{t^5}{5}$$

we get the derivative

$$\frac{\mathrm{d}f}{\mathrm{d}t} = t^4$$

which is what we want!

Non-uniqueness of anti-derivatives

One of the biggest issues with anti-derivatives is that they are not entirely unique. This is due to the fact that

$$\frac{\mathrm{d}}{\mathrm{d}x}C = 0$$

for *any* C. This means there are infinitely many things that disappear when we take derivatives; going backwards this means that if we see a 0, there are infinitely many possibilities for where it came from. This is better illustrated through example.

We saw above that if we were given

$$\frac{\mathrm{d}f}{\mathrm{d}t} = t^4$$

then

$$f(t) = \frac{t^5}{5}$$

What if instead we examine

$$f(t) = \frac{t^5}{5} + 10$$

then

$$\frac{\mathrm{d}f}{\mathrm{d}t} = t^4$$

Our original criteria is still satisfied, therefore $f(t) = t^5/5 + 10$ is an equally valid solution. This comes from the fact that our derivative condition can be written as

$$\frac{\mathrm{d}f}{\mathrm{d}t} = t^4 + 0$$

and so we need the anti-derivative of t^4 and the anti-derivative of 0, the latter of which has infinitely many possibilities.

In fact, we can write +0 onto *any* function that we want. Adding 0 doesn't change anything. So we always have a constant term that must be added, but we have no way of knowing just from the derivative what that constant term should be. Therefore, we add it as an undetermined constant, C.

The true anti-derivative to

$$\frac{\mathrm{d}f}{\mathrm{d}t} = t^4$$

is

$$f(t) = \frac{t^5}{5} + C$$

6.4 FUNDAMENTAL THEOREM OF CALCULUS REVISITED

First we will remind ourselves of the statement of the fundamental theorem of calculus. If

$$F(y) = \int_a^y f(x)\mathrm{d}x$$

then

$$f(x) = \frac{\mathrm{d}F}{\mathrm{d}x}$$

In our above example using the fundamental theorem of calculus, the function and a were strategically chosen so that the potential non-uniqueness of the solution was hidden.

An integral has a second property that must be satisfied that we touched on in passing: **the area under a single point must be** 0. Directly under a point there is no width, therefore no area. This means that

$$F(a) = \int_a^a f(x)\mathrm{d}x = 0$$

In our above example, this works out fine since $\sin(0) = 0$, but generally speaking, unless we include a C in our anti-derivative, this won't be true.

For example, let's look at the integral

$$\int_0^y \sin(x)\mathrm{d}x$$

Naively, if we were to proceed as above, we might say $F_0(y) = \cos(y)$[14], which is a safe start. The problem is that $F_0(0) = 1$. This is equivalent to saying

$$F_0(0) = \int_0^0 \sin(x)\mathrm{d}x = 1$$

which can't be true!

If instead, we used the fact that the anti-derivative is not unique, we would say that

$$F(y) = \cos(y) + C$$

and we know, by the property

$$\int_a^a f(x)\mathrm{d}x = 0$$

that, for our example, $F(0) = 0 = \cos(0) + C$. If we solve for C, we see that $C = -1$. The function that satisfies our integral is thus

$$F(y) = \cos(y) - 1$$

It is not a coincidence in this case that C takes the value $C = -F_0(a)$.

[14] We call this F_0 since it is the anti-derivative with the constant term equal to 0.

Theorem 6.4.1. *This leads us to the second part of the Fundamental Theorem of Calculus. If*

$$F(y) = \int_a^y f(x)\mathrm{d}x$$

and b is a point such that $b > a$, then

$$\int_a^b f(x)\mathrm{d}x = F(b) - F(a) = F_0(b) - F_0(a)$$

Sometimes we right the right-hand side difference as

$$F(b) - F(a) = F(x)\Big|_a^b$$

This says that the area under a function between two points a and b is just the difference between their anti-derivatives.

Defining the canonical anti-derivative

I've snuck in this idea of $F_0(x)$ above. I'm going to coin the term **canonical anti-derivative** here. If we define the canonical anti-derivative as the anti-derivative with $C = 0$, then the two parts of the fundamental theorem of calculus are interchangeable. We can state that

$$\int_a^b f(x)\mathrm{d}x = F_0(b) - F_0(a)$$

covers the case where b is a number, or $b = y$, a variable.

6.5 PROPERTIES OF INTEGRALS

Linearity

An integral, like the derivative, is what's called a **linear operator**, and therefore has two properties standard among all linear operators.

Additivity

$$\int_a^b f(x) + g(x)\mathrm{d}x = \int_a^b f(x)\mathrm{d}x + \int_a^b g(x)\mathrm{d}x$$

Scalar multiplication If k is a number then

$$\int_a^b kf(x)\mathrm{d}x = k\int_a^b f(x)\mathrm{d}x$$

These two properties are often written together as the property of linearity

$$\int_a^b kf(x) + ng(x)\mathrm{d}x = k\int_a^b f(x)\mathrm{d}x + n\int_a^b g(x)\mathrm{d}x$$

Negative integrals

Integrals give the net area under a curve; areas above the x-axis are considered positive areas, and areas below the x-axis are considered negative areas. In this sense, the area under $\sin(x)$ from 0 to 2π is exactly zero because the positive area and the negative area cancel each other. This concept of negative areas is important because the fundamental theorem of calculus allows for a property that leads to negative integrals.

This leads to the property

$$int_a^b f(x)\mathrm{d}x = -\int_b^a f(x)\mathrm{d}x$$

The exercises will help you understand how this follows from the fundamental theorem of calculus[15].

Additivity in the limits

Integrals also have a second type of additivity that again follows easily from the Fundamental Theorem of Calculus.

If c is a point such that $a < c < b$ then

$$\int_a^b f(x)\mathrm{d}x = \int_a^c f(x)\mathrm{d}x + \int_c^b f(x) mathrm dx$$

This also works with subtraction in the sense that

$$\int_c^d f(x)\mathrm{d}x = \int_a^b f(x)\mathrm{d}x - \int_a^c f(x)\mathrm{d}x - \int_d^b f(x)\mathrm{d}x$$

while the formula may look intimidating, figure 6.6 summarizes this.

[15] You may even already see it.

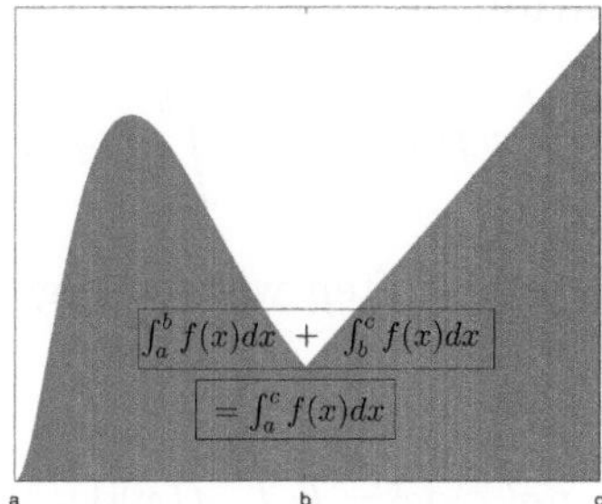

Figure 6.6: We can split an integral into two integrals and add the two integrals up together. This is particularly useful when we have a sharp corner, or discontinuity in our function at a certain point.

6.6 INTEGRATION BY PARTS

The above properties allow us to deal with the addition of functions. We also have a technique for dealing with specific types of products of functions.

The important thing to note is that if we have two functions, $f(x)$ and $g(x)$

$$\int_a^b f(x)g(x)\mathrm{d}x \neq \int_a^b f(x)\mathrm{d}x \int_a^b g(x)\mathrm{d}x$$

Just like with differentiation, multiplication is a little more complicated than addition.

Let's revisit the product rule

$$\frac{\mathrm{d}}{\mathrm{d}x}(f(x)g(x)) = f(x)\frac{\mathrm{d}g}{\mathrm{d}x} + \frac{\mathrm{d}f}{\mathrm{d}x}g(x)$$

We can integrate both sides of this equation

$$\int_a^b \frac{\mathrm{d}}{\mathrm{d}x}(f(x)g(x)) = \int_a^b \left(f(x)\frac{\mathrm{d}g}{\mathrm{d}x} + \frac{\mathrm{d}f}{\mathrm{d}x}g(x)\right)\mathrm{d}x$$

We can use the fundamental theorem of calculus on the left-hand side, and we can separate the right-hand side using our linearity

$$f(b)g(b) - f(a)g(a) = \int_a^b \left(f(x)\frac{\mathrm{d}g}{\mathrm{d}x}\right)\mathrm{d}x + \int_a^b \left(\frac{\mathrm{d}f}{\mathrm{d}x}g(x)\right)\mathrm{d}x$$

If we rearrange this formula, we can express one of our unknown integrals in terms of the other

$$\int_a^b \left(f(x) \frac{\mathrm{d}g}{\mathrm{d}x} \right) \mathrm{d}x = f(x)g(x)\Big|_a^b - \int_a^b \left(\frac{\mathrm{d}f}{\mathrm{d}x} g(x) \right) \mathrm{d}x$$

This formula is known as **integration by parts**.

It may seem useless; we are switching out one integral for another but sometimes one of the two integrals is easier to solve than the other. As an example consider

$$\int_a^b x \log(x) \mathrm{d}x$$

While this integral isn't computable as is, if we use integration by parts then we will see the resulting integral much easier.

In this case, we want

$$f(x) = \log(x)$$
$$\frac{\mathrm{d}g}{\mathrm{d}x} = x$$

then

$$\frac{\mathrm{d}f}{\mathrm{d}x} = \frac{1}{x}$$
$$g(x) = \frac{x^2}{2}$$

Notice in this case we can use the canonical anti-derivative since the constant terms will all be contained in the $f(x)g(X)\Big|_a^b$ term.

Since we have all four necessary terms, we can now plug these into the formula for integration by parts

$$\int_a^b x \log(x) \mathrm{d}x = \frac{x^2 \log(x)}{2}\Big|_a^b - \int_a^b \frac{x^2}{2} \frac{1}{x} \mathrm{d}x$$
$$\int_a^b x \log(x) \mathrm{d}x = \frac{x^2 \log(x)}{2}\Big|_a^b - \int_a^b \frac{x}{2} \mathrm{d}x$$

The integral

$$\int_a^b \frac{x}{2} \mathrm{d}x$$

is far more accessible and can be computed, this will allow us to get an expression for the integral that we *couldn't* solve

$$\int_a^b x\log(x)\mathrm{d}x = \left.\frac{x^2\log(x)}{2}\right|_a^b - \left.\frac{x^2}{4}\right|_a^b$$

Notice that if we had chosen differently,

$$f(x) = x$$
$$\frac{\mathrm{d}g}{\mathrm{d}x} = \log(x)$$

we would immediately get stuck because we would need the anti-derivative of $\log(x)$, which is not easy[16]. In general, we want to choose $f(x)$ in the following order[17]:

1. Logarithms
2. Inverse trigonometric functions[18]
3. Algebraic[19]
4. Trigonometric functions[20]
5. Exponential functions[21]

As a simple example, but useful quantity, we can use this to compute the integral of $\log(x)$. Let's look at the integral

$$\int_a^b \log(x)\mathrm{d}x$$

Any function can be interpreted as being multiplied by 1; we could write this integral as

$$\int_a^b \log(x)\mathrm{d}x = \int_a^b 1\cdot\log(x)\mathrm{d}x$$

[16] This itself requires integration by parts.
[17] It would be great if the order lent itself to a catchy acronym, but alas...
[18] We haven't really spoken much about these.
[19] Terms of the form x^n
[20] sin and cos mostly
[21] e^x or equivalent

By our ordering, we would choose

$$f(x) = \log(x)$$
$$\frac{dg}{dx} = 1$$

which means that

$$\frac{df}{dx} = \frac{1}{x}$$
$$g(x) = x$$

Using integration by parts, we can then write

$$\int_a^b \log(x)dx = x\log(x)\Big|_a^b - \int_a^b x\frac{1}{x}dx$$
$$\int_a^b \log(x)dx = x\log(x)\Big|_a^b - x\Big|_a^b$$

Using the fundamental theorem of calculus[22], we can say that the anti-derivative of $\log(x)$ is $x\log(x) - x + C$.

6.7 SUBSTITUTION

It turns out that the dx in an integral and the denominator of $\frac{d}{dx}$ actually mean the same thing, a small change in x. Let's see how we can use this fact to simplify integrals.

Consider the function

$$y = \sqrt{r^2 - x^2}$$

which represents a half circle. Computing the area under this curve

$$A = \int_{-1}^{1} \sqrt{r^2 - x^2}dx$$

is not simple, because the anti-derivative of $\sqrt{r^2 - x^2}$ is not simple, but it can be made simple.

We can reframe this problem. Instead of looking at infinitely many, infinitely thin vertical rectangles at different x's, we can look at infinitely many, infinitely thin *radial* rectangles at different *angles*, θ, as in figure 6.7, and θ must run from 0 to π[23]

[22] In a bit of an inverted way

[23] We measure angles in radians in math. 360 degrees = 2π radians.

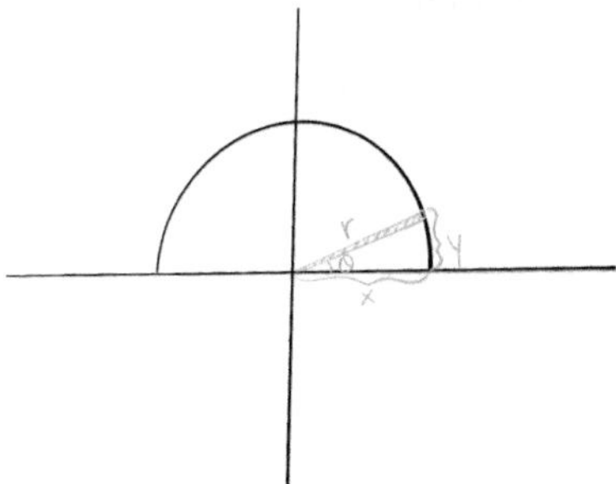

Figure 6.7: We can look at our area as a series of rectangles as different angles instead of at different x positions

The question we must ask ourselves then is *how is x related to θ?* Our old friends Pythagorean theorem and our trigonometric identities can help. If we have a right angle triangle with a hypotenuse, r, a base, x and a height, y and angle θ as in figure 6.8 then

$$\cos(\theta) = \frac{x}{r}$$
$$x = r\cos(\theta)$$

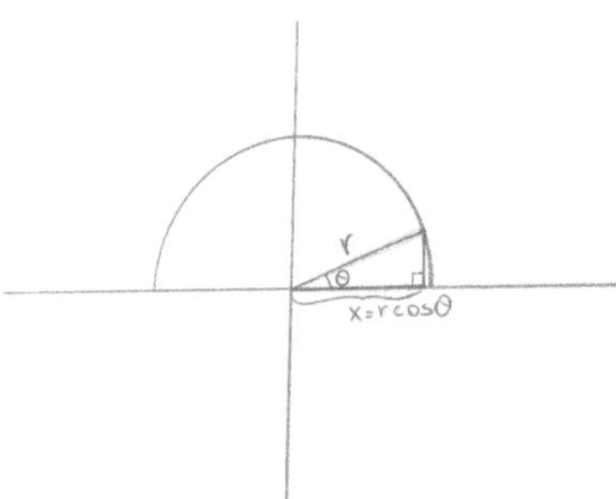

Figure 6.8: Trigonometry helps us relate angles to distances.

Great, we can write x as a function of θ. We are halfway to turning our integral into an integral in θ instead of x. If we make this substitution, we get

$$A = \int_0^{\pi} \sqrt{r^2 - r^2\cos^2(\theta)}\mathrm{d}x$$

We still have an x in the form of $\mathrm{d}x$ that we need to get rid of. Ideally, we need a *mathrmdθ* in its place.

We can get there by determining how x **changes** with θ. In other words, by computing

$$\frac{\mathrm{d}x}{\mathrm{d}\theta}$$

we can find the equivalent change in θ, $\mathrm{d}\theta$, for a change in x, $\mathrm{d}x$. Since we know how x is related to θ, this is as simple as taking a derivative!

$$x = r\cos(\theta)$$

$$\frac{\mathrm{d}x}{\mathrm{d}\theta} = -r\sin(\theta)$$

Solving for $\mathrm{d}x$ then gives

$$\mathrm{d}x = -r\sin(\theta)\mathrm{d}\theta$$

Since we now have an expression for $\mathrm{d}x$, we can substitute this into our integral and now everything is in terms of θ!

$$A = \int_0^{\pi} -r\sin(\theta)\sqrt{r^2 - r^2\cos^2(\theta)}\mathrm{d}\theta$$

Using the fact that $\sin^2(\theta) + \cos^2(\theta) = 1$, we can simplify this to

$$A = \int_0^{\pi} -r^2\sin^2(\theta)\mathrm{d}\theta$$

and using the trigonometric identity

$$\sin^2(\theta) = \frac{\cos(2\theta) + 1}{2}$$

we get

$$A = \int_0^{\pi} -r^2\frac{\cos(2\theta) + 1}{2}\mathrm{d}\theta$$

which yields

$$A = \frac{\pi r^2}{2}$$

This method would, and should, anger any purist of mathematics. Based on our definitions of a derivative and integral, these *aren't* the same, and we did *doesn't* work. This is actually a consequence of the fundamental theorem of calculus and the chain rule. The proper rule is that

$$\int_a^b f(x)\mathrm{d}x = \int_{y(a)}^{y(b)} g(y)\frac{\mathrm{d}x}{\mathrm{d}y}\mathrm{d}y$$

for the substitution $x = x(y)$. In our case $y = \theta$.

Treating the infinitesimal quantities $\mathrm{d}x$ and $\mathrm{d}y$ as autonomous entities that can move freely is more intuitive.

PRACTICE PROBLEMS

1. Overall, in the last 70 years, people have been living longer. The annual death rate of individuals from 1950 to 2019[24] can be modelled as

$$d(t) = A_0 * \mathrm{e}^{-kt}$$

where $t = 0$ is the year 1950.

In order to determine the number of people that died in a given year, we can multiply this rate by the global population, which is another model since it is continually growing. For this time period, a linear model works decently well:

$$p(t) = gt + P$$

where $p(0) = P$ is the world population in 1950. Using the above information, write out the integral that would give us the total number of deaths between 1950 and 2019.

2. Use integration by parts to solve the above integral in terms of your parameters

3. Using the values

$$\begin{aligned} A_0 &= 0.02015 \\ k &= 0.01418 \\ g &= 75029521 \\ P &= 253631149 \end{aligned}$$

and your above results, calculate the total number of people who have died between 1950 and 2019.

4. Consider the integral

$$\int_0^x \sin(\theta)\cos(\theta)\mathrm{d}\theta$$

choose an $f(\theta)$ and $\frac{\mathrm{d}g}{\mathrm{d}\theta}$ and write out both sides of integration by parts definition for the above integral

---

[24] After this, things get tricky since the world started to get weird.

5. From the above result, make the substitution

$$y = \int_0^x \sin(\theta)\cos(\theta)d\theta$$

on both sides of your equation.

6. Solve your above answer for y.

7. What is

$$\int_0^x \sin(\theta)\cos(\theta)d\theta$$

8. Repeat the above steps except use the opposite functions for integration by parts; in other words if you chose $f(\theta) = \sin(\theta)$ above then choose $f(\theta) = \cos(\theta)$. Use the steps above to solve the integral. Does your answer change? Did you expect it to? Why or why not?

9. What happens if we try to use a substitution to compute the integral of a full circle instead of a half-circle? Why do you think this happens? Can you use your area of half a circle to get a full circle?

10. If the integral of a $1 - D$ function gives an area, what do you think the integral of a $2 - D$ function gives?

11. A cylinder can be seen as infinitely, many infinitely thin circles piled on top of each other from $z = 0$ to $z = h$. Use the area of a circle formula to write an integral for the volume of a cylinder of height, h.

12. What if instead of the radius remaining constant for $z = 0$ to $z = h$, the radius, r changed like $r(z) = h - z$, so that we produced a cone. Use a substitution and an integral to compute the volume of the cone.

13. Write the integral that would give the volume of **half**[25] a cylinder of height h.

14. Substitute the area in the above question for the integral from which it came. This is called a **double integral**. A $2 - D$ equivalent of the integral.

[25] Half as in half a circle, not half the height

II

Applied Stats & Data Science

Some Context to Anchor Us

DATA VS. INFORMATION VS. KNOWLEDGE VS. WISDOM

The distinction between data, information, knowledge, and wisdom is of mostly philosophical relevance but I am including it here because I think it can give direction and purpose to what is to come.

Russell Ackoff used his wisdom to make these distinctions[26]. I will summarize his arguments here so that you needn't go and read his article (but you can if you like[1]). The short answer is the distinction between the four concepts is one of usefulness.

DATA ITSELF IS JUST A COLLECTION OF SYMBOLS. It's scratches on paper and inherently meaningless. Even if we can interpret the symbols, staring at a table of data is still, for the most part, without meaning. I like to think of this as the difference between being able to recognize your letters and being able to read a word. If I can only recognize my letters, I can identify that something is being encoded in writing, but without some extra skills[27]. I am unable to do more than just identify the existence of what's in front of me.

INFORMATION IS THE NEXT STEP; IT IS DERIVING SOMETHING USEFUL FROM DATA. It is drawing conclusions, inferences, and context from data. Knowledge uses data to answer basic questions[28]. Information is descriptive, just like data, but more concise[29]. Knowledge answers questions like *How?*. How is it that COVID-19 spreads through a population? When seeking knowledge, we look for patterns in our information. With

[26]He pioneered what is known as the DIKW model.

[27]Namely reading

[28]Who? What? When? Where?

[29]When we take a census, we gather data. When we say the mean household size in Canada is 2.5 people, this is information derived from the data.

knowledge we can make decisions. From knowledge comes understanding; an appreciation for *why* things happen. Why the patterns we see form in the first place.

WISDOM IS MUCH HARDER TO DEFINE AND IS, GENERALLY, STILL DEBATED AND THOUGHT ON. Wisdom is the synthesis of all of the above into your state of being. Wisdom allows you to take knowledge learned from information that was derived from data along with your experiences and not only allow you to make decisions, but make the *right* decision. It is seeing trees, and recognizing you are in a forest.

IN THIS COURSE, WE ARE TO WORK ON THE FIRST TWO CONCEPTS: DATA AND INFORMATION. I am making you read this little summary on the DIKW model because context is always important. Here we will learn tools to transform data into information. These are tools you can take without into whatever discipline you choose and you can use the specific information of your chosen field along with the tools from this course to gain knowledge. The hope is that one day this will lead you to positions where you are both knowledgeable and wise, places where you can not only make decisions to direct society, but also make the right and good decision.

Math Versus The World

MATH IS PERFECT; LIFE IS NOT

THE UNDERSTANDING OF OUR WORLD STEMS FROM A SINGLE, SIMPLE PRINCIPLE THAT HAVE VAST, COMPLICATED, AND SOMETIMES DIRE CONSEQUENCES. It is the principle that for every cause, there is an effect. Effects do not come before causes, and things are related to each other by the way they effect each other. Sometimes we can undo an effect, sometimes we can't[30]. Sometimes there are long chains between the cause and the effect we can see.

Take as an example the act of getting sick. The cause is often breathing in infected air[31], and the visible effect is that you get sick. In between that cause and visible effect, it takes a lot of machinery to get you sick. A virus has to infect a cell, and then that cell has to change its machinery to be able to produce virus, then those new viruses have to evade your immune system and go on to infect other cells. All of this happens before you feel anything at all.

Math is perfect[32]. It is perfect because it follows some very simple rules and always produces the same results. If you can describe a physical[33] process through a mathematical formula, then you in theory have perfect knowledge of how that process happens. Take for instance these equations describing the trajectory of thrown object.

$$x(t) = v_x t + x_0$$

$$y(t) = \frac{at^2}{2} + v_y t + y_0$$

For any time, t, we can determine the position of our object. We put in a number, we get back a number. Better yet, we can put in *any* number and we will *always* get back a number. Of course there are things

[30] Hence the *dire* qualifier above

[31] Or licking something you probably shouldn't be licking

[32] As perfect as it can be, anyway

[33] Social, chemical, biological, evolutionary...

we need to know in order to do this. If we plug in say, $t = 2$, we are left with the following:

$$x(2) = 2v_x + x_0$$
$$y(2) = \frac{2^2 a}{2} + 2v_y + y_0$$

In order to get a position, we need to know a few things:

- What is v_x? *It is the speed in the x-direction at time $t = 0$. This is something we can measure.*
- What is v_y? *It is the speed in the y-direction at time $t = 0$. This is something we can measure.*
- What is a? *It is the acceleration of our object in the y-direction. For most objects (like a ball, bullet, arrow, etc.) this is the acceleration due to gravity. It's a known value*[34].
- What is x_0? *It is the starting x-coordinate from which we throw the object.*
- What is y_0? *It is the starting y-coordinate from which we throw the object.*

These are all values we have to measure and the truth is we can never ever measure things *perfectly*[35]. Each time we throw a ball, even if we try our hardest to match the exact position and the exact initial speed we will have some *variation*. Every experiment we do is slightly different. When we record our measurements, we are never truly on one curve; each experiment gives us a piece of data from a different curve. What we end up with is a bunch of points that tell us something about the family of curves.

ANOTHER WAY THAT MATHEMATICS AND THE REAL WORLD DIFFER, IS THAT IN THE GRAND SCHEME OF THINGS, WE DON'T KNOW HOW TO DO A LOT OF MATH. We only truly know how to deal with *linear* things[36]. We can say some things about non-linear things, but many of our tools are just fancy ways of straightening things out. The less

[34] $a \approx 9.81m/s^2 = g$

[35] In truth, I don't think perfection exists in the real world. It's a concept relegated to the abstract and fictitious.

[36] That is, things that look like straight lines when we graph them

straight things are, the fewer mathematical tools we have to deal with it[37]. So when we *model* real-world phenomena with mathematics, one of the things we do is *simplify the situation*. Our model above of throwing an object is extremely simplified. We take into account gravity and the act of throwing, but otherwise we assume nothing else exists. We have no wind, no air resistance, the object isn't spinning, we don't even account for the shape of the object!

All of this to say that mathematics and the real world are never going to match up perfectly, and that's ok. I like to think of statistics as a bridge between mathematics and the real world. The tools of statistics can take all of these points that are related by unknown quantities and see through the unknowns to glean important information. Statistics can reveal what is hidden, show us where our simplifications fail, and how we can be certain of things in a world where nothing is perfect[38].

---

[37] Generally, of course

[38] Or perfectly accurate

CHAPTER 7

Data and Summary Statistics

7.1 WHAT IS DATA?

FIRST AND FOREMOST, IT'S PLURAL. SO THE TITLE OF THIS SECTION SHOULD BE "WHAT ARE DATA?" This in itself is an important distinction that is often lost[1]. Data is the plural form of datum. Data is a **collection of observations**. These notes are data; the collection of characteristics that make you *you* is data.

Data comes in all shapes, sizes, and forms, but we can very broadly break up data into two categories: qualitative (or categorical) and quantitative (or numerical).

Categorical Data

This is, arguably, the broader category of data. Categorical data is exactly what it sounds like. Each piece of datum fits into one and only one category. These categories and data have labels which can be words, descriptions, or numbers.

Ordinal data[2] can be ordered; think a store asking to rate your experience on a scale of 1 to 5 or rating a movie on a scale of *bad*, *okay*, *good*, or *great*.

Nominal data necessarily cannot be ordered; think of a survey asking every person for their hair colour: what comes first? Black? Brown?

[1]Important enough that I made, and called out, my own typo instead of fixing it

[2]A subclass of categorical data

DOI: 10.1201/9781003265405-7

Blonde? Chestnut? Red[3]? As an example, here is a table of the majors of students from the first time I taught a class using this half of the book[4].

Can we order these subjects? I know many people would like to, but we can't. An ordering should be *natural* and *objective*. We can all agree that we need to add something *more* to 2 to get 3[5]. Therefore, we can rationally conclude 3 is bigger than 2. Same goes with *bad* versus *good*. We have to add something[6] to move from bad to good. Good is greater than bad.

MATHEMATICALLY, CATEGORICAL DATA IS SEVERELY LIMITED. When it comes to **nominal data**, we can't perform *any* of our basic mathematical operations. In table 7.1, what does Psych + Commerce mean? What is Arts divided by CS? These operations are non-sensical. Since a lot of mathematical and statistical analysis relies, at its core, on our most basic mathematical operations, our analytical tools for categorical data are limited[7].

Ordinal data is a little bit more malleable by virtue of its natural ordering, but again our basic mathematical operations are mostly nonsensical. What is good times bad? What is fine subtracted from great? The answer to both of these questions is *nonsense*. Even ordinal data that looks like we *should* be able to do math to it lacks some interpretability properties.

Back to our store that asked for a rating on a scale from 1 to 5. There is a natural ordering here, since we know that 1 comes before 2 comes before 3, and so on, but when we ask, is a 5 *truly an experience that is* five *times better than a* 1*?* If we pause on this question, we'll see that it's not necessarily an easy question to answer. It *could* be that a 5 is five times better than a 1, but by what measure? In what units? Is it at all quantifiable?

Moreover, we could ask, what is the significance of $1 + 5$? Can these ratings be meaningfully[8] added together? Is an experience of a 1 and an experience of a 5 equivalent to an experience that is better than a 5[9]? The answer here is a resounding *No*. The numbers in this case are

[3]You may have some personal opinions about this, but *that* kind of ordering doesn't count. It's not a *meaningful* ordering.

[4]Winter 2021

[5]We need to add 1.

[6]Quality of some kind, generally

[7]For now, we will develop tools later to *translate* categorical or nominal data into something useable.

[8]This is the key here. They are numbers, so they can be added that much is true, but does the result make sense?

[9]An ever elusive 6; an effort of 120% that this dystopian corporatocracy asks its employees to give.

Table 7.1: Harking back to the discussion of the DIKW model, you may be asking, why are all the arts lumped together, yet the sciences are finely differentiated? This is where wisdom comes into play: the goal was to get a vague idea of the mathematical background of the cohort. This data is already half converted to information in your mind because I gave you context to this list of words: it is the various majors of the W21 cohort of my intro to data science course.

Commerce	Psych	Biology	Commerce	Arts
Biology	Arts	Math	Unknown	Physics
Biology	Biology	Unknown	Arts	Unknown
Psych	Biology	Unknown	Psych	CS
Commerce	Biology	Aviation	Psych	Biology
Biology	Arts	Chemistry	Biochemistry	Biology
Aviation	Biology	Math	Math	CS
CS	Biology	Unknown	Unknown	Biology
Unknown	Unknown	Math	Unknown	Psych
Psych	Biology	Commerce	Biology	Psych
Psych	Psych	Psych	Biology	
Commerce	Unknown	Commerce	Commerce	
Econ	Psych	Unknown	CS	
Arts	Psych	Aviation		
Psych	Psych	CS		
Biology	Biochemistry	Biology		
CS	Biology	Sociology		
Aviation	Psych	Commerce		
Arts	Biology	Psych		
Biology	Commerce	Unknown		

nothing more than labels that can be ordered. A rating scale from 1 to 5 is functionally equivalent to a rating system from *bad* to *great* with steps of *poor*, *meh*, and *good* in between.

This is an important distinction that will hopefully become more clear[10] when we discuss the other overarching class of data: numerical data.

Numerical data

Numerical data allows for the greatest amount of analysis. It comes in two flavours as well: discrete data and continuous data.

[10]If not already

Table 7.2: My coffee intake in the week leading up to Christmas (20–25 December 2020)

Day	Coffees
Monday	1
Tuesday	3
Wednesday	1
Thursday	4
Friday	3

DISCRETE DATA IS SUPERFICIALLY SIMILAR TO ORDINAL DATA. The biggest difference is that discrete data is not only represented by ordered numbers, but the numbers are *mathematically relevant.*

Let's look at a comparative example to try and sort through this. We may be collecting survey data with the following question:

Rate your experience at S-Mart.

- ☐ 1 – Awful; too many undead
- ☐ 2 – Bad; weird man with chainsaw hand made me uncomfortable
- ☐ 3 – Meh
- ☐ 4 – Fine
- ☐ 5 – Excellent

In this case, we have numbers and we can figure out some statistics[11]. What we *can't* do is say things like an *awful* plus a *bad* is basically a *meh.* These numbers are not mathematically relevant. We cannot perform mathematical operations on them; the numbers are just labels to highlight the *natural ordering* of the sentiments captured by the survey.

Meanwhile, if I record how many coffees I drink each day for a week, I can do some math! Here is a table for the week of 20–25 December 2020.

I can actually add these numbers! If I do, we will see that over a week, I drank 12 coffees. I can turn these numbers into proportions and say that I drank 25% of my weekly coffees on Thursday; one of those coffees each day is just to de-monsterize myself after waking up and allow me to

[11]We'll discuss exactly which statistics we can get from ordinal data in the following sections.

be able to function around other people. We can subtract 1 from each day to determine the extraneous coffees I drank. Math.

Could we subtract 1 from each of our rankings of S-Mart and have them still make sense? The answer is no, because the numbers are arbitrary. A 5 − 1 translates to an *excellent* − *awful*; does that or even *should that* equal fine? Who is to say? Math doesn't work here because the numbers are not mathematical in nature[12]. We are not counting anything. We are *using* numbers so that we can *kind of* use *some* math, sometimes[13].

CONTINUOUS DATA IS PROBABLY THE EASIEST TO UNDERSTAND. It is numbers the way we are used to thinking about numbers. It typically represents measurements. Continuous data assumes that there are numbers *in between* the numbers we have measured[14]. For instance, if we measure the heights of individuals, we may see that person A has a height of 180 cm and person B has a height of 172 cm. In theory – even though we haven't observed them – heights of 173 cm, 174 cm, etc. can and should exist. If we are extremely accurate in our measurements, then 172.000001 cm should exist. Continuous data means that no matter how close two observed numbers exist to one another, there is theoretically a number that may exist between them. Compare this to the case of coffees drunk[15]. I'd be a pretty wasteful person if I didn't finish my coffees, and I like to make a conscious effort of not being wasteful. So, for me, I can have 2 coffees or 3 coffees but 2.5 coffees is morally out of the question. There is nothing that exists between 2 and 3; this data is not continuous.

The key to understanding the different types of data is a lot to do with **context**. Table 7.3 shows some context-less data. Without context, it's impossible to determine what type of data this is. This data could be S-Mart reviews on their new scale:

- ☐ 1 – Awful; too many undead
- ☐ 2 – Bad; weird man with chainsaw hand made me uncomfortable
- ☐ 3 – Meh
- ☐ 4 – Fine

[12] In the words of King Theoden, "You have no power here."

[13] Don't worry if this doesn't make sense right now; we only need to keep this stuff in the back of our mind. Once we learn about our tools for analysing data, the data types will become more obvious.

[14] At least in theory

[15] Drank? Drinked? Dranked? Lucky this isn't an English class.

Table 7.3: A table of ambiguous data. It's hard to tell if this data is categorical ordinal, numerical discrete, or even numerical continuous without some contextual clues.

Event	Observation
1	4
2	3
3	6
4	2
5	3
6	2
7	4

☐ 5 – Excellent

☐ 6 – Corporate perfection

It could be litres of coffee drank, where we just happened to make integer measurements, but theoretically we *could* get any number as a measurement. Less probable, but still possible, this could be a roll call for some extra-terrestrial school where children are named using the symbols we use for numbers, which would make this data nominal.

In truth, this is me rolling a die six times and recording the results. A die can only take the values 1, 2, 3, 4, 5, or 6. This is discrete data. The important takeaway here is that data is context sensitive. It's important to always have proper context for your data as it will tell you what kind of analysis is possible and sensical.

A lot of ambiguity can be fixed just from **properly naming the columns of your data tables.** We can see in table 7.4 that simply renaming our columns makes the context and data type obvious.

Representing Data

WE NEED TO AGREE ON HOW WE ARE GOING TO REPRESENT DATA MATHEMATICALLY. In order to do this, let's first agree that no matter what we are working with, one piece of datum is going to be represented by the letter X_i, where the i is a number between 1 and the number of pieces of data we have. For the data we have in table 7.1, X is a placeholder for a major. If we number the entries in the table starting in the top left corner and then moving first down and then to the right, we

Table 7.4: Simply changing the column headings makes our data significantly less ambiguous, assuming we are familiar with the idea of rolling a six-sided die.

Roll	Die Value
1	4
2	3
3	6
4	2
5	3
6	2
7	4

get that

$$\begin{aligned} X_1 &= \text{Commerce} \\ X_2 &= \text{Biology} \\ X_3 &= \text{Biology} \\ X_4 &= \text{Psych} \\ &\vdots \\ X_{83} &= \text{CS} \end{aligned}$$

The i goes up to 83 because there are 83 students.

If data comes **in a pair** as in table 7.2, then we represent each pair as (X_i, Y_i). So the first entry in table 7.2 can be written as

$$(X_1, Y_1) = (\text{Monday}, 1).$$

It's important that we get used to using placeholders[16] to represent data because a lot of our statistics will hinge on the existence of an arbitrary data set that we know nothing about other than the fact that it is made up of individual datum.

Working Representation of Data

In this book, the data we work with will come in **tables**. Tables are made up of **rows** and **columns**. Rows go this way $\leftrightarrow$, and columns go this way $\updownarrow$. Generally, when dealing with tabular data, **one row represents one set of measurements with respect to an independent variable**.

[16] Better known in math as variables

Rows can represent measurements taken at different places, at different times, during different experiments, or from different people. In contrast, **one column represents a particular measurement at different instances of the independent variable**. Tabular data can have many rows and many columns.

It would be absolutely wonderful if all data in the real world were provided in the aforementioned format. We have to note very early on that while tabular data is ideal for processing[17], it is *very, very far from the norm.* Typically, data is messy and gross and given in ways that make little to no sense. I wish there was a set of steps for going from gross data to good data, but there isn't. Each and every data set is messed up in its own unique way; much like people. Our only hope is to learn through experience and to recognize what workable, clean data looks like. If we know where we are starting from and we know where we should end up, our experience and tools will help us navigate a path between the two.

7.2 DATA IN PYTHON

Python is the language we will use to communicate with a computer because when it comes to data and when it comes to data and statistics, a lot of the work is simple calculations, but we have to do *a lot* of them. It can quickly become tedious to do by hand, and so we leave the tedious work for a computer[18].

Since it's very likely you have no idea what Python is[19], which is absolutely fine, let's go through the anatomy of a Python file in figure 7.1.

A crash course in Python is given in Appendix A.

Python uses tabular data representation where rows are given by an **index** and a column is described by its **header** (or label). In order to work with data in Python, we will be using the library `pandas`. This library allows us to create or import data into what is called a **data frame** that can be used and manipulated for analysis.

Whenever you want to use the pandas library, you need to write at the top of your Python file

```
import pandas
```

[17]In most cases

[18]By virtue of their names, while we may see them as internet machines or things to play games on, a computer's main purpose is to compute and to compute quickly.

[19]Yes, it's also a type of snake but not that python.

Line numbers help
us identify where in our
code problems might exist
These are libraries
They are non-standard
things we add to Python
to extend capabilities
We can define functions (much like
mathematical functions) that create
reusable code!
Code inside a function
only gets executed if we
call the function
Any code that is all the way to
the left is the instructions the
computer will run

```
import numpy as np
from scipy.optimize import curve_fit
import matplotlib.pyplot as plt
from scipy import stats
import pandas as pd
import csv
import datetime

def buildIncidence(file,province):

    case_data = pd.read_csv(file)

    case_data = case_data.loc[case_data['Entity']==province]

    case_data['Date'] = pd.to_datetime(case_data['Date'], dayfirst=True)

    case_data = case_data.sort_values(by=['Date'])

    total_cases_per_day = case_data[['Date','Total','Daily']]

    province = province.replace(' ','')

    total_cases_per_day.to_csv(province+'Data.csv',index=False,header=True)

provinces = ['United Kingdom', 'Bangladesh',]

for pr in provinces:
    buildIncidence('totalvsconfirm.csv',pr)
```

Figure 7.1: This is what a typical Python file looks like. Don't worry if you don't understand what this file is doing; you're not supposed to. It is just to show you what a file looks like so that if someone puts one in front of you, you can say Hey! That's Python!

More commonly, we use the command

```
import pandas as pd
```

so that we don't need to write out pandas every time we want to use something from the library. Other common libraries that we will use in this course are the following:

```
import numpy as np
from scipy import stats
```

These provide us the necessary math tools to analyse data.

Python's `pandas` can natively import all kinds of different file types. Some common file types for data are the following:

- Comma-separated values (.csv)
- Excel (.xlsx)

- JavaScript Object Notation (.json)
- Word (.docx)
- SQL (.mdf)
- Webpages (.html)

We can import a file with commands like `pd.read_*`, where $*$ is replaced with the file type. If we download `basicdatafile.csv` from the file repository, we can read the data in this file into Python with the command

```
data = pd.read_csv( 'basicdatafile.csv')
```

Figure 7.2 is what it looks like, when we read the data into Python and then print it to the screen.

```
>>> import pandas as pd
>>> data = pd.read_csv('basicdatafile.csv')
>>> print(data)
   Experiment Number  Measurement1  Measurement2  Measurement3
0                  1        0.6429        0.3543        0.2397
1                  2        0.5504        0.3540        0.8920
2                  3        0.7140        0.0763        0.5181
3                  4        0.8641        0.1738        0.9469
4                  5        0.5546        0.9417        0.2760
5                  6        0.3365        0.6365        0.3569
6                  7        0.7846        0.8205        0.3481
7                  8        0.2579        0.5706        0.2567
```

Figure 7.2: Our data frame has four columns of data: Experiment Number, Measurement1, Measurement2, and Measurement3. Notice there is one unlabelled column on the left that starts at zero. This is the **index** *of each row of data. Python starts counting at zero.*

We set our command equal to something (in this case `data`). Now every time we want to access the data, we just refer to `data`. We can get all data in a column by referencing the name of the column as in figure 7.3.

We measure *rows* by the *index* number. Remember, rows come first, then columns[20]. So if we want to specify a particular cell, we do so with `DATANAME[ROWNUMBER][COLUMNNAME]`. In our data set, we can get the first value in the column `Measurement1` with the command `data[0]["Measurement1"]`.

[20] We have to enter a house before we can go up the stairs.

```
>>> data['Measurement1']
0     0.6429
1     0.5504
2     0.7140
3     0.8641
4     0.5546
5     0.3365
6     0.7846
7     0.2579
8     0.6029
9     0.7808
10    0.9029
11    0.6342
12    0.3098
13    0.5262
14    0.9454
15    0.9456
16    0.1792
17    0.6000
18    0.4999
19    0.5332
Name: Measurement1, dtype: float64
```

Figure 7.3: When we type `NAME['COLUMN NAME']`*, we access a particular column in our data frame. In this example, we access the column* `Measurement1` *of the* `data`.

Say it again, for those in the back.

Importing and accessing data is the first step in analysing data. Again, I need to reiterate this point because it's important: there is no one standard way for data to be provided. Some key questions to ask when you've been provided with a data set are the following:

- Is everything in this data set necessary?
  - Strip out the unnecessary information.
- What is the **dimension** I would like to compare?
  - Make this your rows[21].

[21] Another way to frame this might be, *what am I taking measurements on?* Generally, this should be your rows.

- Can this data be easily organized as a table?
  - If not, look for an easier data format[22].

The most important thing is not to throw your hands up in the air and give up. Sometimes solutions aren't elegant. Sometimes data is so bad it's actually easier to retype it by hand than try and work with it.

7.3 SUMMARY STATISTICS

When we're given data and are able to identify the type of data, there are a few tools that we have at our disposal for describing the data set as a whole. Many statistics, as mentioned above, will only really work on numerical or ordinal data. We'll start though, with something that can always be calculated.

Mode

Works with categorical and numerical data; *i.e.* all data types.

The **mode** of a data set is **the value which appears most frequently in the sample**. In the data shown in table 7.1, the mode is Biology. In table 1.2, the mode (of the **dependent variable**) is 3. In table 7.2, we don't really care about the days of the week for anything more than giving us a way to order our data[23]. Time marches on, and there is absolutely nothing we can do about that fact. It is **independent** of influence; time is a cause, not an effect.

To find the mode of data by hand, we just keep a tally of each possible outcome and then pick the largest tally. Easy peasy.

Hold up. HOLD. UP. What happens when we have continuous data? Let's look at table 1.4: there is no mode in wind speed! Every value is different! In order to get around this, we can **discretize** our data by **binning** it. When dealing with continuous data, it's sometimes useful to group similar values into *intervals* and then count how many measurements fall into each interval. We can do this with wind speed.

The **modal class** is the interval 3–6. This means that if we pick a wind speed at random from our data set, that measurement is most likely

[22] JSON is visually nice if a human needs to look at data.

[23] This is called time series data.

Table 7.5: Some simulated wind and temperature data

Day	Wind Speed	Temperature
1	3.21	17
2	15.3	2
3	19.3	3
4	4.12	10
5	0.00	3
6	6.19	9
7	3.01	12
8	12.2	11
9	16.2	4
10	5.89	7
11	0.89	7

Bin	Count
0–3	2
3–6	4
6–9	1
9–12	0
12–15	1
15–18	3
18–21	1

to be between 3 and 6. We can calculate a **crude mode** by taking the centre value of our *modal class*. In this case, our crude mode is 4.5. We get this number from the equation

$$mode_{crude} = 3 + \frac{6-3}{2} = 4.5.$$

In general terms, we take the lower bound of our modal class[24] and add the value in the middle of our modal class[25].

Alternatively we can calculate the **mode for grouped continuous data**. This is given by the following formula:

$$M_0 = L + \frac{f_1 - f_0}{2f_1 - f_0 - f_2} w \tag{7.1}$$

[24] 3, in this case

[25] 4.5 is exactly halfway between 3 and 6.

Let's explain the terms:

- L is the *lower bound* of the *modal class.*
- f_1 is the *count*[26] of the *modal class.*
- f_0 is the *count* of the *class before* the modal class.
- f_2 is the *count* of the *class after* the modal class.
- w is the *width* of the classes.

In our example, we have the modal class as 3–6. The class before it is 0–3, and the class after it is 6–9. Now let's plug in all our numbers:

- $L = 3$
- $f_1 = 4$
- $f_0 = 2$
- $f_2 = 1$
- $w = 3$

Therefore,

$$M_0 = 3 + \frac{4-2}{8-1-1}3$$
$$M_0 = 4$$

Why is this a better estimate of the mode? Because it takes into account that the data in the class before the modal class has more values in it than the class after the modal class. We take into account the fact that the data seems to be skewed towards the lower end of the interval.

Median

Works with categorical ordinal or numerical data[27].

The **median** of a data set is **the middle value when the data is ordered smallest to largest**. Once we order the data from smallest to largest, the median is the value that appears in the middle. For our wind

[26]Also called the frequency, hence the f

[27]In other words, any data that has a natural ordering

speed data, the ordered values look like this: we see here that 5.89 falls directly in the middle of the ordered set. We say that the median wind speed is 5.89.

0.00	0.89	3.01	3.21	4.12	**5.89**	6.19	12.2	15.3	16.2	19.3

If we have repeated values, we write a value as many times as it appears. In the same table, we have temperature data. The ordered list of values is the following:

2	3	3	4	7	**7**	9	10	11	12	17

The median temperature in this data set is 7.

This also works on categorical ordinal data. Consider the following data which shows the birth month of random people[28]. Listing the months in order gives the following:

1	January	11	March	21	August
2	January	12	April	22	September
3	January	13	April	23	September
4	January	14	April	24	September
5	February	15	**May**	25	October
6	February	16	**May**	26	October
7	February	17	May	27	October
8	February	18	May	28	November
9	March	19	June	29	December
10	March	20	August	30	December

In this case, since we have an even number of data, we have **two** medians. Luckily, in our data, they are both the same value, so we can say that the median birth month is May. If this weren't the case, we would have to report *both* possible median values for our categorical data.

[28]These are synthetic people, each assigned a birth month using data from [4] to determine birth month.

ID	Birth Month	ID	Birth Month	ID	Birth Month
1	April	11	August	21	January
2	December	12	March	22	May
3	January	13	April	23	February
4	March	14	October	24	November
5	January	15	October	25	October
6	February	16	February	26	June
7	January	17	September	27	May
8	August	18	March	28	February
9	May	19	September	29	December
10	May	20	April	30	September

When dealing with numerical data and an even number of data points, we can take the value in the middle of our two candidates as our median, for example.

Mean

Works with categorical ordinal or numerical data.

The mean is probably the summary statistic we are all most familiar with; particularly the **arithmetic mean**.

For N data points of the form X_i (where $1 \leq i \leq N$), the **arithmetic mean** is given by

$$m_A = \frac{1}{N}\sum_{i=1}^{N} X_i$$

This is the most common type of mean and is the most applicable to data. Almost all quantitative data can be added, and thus this mean makes sense most of the time.

For our example above, the *arithmetic mean* of our wind speed is

$$m_A = \frac{0.00 + 0.89 + 3.01 + 3.21 + 4.12 + 5.89 + 6.19 + 12.2 + 15.3 + 16.2 + 19.3}{11}$$
$$= 7.8464$$

In a lot of situations[29], a different kind of mean is more useful. The formula for the **geometric mean** is given as follows:

$$m_G = \left(\prod_{i=1}^{N} X_i\right)^{1/n}$$

[29]We will discuss later when it's best to use which mean.

If we calculate the *geometric mean* of our wind speed, we get

$$m._G = 0.$$

It should be clear and obvious that if our data has any zeros in it, the geometric mean isn't useful.

The third type of mean is the harmonic mean. We calculate the **harmonic mean** by

$$m_H = \frac{N}{\sum_{i=1}^{N} \frac{1}{X_i}}$$

The harmonic mean of our wind speed data above is given by

$$m_H = 0.$$

Again, because the zero value in our data makes the harmonic mean unusable.

Which mean to use and when

I hate to do this so early on, but it will be something that will have to be said a lot: *the correct mean to use depends almost entirely on the context of the data and is usually a judgement call.* There are a few guidelines that might help you choose though:

The arithmetic mean is best used when

- The data is additive.
- The data is independent. (One value does not depend on the other values.)
- Values are in a similar range.
- You have a large sample size.
- There are not many *outliers.*
- There are zeros in your data set.

The geometric mean is best used when

- The data is multiplicative.
- You are averaging over things with very different scales.

- When averaging ratios.
- Averaging over data that has different units.
- Almost always when averaging compound interest rates.

The harmonic mean is best used when

- When averaging rates (data made up of the ratio of two independent variables, like speed).
- Data does not have zeros in it.

It's important to remember that *these are guidelines only.* The best mean to use will depend on the context of your data, the questions you are trying to answer, and what you are using the mean for.

For the particular data set used in 7.5, the arithmetic mean is best even though wind speed is in fact a rate. This is because I made up all the values. While I was sitting away typing and wondering about what the best example for this section would be, I made a table and I mashed my keyboard until random numbers popped out. These numbers are not related to each other, or any other thing. Context is everything.

7.4 ETHICAL AND MORAL CONSIDERATIONS: PART 1

I can say one thing for sure: you might be tempted to remove zeros from your data in order to use the geometric or harmonic mean. Removing raw data is almost never advised. For instance, if we removed our zero from the wind speed, we would no longer be taking the average wind speed over our 11 days, but would instead be taking *the average wind speed over the last ten windy days.* It's a subtle difference, but it is a difference. You are *changing the context* of the data and of your analysis. This is fine as long as you are clear and open and transparent about what you are doing. Unfortunately, many people will remove data points out of convenience with little regard to how it is affecting the context.

Deeming data important or unimportant can lead to shaky arguments to remove data because it doesn't *fit* what you think should be happening. Often[30] this isn't done with malicious intent, but out of passion and determination. That doesn't make it any less harmful though. A good baseline is to just not remove data unless you have very good reason[31].

[30]But not always

[31]In our wind speed example, perhaps we want to remove the zeros because the machine broke on those days and was reading out zero when it shouldn't have been.

Even then, it's best to note that it has been *excluded from analysis* but should remain present in the raw data. There is almost no reason ever to alter raw data.

7.5 MEAN VS. MEDIAN VS. MODE

The mean, median, and mode are measures of the **central tendency** of data. It measures, roughly speaking, an equivalent constant value that would produce the same outcomes. Another way to think about the central tendency is, *if I were to make a measurement right now, what is the most likely value I am to see?* Since data comes in many different shapes, sometimes one measure of central tendency will tell you more than another.

When a data set is **symmetrical** and **unimodal**, the **mean** is most accurate.

When a data set is **asymmetrical** and **unimodal**, the **median** is most accurate.

If a data set is **bimodal**, both the mean and median will be woefully inaccurate. The **mode** is the only measure that has any hope of capturing the central tendency[32].

7.6 VARIANCE AND STANDARD DEVIATION

The central tendency of a data set doesn't tell us everything about a data set. For example, take the data sets in figure 7.5.

We can get a sense of how *spread out* a data set is by computing the **variance**. The **variance** measures how far numbers tend to deviate from the central tendency. In figure 7.5, (b) has the smallest variance and (d) has the largest[33].

Before we talk about variance, it is at this point where *how* our data is collected starts to affect how we calculate things. So, before figuring out how to calculate the variance, let's take a slight detour.

A population vs. a sample

When we collect data, one of the first things we have to decide is, *where are we going to collect data from?* For instance, let's say we wanted to calculate the average GPA of Mount Allison University students. There's

[32]Tendencies, in this case

[33]Although it is hard to see, because the x-axis only runs from -5 to 5 to highlight the shape of the data. (c) and (d) both spread much farther in both directions, with (d) spreading the most.

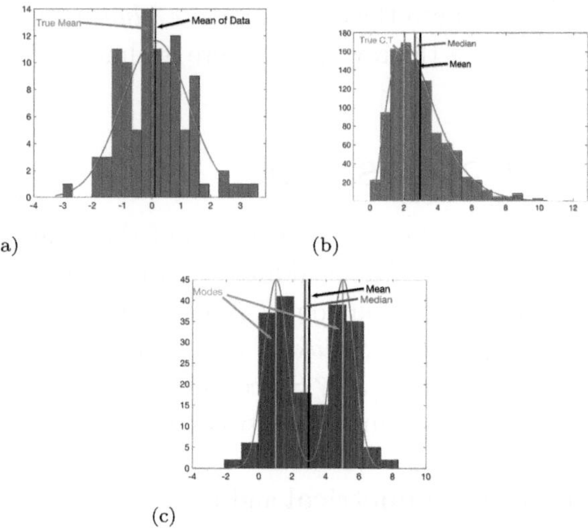

Figure 7.4: The mean of unimodal data, (a), is usually very accurate when capturing the central tendency. The median is a better indicator of central tendency for a skewed data set, (b), than the mean. The mode is best for capturing the central tendencies of something that has more than one peak, (c).

two ways we can go about this: (1) we can ask every single student at Mount Allison University what their GPA is or (2) we can post up at a spot on campus and only ask the students who happen to pass by us. This first would give us **population** data as we have data for *all* Mount Allison University students, whereas the second would give us **sample** data. We don't ask everyone about their GPA, but we ask enough random students that we believe we can *infer* something about the larger population from our sample.

As with just about everything in this course, what is a population and what is a sample entirely depends on your *context*. If we are looking to gain information[34] about current Mount Allison University students, then the current student body is the population. If instead we wished to study Mount Allison University students' *past and present*, then our data set involving only the current students is just a sample[35].

We have to be careful when talking about samples and populations. A sample of a population introduces error, because we are not accounting for all possibilities within the population. We are necessarily leaving

[34] And, potentially, eventual knowledge

[35] And not a very good sample at that as it is heavily biased; we'll talk about this in the next episode of ethical considerations.

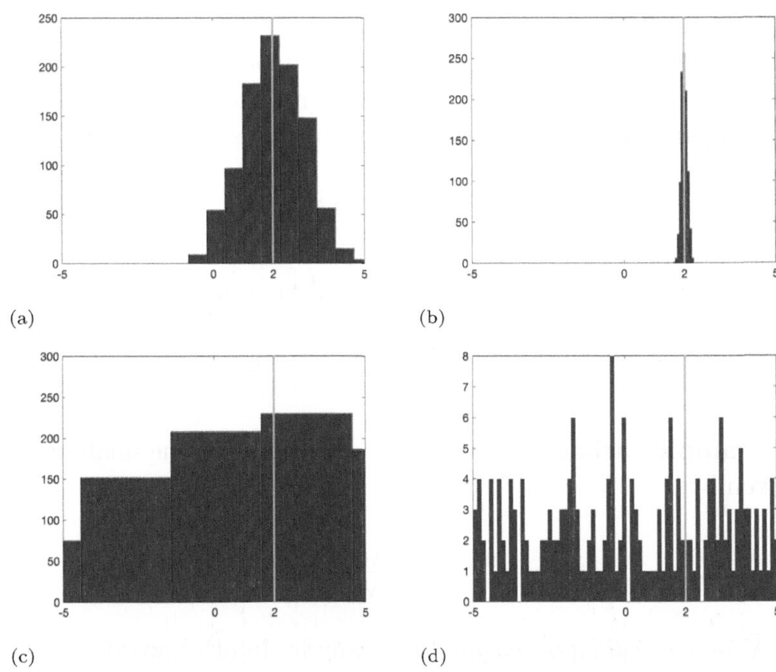

Figure 7.5: All four of these data sets have the same mean and median, yet they are all wildly different. Three of the four even have the same mode! The central tendency alone can't tell us everything about a data set.

information out. That being said, it's often completely unrealistic[36] to poll an entire population. If we are studying, say, sheep instead of people, it would be very difficult and prohibitively expensive[37] to take a measurement on *all* the sheep. Instead, we take a sample. While this introduces what's called a **sampling error**[38] if we ensure our sample is both the representative of the population as a whole and large enough, we can make the sampling error very small.

For a simple calculation like the mean, median, or mode, the formula doesn't change between a sample and a population. Again, whether the mean you are speaking about is a sample mean or a population mean depends on the data itself and the context surrounding it.

[36]Or even straight up impossible

[37]In both a financial and temporal sense

[38]This is the difference between the statistics of your sample and the same statistics on the population.

Back to the variance

The variance on the other hand *does* change depending on whether you are speaking of a sample or a population; but it doesn't change a lot.

The **population variance** of N data points is given by

$$V_p = \sigma^2 = \frac{\sum_{i=1}^{N}(X_i - \mu)^2}{N} \tag{7.2}$$

where μ is the *arithmetic mean* of the population. In other words,

$$\mu = \frac{\sum_{i=1}^{N} X_i}{N}.$$

The **sample variance** is almost the same but with one small tweak[39]. It is given by

$$V_s = s^2 = \frac{\sum_{i=1}^{N}(X_i - \bar{X})^2}{N-1} \tag{7.3}$$

where $\bar{X}$ is the *arithmetic mean* of the sample. In other words,

$$\bar{X} = \frac{\sum_{i=1}^{N} X_i}{N}.$$

Notice, the only practical difference here is the $N-1$ in the denominator instead of N.

If you read it again, you may notice that I slipped in a squared letter in between the variance and the formula for the variance. You may be asking yourself why I did this.

The variance is nice because it is mathematically easy to work with[40], but this comes at the cost of the variance lacking interpretability. We can see that this becomes cumbersome to interpret just by looking at the units of the quantities in our formula. The variables X_i and $\bar{X}$ (or μ) have units of whatever our measurement is. N is just a number. When we add (or subtract) two quantities with the same units, the units do not change[41]. When we divide by a number, we don't change units[42]. When we square a number, we change the units[43]. So our variance does

[39] And we use different symbols to help keep things organized.

[40] In a more math-intensive stats course, you will see why.

[41] 4 apples - 1 apple = 3 apples. Same units.

[42] $4\ apples/2 = 2\ apples$. Same units

[43] $(3 \text{ apples})^2 = 9 \text{ apples}^2$

not have the same units as our mean or our data. This makes it hard to compare the two.

Enter the **standard deviation**: the **square root of the variance**. The standard deviation has the same units as the mean and the data, and so we can easily use it to compare with the data. This is why there are those symbols in the above formulas. The **standard deviation of the population** is denoted as σ, and σ^2 is the population variance. The **standard deviation of a sample** is denoted s, and s^2 is the sample variance.

The standard deviation allows us to set limits around our mean.

7.7 ETHICAL AND MORAL CONSIDERATIONS: EPISODE 2

WHY IS THIS DIFFERENCE BETWEEN A SAMPLE AND A POPULATION SO IMPORTANT? A sample necessarily has holes in the data; it is by definition missing something. Depending on what it is missing, we can manipulate data in such a way that things *sound* true, without actually being true. It is important to be open, clear, and transparent[44] about *how* data was collected and on what groups. This will help us set reasonable bounds on the conclusions that we try to draw from our data.

A common example of biased sampling is through self-reporting. As a fairly on-the-nose example, let's say that we have a new vaccine for a population of cows[45]. Let's also imagine that our hyper-intelligent and communicative cows can choose whether or not they want this new vaccine. If we wish to know roughly how many cows at a farm have received the vaccine, one way to do that is to set up a poll which the cows can voluntarily respond to.

Upon looking at our survey data, we notice that of the 500 responses, 498 cows said they received the vaccine. *Amazing!* We think 99.6% *of cows are vaccinated!* This is when our resident moralist[46] turns to us and says *we can only say that* 99.6% *of cows* **on this farm** *are vaccinated.* Always willing to listen to our resident moralist, we refine our conclusion to reflect our sampling.

We still have a problem though. The data is self-reported. Much like in humans, choosing not to get vaccinated is a fringe, ostracizing opinion in cows. We should ask, *could this affect how many cows decide to*

[44]Notice a theme in this series of ethical and moral considerations?

[45]The root word of vaccine is the Latin *vacca*, which means cow. This is because the first vaccine was derived from cows!

[46]A dog named Brusco, if you're curious what this wild world in my head looks like.

self-report their vaccination status? The answer may not be a resounding *yes*, but at the very least we cannot confidently say *no*. This is enough to make our conclusion misleading. It's possible that many did not answer our survey because they know they would answer in the negative.

Upon further investigating, we see that the farm has a total population of 624 cows. This should further call into question the results of our survey; only $\approx 80\%$ of our cows participated in the survey! Given the implicit bias of self-reported data, we would have to have some extremely strong arguments to infer things about our **population** based on our **sample**. In truth, this data is all but meaningless due to the societal contention when it comes to vaccines. The only honest conclusion we can make from this data is that *at least* 79.8% *of our farm is vaccinated.*

How could we improve our sampling? Generally, random samples are the gold standard. If a sample is truly random, then we can be fairly certain that any variation at an individual level is not affecting the relationship we wish to study. These variations appear in the data, well, *randomly* and not *systematically*.

7.8 AN EXAMPLE

Here is a data set. It is data related to U.S. politicians.

We can import this data set into Python with the following bit of code:

```
poli_data = pd.read_csv('dataset.csv').
```

Now, whenever we would like to refer to the entire data set, we need only reference `poli_data` in Python. If we `print` our data set, we can get some information about it.

A question we might want to ask about this data is, *what is the average age of U.S. politicians?* These ages are independent of each other[47], and these are not rates, so using the *arithmetic mean* is sufficient. Since the arithmetic mean is the most common, when we call the `mean` function from `numpy`, we get back the arithmetic mean.

We can get the age column with the command

```
poli_age_mean = np.mean(poli_data['Age'])
```

Of course, we could also calculate the geometric and harmonic means of this data. The commands for these means are in the `stats` library of the `scipy` package.

[47] One politician's age, arguably, doesn't depend on another's.

US Politicians

Name	Sex	Birthplace	Age	Political_party
Daniel Squadron	male	United States of America	41.0	Democratic Party
Terry Kilgore	male	United States of America	59.0	Republican Party
Anitere Flores	female	United States of America	44.0	Republican Party
Kenneth Mapp	male	United States of America	65.0	independent politician
Xochitl Torres Small	female	United States of America	36.0	Democratic Party
Leonard Lance	male	United States of America	68.0	Republican Party
Tom Rooney	male	United States of America	49.0	Republican Party
Tom Malinowski	male	United States of America	55.0	Democratic Party
Blake Farenthold	male	United States of America	58.0	Republican Party
Scott Tipton	male	United States of America	64.0	Republican Party
Marc Veasey	male	United States of America	49.0	Democratic Party
Yvette Clarke	female	United States of America	55.0	Democratic Party
Bill Keating	male	United States of America	68.0	Democratic Party
Pam Faris	female	United States of America	63.0	Democratic Party
Jim McGovern	male	United States of America	60.0	Democratic Party
Karyn Polito	female	United States of America	54.0	Republican Party
Tammy Baldwin	female	United States of America	58.0	Democratic Party
Ted Budd	male	United States of America	49.0	Republican Party
Laura Bush	female	United States of America	74.0	Republican Party

Figure 7.6: A data set of U.S. politicians and some metrics (age, birthplace, political party, and sex). Why U.S. politicians? Because it was easier to get than Canadian politicians. The **headers** *should be descriptive and tell us exactly what we are measuring. In this data set, each row represents a politician. We have three (birthplace, political party, and sex) categorical, nominal data columns. The age column is numerical discrete data. This data represents a sample because the politicians in the data set are taken from Twitter. Not all politicians are on Twitter, and some may be on Twitter twice!*

```
poli_age_gmean = stats.gmean(poli_data['Age'])#Geometric mean
poli_age_hmean = stats.hmean(polidata['Age']) #Harmonic mean
```

If we try this, we will get some weird results. First thing you might notice if you try to print the means to the screen is that the geometric mean is `NaN`. This stands for *Not A Number*. It is a special character in data analysis that shows up when the computer expects a number and receives something that is, well, not a number. A nan has some special properties in that, if x is a number, then

$$nan + x = nan$$
$$nan - x = nannan \cdot x = nan$$
$$x/nan = nan.$$

It basically turns numbers into `nan`'s. It's kinda like a zombie virus in that if it interacts with a number it infects them with `nan`.

```
                      Name Twitter_username  ... Instagram_username    Political_party
0        A. Donald McEachin      RepMcEachin  ...         repmceachin   Democratic Party
1          Aaron Michlewitz   RepMichlewitz  ...                 NaN   Democratic Party
2              Aaron Peskin     AaronPeskin  ...          apeskin52   Democratic Party
3                Aaron Peña       AaronPena  ...                 NaN   Republican Party
4              Aaron Schock     aaronschock  ...         aaronschock   Republican Party
...                     ...             ...  ...                 ...                ...
2509          Yvette Clarke RepYvetteClarke  ...    repyvettedclarke   Democratic Party
2510         Yvette Herrell yvette4congress  ...                 NaN   Republican Party
2511        Zephyr Teachout  zephyrteachout  ...      zephyrteachout   Democratic Party
2512            Zoe Lofgren    RepZoeLofgren  ...      repzoelofgren   Democratic Party
2513          Zoltan Istvan   zoltan_istvan  ...       zoltan_istvan  Libertarian Party

[2514 rows x 10 columns]
```

Figure 7.7: This is what the data set looks like in Python when we `print` *it to the Shell. We get a sample of the rows and columns, and we also see how many rows our data set has and how many columns it has.*

By trying to compute the geometric mean, we see very clearly that the issue with our data set is that it contains a `nan`. If we instead try to compute the *harmonic* mean, we get an error that seems to imply that we have ages that are either zero, or possibly negative[48].

It's important that the data is **clean** before we work with it. In general, we want to avoid problems like `nan` in a data set. Cleanliness is, like everything else, context sensitive. We can do a blanket removal of *rows* with the command `poli_data.dropna()`. The command `dropna` works *on* the data set, and so it uses this weird .notation, when we use this to create a clean data set with the command `poli_data_too_clean = poli_data.dropna()`. When we print this onto the screen, we get the following:

In the data set, there are *many* `nan`'s and not all of them appear in the Age column. Do we want to disqualify entries for having `nan`'s in columns we don't care about? Probably not. We can modify our `dropna` to only drop `nan` if it exists in a specific column[49].

```
poli_data_clean = poli_data.dropna(subset=['Age'])
```

This allows us to only drop a `nan` if it appears in the Age column. When we do this, we get this data set.

Now, we can calculate all the means! When we do, we get the following results:

[48] But in actual fact, the error is caused by a nan.

[49] Or columns

```
                  Name Twitter_username  ... Instagram_username    Political_party
0     A. Donald McEachin       RepMcEachin  ...         repmceachin  Democratic Party
2            Aaron Peskin       AaronPeskin  ...           apeskin52  Democratic Party
4            Aaron Schock       aaronschock  ...         aaronschock  Republican Party
5         Abby Finkenauer         Abby4Iowa  ...           abby4iowa  Democratic Party
6      Abigail Spanberger    SpanbergerVA07  ...       repspanberger  Democratic Party
...                   ...               ...  ...                 ...               ...
2508         Yvette Clarke      YvetteClarke  ...    repyvettedclarke  Democratic Party
2509         Yvette Clarke   RepYvetteClarke  ...    repyvettedclarke  Democratic Party
2511       Zephyr Teachout    zephyrteachout  ...      zephyrteachout  Democratic Party
2512            Zoe Lofgren     RepZoeLofgren  ...       repzoelofgren  Democratic Party
2513         Zoltan Istvan     zoltan_istvan  ...       zoltan_istvan Libertarian Party

[1007 rows x 10 columns]
```

Figure 7.8: This data set doesn't have any `nan`*'s in it, but there's a reason we called it too clean. If you notice we only have 1007 rows! We've lost more than HALF of our data! As a general rule, if the data exists, we should use it in some way.*

```
                  Name Twitter_username  ... Instagram_username    Political_party
0     A. Donald McEachin       RepMcEachin  ...         repmceachin  Democratic Party
1       Aaron Michlewitz     RepMichlewitz  ...                 NaN  Democratic Party
2            Aaron Peskin       AaronPeskin  ...           apeskin52  Democratic Party
3              Aaron Peña         AaronPena  ...                 NaN  Republican Party
4            Aaron Schock       aaronschock  ...         aaronschock  Republican Party
...                   ...               ...  ...                 ...               ...
2509         Yvette Clarke   RepYvetteClarke  ...    repyvettedclarke  Democratic Party
2510        Yvette Herrell  yvette4congress  ...                 NaN  Republican Party
2511       Zephyr Teachout    zephyrteachout  ...      zephyrteachout  Democratic Party
2512            Zoe Lofgren     RepZoeLofgren  ...       repzoelofgren  Democratic Party
2513         Zoltan Istvan     zoltan_istvan  ...       zoltan_istvan Libertarian Party

[2491 rows x 10 columns]
```

Figure 7.9: For our question, we only care about age data. So, we only remove a `nan` *if it appears in this column. When we do this, we see that we only lose 13 entries. This is way better than losing over half!*

We can even calculate the standard deviation using Python! In Python, using `numpy`, the default standard deviation is the *population standard deviation.* Moreover, it is always the arithmetic standard deviation. It's important to remember these facts when we are using the command.

In Python, we can calculate the population standard deviation with

```
sigma = np.std(our_data)
```

where we obviously replace `our_data` with the data on which we are calculating the standard deviation.

If we would like to instead calculate the *sample standard deviation,* we have to add a modifier parameter to our command:

```
s = np.std(our_data, ddof=1).
```

Table 7.6: The three different means for our sample data set of U.S. politicians. Notice that the arithmetic mean is the biggest, followed by the geometric mean, followed by the harmonic mean. This isn't a coincidence; this will always be the case (unless they're all equal).

	Arithmetic Mean	Geometric Mean	Harmonic Mean
U.S. Politician Age	58.74	57.32	55.81

Table 7.7: The arithmetic means and standard deviation for our sample data set of U.S. Politicians

	Arithmetic Mean ± Standard Deviation
U.S. Politician Age	58.74 ± 12.59

For our data, since we have used `dropna()`, we have lost rows. So even if we did start out with a population's worth of data, we now have a sample of our data. When we calculate the standard deviation, we get the following:

The standard deviation tells us **the limits between which most of the data falls**. You can see in figure 7.10 all the data points (that are not `NaN`), the mean, and the area between $\bar{x} - s$ and $\bar{x} + s$.

How much of the data falls within this area? Using the commands

```
poli_age = poli_data_clean['Age']
num_of_pts_in_area = len(poli_age.loc[(poli_age>m-s)&
      (poli_age<m+s)])
total_points = len((poli_data_clean['Age']))
fraction_of_points_in_area = num_of_pts_in_area/total_points
```

we can see that about 67% of the points fall within the area. This number is not arbitrary.

7.9 THE EMPIRICAL RULE

Like the last sentence in the last section said, the fact that 67% of the data falls within the limits of one standard deviation is not random. This is called **the empirical rule**. It states the following:

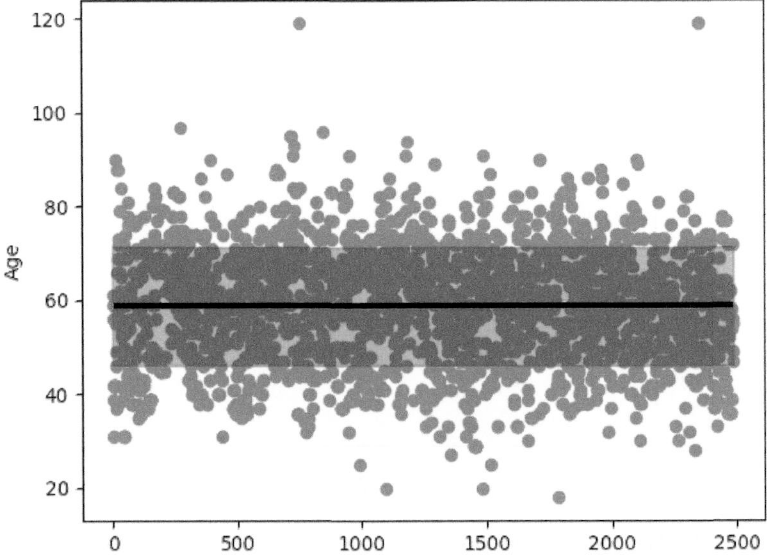

Figure 7.10: The points are all our data, the black line is the arithmetic mean and the shaded area is the range set by the standard deviation.

Definition 7.9.1. If data is normally distributed (i.e. is bell shaped) and $\bar{x}$ is the mean of the data and σ is the standard deviation, then approximately 68% of the data will fall within $\bar{x} \pm \sigma$, 95% of the data will fall within $\bar{x} \pm 2\sigma$ and 99.7% of the data will fall within $\bar{x} \pm 3\sigma$.

Figure 7.11 shows how the politician age data is approximately "bell shaped," and figure 7.12 shows overlays of areas of $\pm\sigma$, $\pm 2\sigma$ and $\pm 3\sigma$. You can see that just about all the data is covered.

But wait! We only have 67% of our data within $\bar{x} \pm \sigma$, not 68%. What gives? Two things:

1. We have a sample, not a population. There is sampling error that may change the data slightly.

2. The 68% is for a *perfect* bell shape. Our data is mostly bell shaped, but not a perfect bell. It looks pretty spot on near the edges, but things get wonky in the middle.

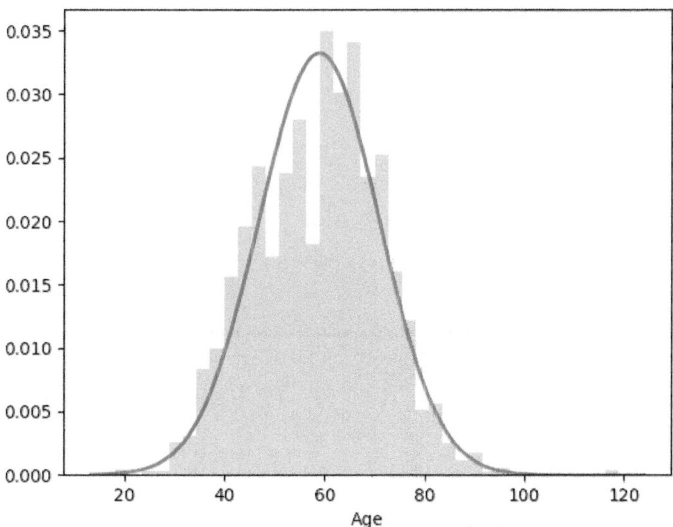

Figure 7.11: Plot showing the fraction of politicians in our data set that fall into each age category, overlaid with a bell curve. Since this data is approximately bell shaped, the empirical rule approximately holds.

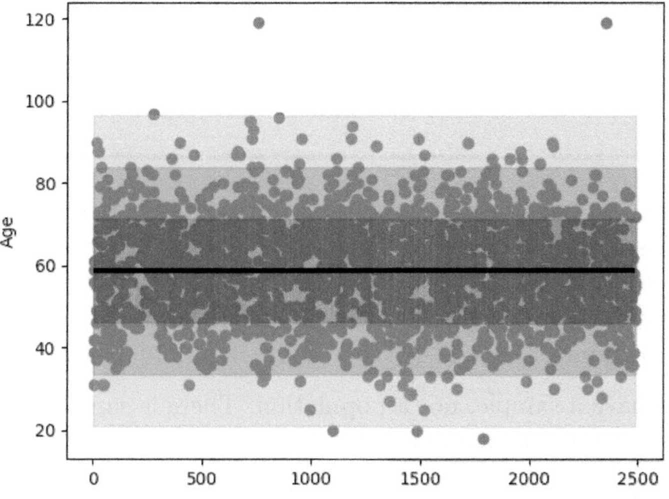

Figure 7.12: Red area is $\pm\sigma$, green area is $\pm 2\sigma$, and the cyan area is $\pm 3\sigma$. We can see that all but six of our data points are captured in these ranges.

We can use the empirical rule to test how bell shaped a data set is. If it roughly matches up for all three categories, it is very bell shaped. You can test this yourself, but our politician age data follows the empirical rule almost perfectly.

YOU MAY HAVE NOTICED SOME WEIRD THINGS IN THOSE GRAPHS. Are there really three U.S. politicians who are 119 years old? Is there really one who is 18? Chances are, without graphing the data, we may have missed these weird anomalies. Before we continue to talk about data analysis, let's spend some time talking about the ways in which we can visualize data.

CHAPTER 8

Visualizing Data

One of the best ways we can summarize data is through pictures. With the right picture, we can see trends, outliers, patterns, and anomalies in data. Data as pictures works because humans are drawn to patterns and colours. Our brains are so good at pattern recognition that we even see them when none should exist[1]. Many conspiracy theories are the result of the need to see patterns where none exist, making connections that aren't there. A more concrete example is the need to make sense of cars by seeing faces (Figure 8.1).

Figure 8.1: The Mazda MX-5 is a very happy car, right?

THERE IS NO ONE RIGHT WAY TO CREATE VISUALIZATIONS OF DATA, but there are most certainly wrong ways. The easiest way to summarize *wrong* is *dishonest*. Any visualization that hides important information or takes advantage of our need to see patterns to imply something that isn't true is wrong. Good data visualization highlights the important parts of a data set: it tells a story in a picture and is honest about its

[1] We do it so much; we have two names for it: pareidolia (for visual patterns) and apophenia, more generally.

DOI: 10.1201/9781003265405-8

context, shortcomings, and strengths. The best data visualization needs very little explanation to tell its story.

We could realistically spend the rest of this book talking about data visualization, but we have so much else to do. We'll look at some simple types of visualization and when/where they are most useful and how we can implement them in Python. That's not to say this is the final word on visualization, or that we always need to represent information in familiar ways.

THERE ARE SOME QUESTIONS WE SHOULD THINK ABOUT DECIDING HOW WE VISUALIZE OUR DATA. They are the following:

- **Is there a central theme to this graphic?** A graphic should tell a story with one central plot. Everything in the graphic should relate and reinforce your point.
- **Is everything in my graphic relevant to my point?** An observer should be able to point to any piece of your graphic and ask, *why is this included?* and you should have an answer beyond *it looks pretty.*
- **Is the graphic misrepresenting the data?** This may be tempting sometimes. Sometimes a half-truth through omission or lining things up a certain way to take advantage of apophenia will make you stand out; it may make your work seem more interesting than it is. Don't do it. Keep it honest; play the slow game.
- **Is this as simple as I can make it without changing the meaning?** No? Simplify.

Likewise, we should also keep in mind a few rules about visualization that, I think, transcend the type of visualization.

- Make thick lines.
- Your fonts should be large, but not overbearing. Generally, I'd aim for two points larger than the surrounding text for any text on a graph or figure.
- Use complementary colours, preferably under a colour palette that is colour blind friendly.

- Related, use colour effectively. Make sure it *means* something.
- Labels. Labels. Labels. Label every axis, every different colour with what it is representing.
- While $3D$ graphs have more data and information in them, they tend to be harder to read and thus effectively convey less information. Often, two side-by-side $2D$ figures are better than one $3D$ plot.

8.1 PLOTTING IN PYTHON

Plotting in Python requires a new package. We already have `pandas` for manipulating data tables, and we have `scipy` and `numpy` for doing math (and stats). We need to install `matplotlib` and `seaborn` in order to visualize data. The first package, `matplotlib`, is the base package for visualization and can do a lot of things natively. You have a lot of control over how things look. The package `seaborn` allows for more advanced types of plots and adds certain quality-of-life elements that were missing from `matplotlib`.

Typically, when we import these packages, we import them as

```
import matplotlib.pyplot as plt
import seaborn as sns
```

8.2 SCATTER PLOTS

Scatterplot graphs are the simplest way to compare two separate columns of data and look for a relationship between the two. For each *row* of data in our data set, we choose data from two columns; let's call these individual datum X_i and Y_i. In two dimensions, these give us a point in planar space[2]: (X_i, Y_i). When we plot these points, we can see the relationship between columns X and Y.

Here is a scatter plot of two columns with hidden names, $mystery_x$ and $mystery_y$. These plots show how we can see trends in data much better than in a table. In figure 8.2(a), we see that the two variables have no relationship; there is no pattern in that data at all. In figure 8.2(b), we see there is a very clear relationship. When we increase $mystery_x$, we (generally) see $mystery_x$ increase as well. In figure 8.2(c), we see that the opposite is true, if $mystery_x$ is small, which generally means $mystery_y$

[2] i.e. on a plane, or a standard $2D$ set of axes

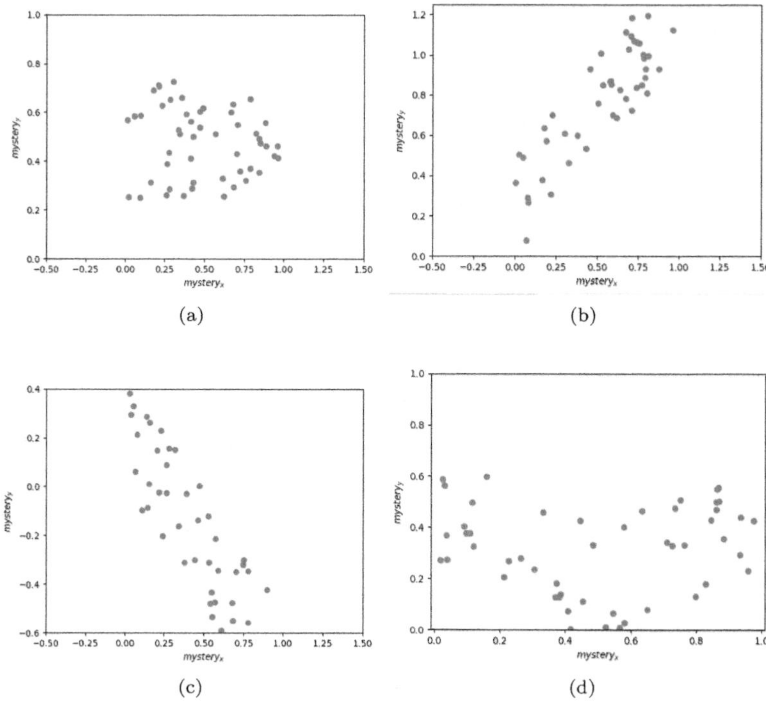

Figure 8.2: Four different qualitative relationships we might see in data. Panel (a) shows two variables X and Y that are not correlated. Panel (b) shows a positive relationship between X and Y; as X increases, Y also increases. Panel (c) shows a negative relationship between X and Y; as X increases, Y decreases. Panel (d) looks much like panel (a), but this is an issue of not having enough data to sufficiently see the relationship between X and Y.

is big and vice versa. What about figure 8.2(d)? This data looks random, much like the first plot, but it's not.

A quick note on sample size

A lot of times in other fields, you will see linear relationships. This isn't necessarily because everything in the world is linear, but because we don't[3] get enough samples to be sure of anything beyond linear relationships. The amount of data we get is almost directly proportional to what we can say about relationships from a strictly statistical point

[3]Or can't

of view. For example, figure 8.2(d) is not as random as it seems. The points that are in fact quadratically related there just isn't enough of them to see this. If instead we plot 5000 points instead of 50 points, the relationship becomes much more clear, like in figure 8.3.

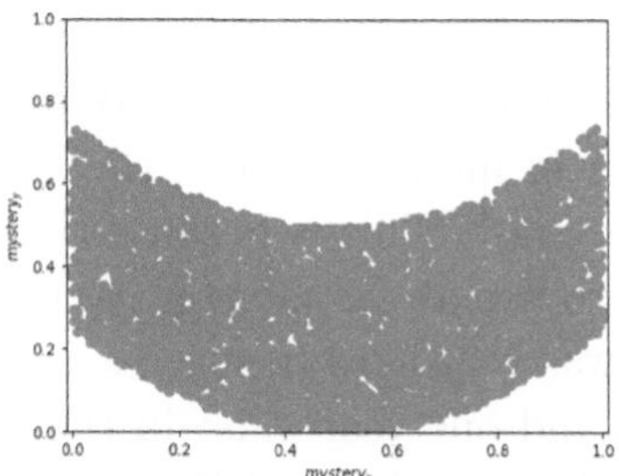

Figure 8.3: When we have a lot of points, it's much easier to see the quadratic relationship between $mystery_x$ and $mystery_y$.

In many of the natural sciences, we can often only experimentally determine linear relationships between two variables because of the complex relationships that exist. To do better requires an understanding of the underlying mechanisms of a system and careful construction of mathematical models.

SCATTER PLOTS CAN ALSO HELP US DETERMINE IF THERE ARE OUTLIERS IN THE DATA AND WHERE THEY ARE. An **outlier** is a point that is very far from the rest of the data, or to put it another way, a point that doesn't follow the general trend of the data.

8.3 OUTLIERS

Outliers occur for one of two reasons:

1. They represent a rare occurrence or event that happened to coincide with when we took a measurement.

2. They may represent an error in data.

Outliers can cause real problems with statistical analysis. We briefly touched on this earlier when looking at the differences between arithmetic and geometric means. The geometric mean is less sensitive to outliers than the arithmetic mean. In a data set with many extreme outliers, the geometric mean may be preferable.

In general, we have to be very careful about removing outliers as any data removed *changes the context surrounding the data.* For instance, if we manage to miraculously collect data on a whole population for a particular measurement and then remove any outliers[4], we no longer have a population.

As an example, we can think of life expectancy. The statistic most commonly used is *life expectancy at birth.* While it is true that people have been living longer in modern times[5], the life expectancy is skewed to the low end in other eras[6]. This is because the statistic measures how long you are expected to live at the moment you are born. Unfortunately, for much of history, it was very hard to survive as a child. Therefore, there are lots of 0 year olds and one year olds[7] in the death tables that bring down the average. In fact, in classical Rome[8], the life expectancy was 20–30 years old. However, if we *change the context of the data* and only count those individuals who survived up to 20 years, then life expectancy rises to $\approx$ 50 years old. This shows you how much childhood mortality can bring down the life expectancy statistic, and how *context is everything.* When we remove childhood mortality, we are removing points from our data set, and it completely changes the context. We are now calculating life expectancy of individuals who survive to adulthood.

Often, unless the outliers are considered *bad* due to measurement error[9] of some kind, **it is poor practice to remove outliers**. Like all data points, outliers give us information about the system we are studying. We should take care to try to understand why the outlier exists, where it came from, and whether it will happen again. Sometimes outlier data can point us to relationships we were not considering. Sometimes, they point to a common behaviour that *looks* rare because of sample size issues. Sometimes, outliers are just rare phenomena we happened to capture. While much of statistics is sensitive to outliers, it is important to try and retain as many as possible and use methods that are robust against outliers. If points are removed for *any* reason[10], you should be clear, honest, and transparent about it. Remember, and this can't be stated enough, removing data points changes the context of your data. **You should also always maintain an original, unmodified copy**

[4]For whatever reason

[5]With the advent of modern medicine

[6]Ancient, classical, medieval, etc.

[7]And other child ages

[8]800BCE–400CE

[9]Often caused by failure of our measuring machines

[10]Including if that reason is just "it's a bad point."

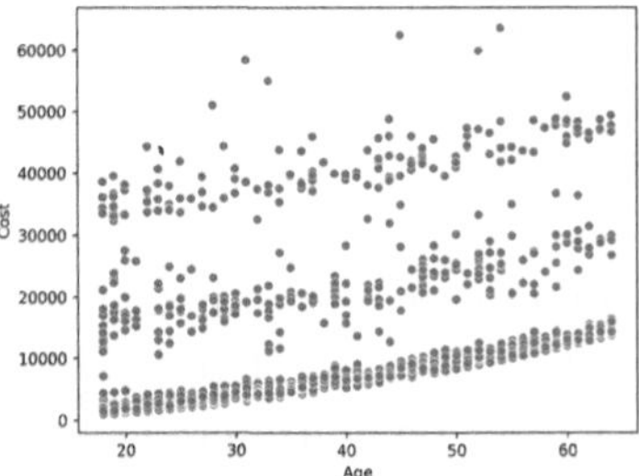

Figure 8.4: Age vs. Cost of Healthcare, notice the data naturally separates into three distinct bands. There is something else going on with this data set!

of the data so that the outliers can be reintroduced and the data can be reanalysed from scratch.

As an example, take a look at this scatterplot of age vs. healthcare costs in figure 8.4. This data set is particularly interesting. By plotting we can see three specific bands that we may not have noticed by looking at the data in tabular form; to prove this, here are the first twenty rows of the data as a table:

Age	Cost
19	16884.92
18	1725.55
28	4449.46
33	21984.47
32	3866.85
31	3756.62
46	8240.58
37	7281.50
37	6406.41
60	28923.13
25	2721.32
62	27808.72
23	1826.84
56	11090.71
27	39611.75
19	1837.23
52	10797.33
23	2395.17
56	10602.38
30	36837.46
60	13228.84

We *could* analyse the data all at once, but it's very clear something else is happening here. Perhaps it is age *and* some other measurement that are actually contributing to healthcare costs. We don't have enough information[11] to determine this, but that isn't the goal of this particular thought experiment.

Again, the goal is to hammer home that what is or isn't considered an outlier is *context sensitive*. If we want to analyse the lowest band, the top most points would surely be considered outliers; more than likely we would filter out the middle band as well. Again, we currently don't have the information to be able to do this automatically, but we can imagine filtering the data points via some criteria if we knew the criteria. Keep in mind that filtering data changes what we are analysing.

No matter which band we are analysing, or even if we are analysing the whole data set, we may consider the three points highest on the y-axis to be outliers as they don't seem to be following any trend we can see in the data. We may want to explore why this is, or adjust our techniques to account for the outliers.

8.4 HISTOGRAMS

Sometimes, our question is not *what is the relationship between two things*, but rather, *what is the probability of a specific outcome?* Or, equivalently, *how often does a specific outcome occur?* In such a case, a **histogram** can be used to display the answers to these questions.

A histogram is visually similar to a bar graph, which is probably the first type of graph many people learn about. While they look similar, they are used for very different purposes. As mentioned above, a histogram is used to show the frequency of specific values in column of data (for continuous variables, we bin the data much like we do when trying to calculate the mode). A bar graph, on the other hand, is used to plot a categorical or discrete variable against a summary statistics[12]. Figure 8.5 shows the difference.

A histogram shows us, roughly, how a variable is distributed among a population. Another way to think of this is while a scatter plot shows us the relationship between two columns of data, a histogram will show us the relationship between the measurements in one column of data. For example, the histogram in figure 8.5 shows us that the library on Netflix is heavily skewed towards recent releases. We can ask ourselves

[11] Right now

[12] Usually

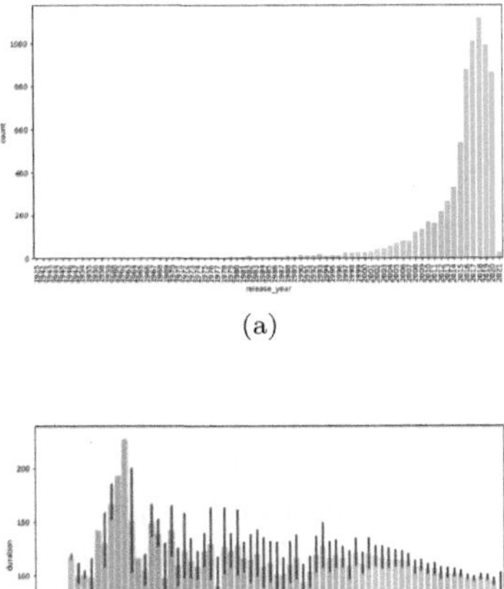

(a)

(b)

Figure 8.5: While they look similar, histograms and bar graphs have extremely different functions. The histogram here shows how many movies/TV shows in the Netflix library were released in a given year. The bar graphs show the mean runtime of movies in the Netflix library by year.

why this might be, and we probably need a lot more data and *context* to answer this. One possibility is that as streaming services became more competitive, Netflix has had to rely more and more on original content. Perhaps this histogram is taking into account the uptick in Netflix original programming in recent years; perhaps it has to do with the licensing agreements for certain things. Netflix usually gets releases that are a few years old, and the streaming agreement may lapse after 5 to 10 years. This could also explain the distribution. It would be interesting to look back in 5 years and see if the distribution has shifted.

Another question we could ask[13] is, *how is the world represented in the Netflix library?* By building a histogram, we get to see how many of movies/TV shows are from a certain country. This is shown in figure 8.6. We can see that the U.S. and India dominate the Netflix library. This shouldn't be surprising India has the largest film industry in the world by number of movies produced per year, and the U.S. has the most popular movies globally.

[13]And then answer, with a histogram

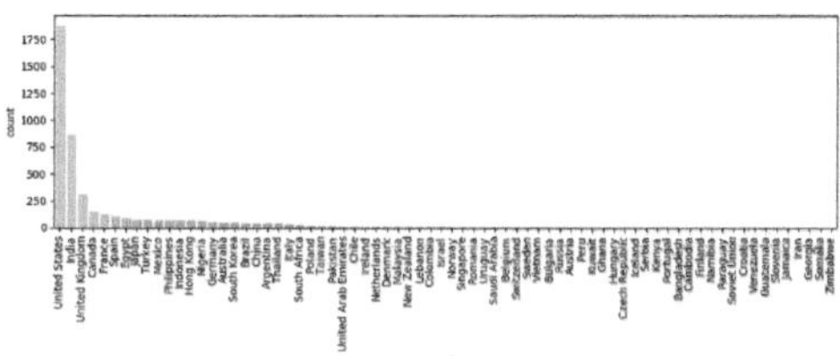

Figure 8.6: While India produces the most movies per year, the USA produces the most TV shows per year. So it is not surprising that these two countries are at the top of the list. Why does the USA seem to dominate the Netflix library if India produces about 3× as many movies per year? There could be many reasons for this, and some of them aren't so pretty: maybe Netflix is pushing away from movies and more towards TV shows, or this could be due to English becoming the lingua franca, or it could be due to xenophobia and subtle racism that leads to other industries around the world getting less exposure.

We can also create histograms for numerical data[14] by binning the data. Figure 8.7 shows the number of movies (we exclude TV shows as they're not measured in minutes) that are of a certain length.

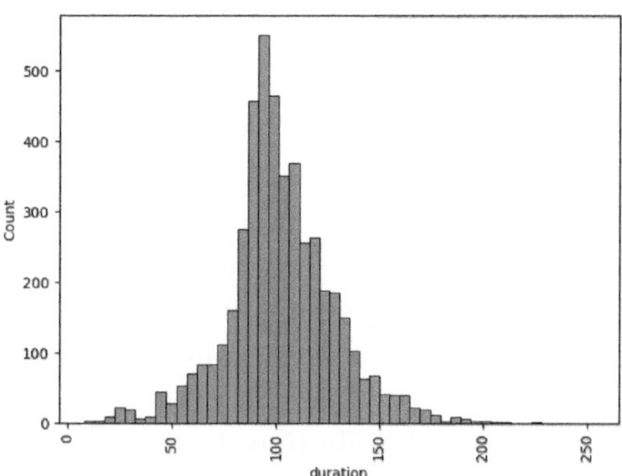

Figure 8.7: This histogram shows us that most movies in the Netflix library are about 90 minutes long. This isn't unique to Netflix; this is the modern "ideal" length of a movie. What is that little bump way down around the 25-minute mark? Short films.

While there are histogram capabilities in `matplotlib.pylot`, the functions in `seaborn` do more and make nicer graphs. We can create a histogram with the following command:

```
sns.histplot(x = x_data)
```

[14]Discrete or continuous

This will plot a histogram with the data; we can use the keyword `bins` to change the number of bins from the default[15].

```
sns.histplot( x = x_data, bins = num_bins )
```

Usually, when dealing with categorical data, `histplot` makes some funny choices that may leave us surprised and confused when looking at the final product. When dealing with categorical data and histograms, it's much nicer to use

```
sns.countplot( x = x_data )
```

This function can be thought of as equivalent to histplot for categorical data. It just makes some alignment issues nicer. I encourage you to play with both and see the difference. I'm confident you'll come to the same conclusion.

Bar Plots

OK, SO THEN WHAT GOOD ARE BAR PLOTS? Bar plots are good **summary plots**. It can draw attention to one aspect of categorical variables that you would like to compare. When we look at a bar plot, we can see that there is *one* value on the y-axis for each categorical variable on the x-axis. In a simple bar plot, we have to compare exactly *one* attribute from each category.

Figure 8.8 shows how a bar plot shows the summary information compared to the raw information of a scatter plot. We can see that the bar plot makes it really easy to compare our favourite summary statistics for duration, filtered by year, graphically. The best part is that *we don't need to compute the means or standard deviations or do any of the filtering manually.* The `seaborn` package understands what the purpose of a bar plot is and does this automatically! The command to generate the above plot is extraordinarily simple:

```
sns.barplot(x=netflix_data['release_year'],
                y = netflix_data['duration']
                )
```

The default is to use the arithmetic mean and to use two standard deviations[16].

[15]The default isn't fixed; Python and seaborn are able to decide what it thinks is best. It's not always correct.

[16]The 95% part of the empirical rule

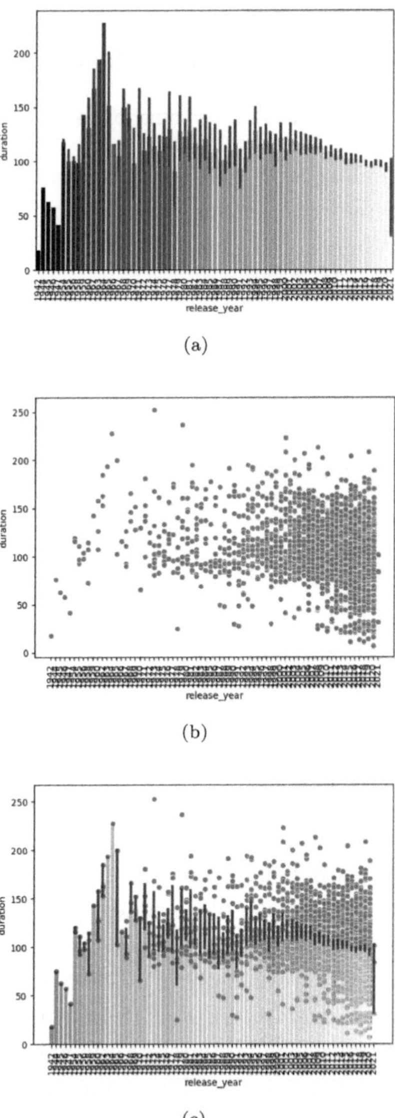

Figure 8.8: This bar plot, (a), shows the release year of a movie on the x-axis and the mean duration of films released in that year. The grey line on the bar represents two standard deviations from the mean. You can think of a bar plot as a summary plot of an equivalent scatter plot (b). This scatterplot above shows all the movies in the (filtered) data set, and plot (c) shows the bar plot (showing the mean and two standard deviations) on top of the scatter plot.

IF YOU DO TRY AND REPRODUCE MY PLOT, YOU MAY NOTICE ONE GLARING DIFFERENCE: THE COLOUR. How did I change the colour of my bars? Moreover, if you plot the bar plot and scatter plot together, yours won't be translucent. How did I do these things? How can you do these things as well[17]?

There are so many ways to modify a plot in Python[18] that it's just not feasible to go through and show you every single thing you can do. What I can do is show you how to read technical documentation that way you have the tools to learn about the different ways to manipulate things yourself.

8.5 THE ANATOMY OF A TECHNICAL DOCUMENT

HOW CAN WE LEARN ABOUT THE DIFFERENT THINGS PYTHON CAN DO[19]? We can read the documentation for the different commands[20] in Python. When you open a help file, it can look quite intimidating, but they're generally pretty easy to follow. Let's take a look at the page below, and we will go through each lettered section individually.

1. This is the command you're looking at.

2. This shows you how to call the command, along with all the possible options that you can put between the brackets. Anything of the form `option_name=` is an *optional argument*. This means that if you do not provide a value, Python will provide a default value automatically. Any arguments within the brackets that do not have this are *required*. If you look at the list of arguments for `seaborn`'s `scatterplot`, you will see that we can run it with a set of empty brackets! There are no required fields[21]!

3. This is a description of what the command is intended to do. Your mileage may vary depending on the package and the command when it comes to how much detail you get here. With `seaborn`, you get a lot of very clear information about how the command `scatterplot` works and also when/why scatter plots are used.

[17] Spoiler alert: yes, you can.
[18] Or any language really
[19] If you answered, *Matt can teach us!* you are sorely mistaken.
[20] Known as functions and/or methods
[21] Give it a try; see what happens. It makes sense.

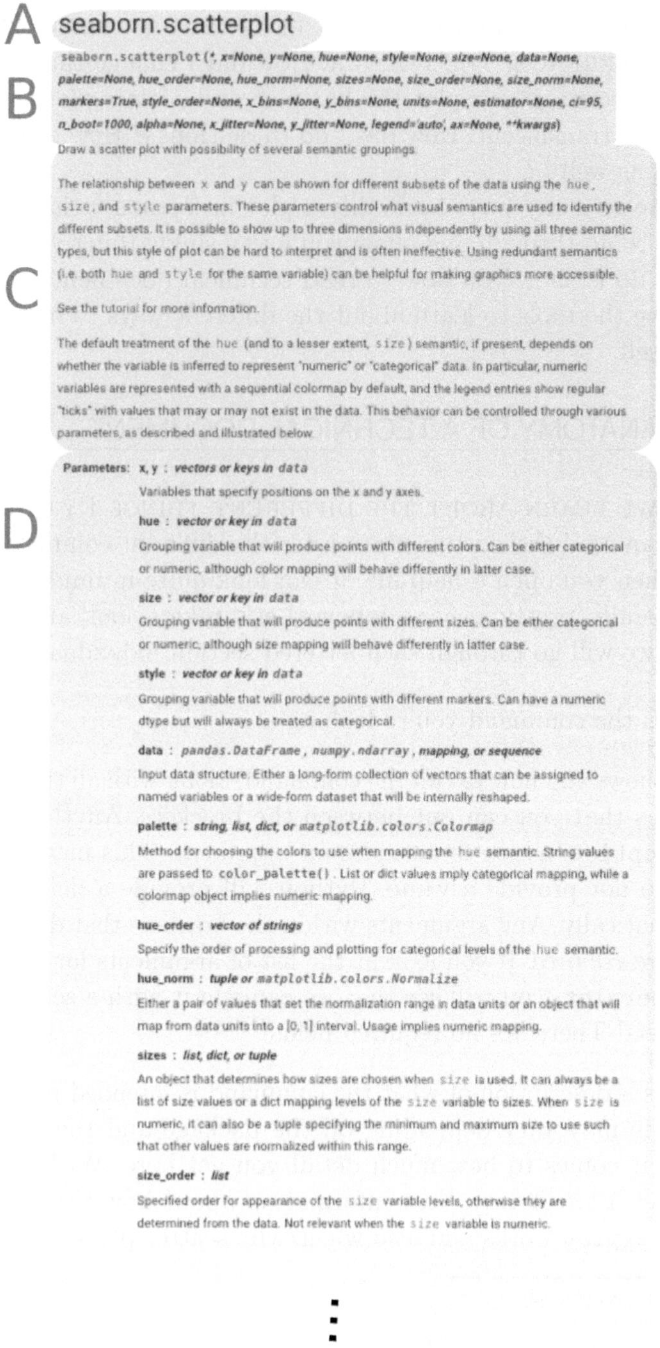

A seaborn.scatterplot

B `seaborn.scatterplot(*, x=None, y=None, hue=None, style=None, size=None, data=None, palette=None, hue_order=None, hue_norm=None, sizes=None, size_order=None, size_norm=None, markers=True, style_order=None, x_bins=None, y_bins=None, units=None, estimator=None, ci=95, n_boot=1000, alpha=None, x_jitter=None, y_jitter=None, legend='auto', ax=None, **kwargs)`

Draw a scatter plot with possibility of several semantic groupings.

C The relationship between `x` and `y` can be shown for different subsets of the data using the `hue`, `size`, and `style` parameters. These parameters control what visual semantics are used to identify the different subsets. It is possible to show up to three dimensions independently by using all three semantic types, but this style of plot can be hard to interpret and is often ineffective. Using redundant semantics (i.e. both `hue` and `style` for the same variable) can be helpful for making graphics more accessible.

See the tutorial for more information.

The default treatment of the `hue` (and to a lesser extent, `size`) semantic, if present, depends on whether the variable is inferred to represent "numeric" or "categorical" data. In particular, numeric variables are represented with a sequential colormap by default, and the legend entries show regular "ticks" with values that may or may not exist in the data. This behavior can be controlled through various parameters, as described and illustrated below.

D **Parameters:** **x, y** : ***vectors or keys in*** `data`

Variables that specify positions on the x and y axes.

hue : ***vector or key in*** `data`

Grouping variable that will produce points with different colors. Can be either categorical or numeric, although color mapping will behave differently in latter case.

size : ***vector or key in*** `data`

Grouping variable that will produce points with different sizes. Can be either categorical or numeric, although size mapping will behave differently in latter case.

style : ***vector or key in*** `data`

Grouping variable that will produce points with different markers. Can have a numeric dtype but will always be treated as categorical.

data : `pandas.DataFrame`, `numpy.ndarray`, ***mapping, or sequence***

Input data structure. Either a long-form collection of vectors that can be assigned to named variables or a wide-form dataset that will be internally reshaped.

palette : ***string, list, dict, or*** `matplotlib.colors.Colormap`

Method for choosing the colors to use when mapping the `hue` semantic. String values are passed to `color_palette()`. List or dict values imply categorical mapping, while a colormap object implies numeric mapping.

hue_order : ***vector of strings***

Specify the order of processing and plotting for categorical levels of the `hue` semantic.

hue_norm : ***tuple or*** `matplotlib.colors.Normalize`

Either a pair of values that set the normalization range in data units or an object that will map from data units into a [0, 1] interval. Usage implies numeric mapping.

sizes : ***list, dict, or tuple***

An object that determines how sizes are chosen when `size` is used. It can always be a list of size values or a dict mapping levels of the `size` variable to sizes. When `size` is numeric, it can also be a tuple specifying the minimum and maximum size to use such that other values are normalized within this range.

size_order : ***list***

Specified order for appearance of the `size` variable levels, otherwise they are determined from the data. Not relevant when the `size` variable is numeric.

⋮

⋮

n_boot : *int*

Number of bootstraps to use for computing the confidence interval. *Currently non-functional.*

alpha : *float*

Proportional opacity of the points.

(x,y)_jitter : *booleans or floats*

Currently non-functional.

legend : *"auto", "brief", "full", or False*

How to draw the legend. If "brief", numeric `hue` and `size` variables will be represented with a sample of evenly spaced values. If "full", every group will get an entry in the legend. If "auto", choose between brief or full representation based on number of levels. If `False`, no legend data is added and no legend is drawn.

ax : *`matplotlib.axes.Axes`*

Pre-existing axes for the plot. Otherwise, call `matplotlib.pyplot.gca()` internally.

kwargs : *key, value mappings*

Other keyword arguments are passed down to `matplotlib.axes.Axes.scatter()`.

E

Returns: `matplotlib.axes.Axes`

The matplotlib axes containing the plot.

F

See also

`lineplot`

Plot data using lines.

`stripplot`

Plot a categorical scatter with jitter.

`swarmplot`

Plot a categorical scatter with non-overlapping points.

G

Examples

These examples will use the "tips" dataset, which has a mixture of numeric and categorical variables:

```
tips = sns.load_dataset("tips")
tips.head()
```

	total_bill	tip	sex	smoker	day	time	size
0	16.99	1.01	Female	No	Sun	Dinner	2
1	10.34	1.66	Male	No	Sun	Dinner	3
2	21.01	3.50	Male	No	Sun	Dinner	3
3	23.68	3.31	Male	No	Sun	Dinner	2
4	24.59	3.61	Female	No	Sun	Dinner	4

Passing long-form data and assigning `x` and `y` will draw a scatter plot between two variables:

```
sns.scatterplot(data=tips, x="total_bill", y="tip")
```

4. This is the parameter list. You are not expected to divide what each of the arguments in B is referring to. This section of a help file gives you a description of each one. It also tells you what it will accept as a value. For example, `x,y` accepts a vector (a column of data) or a key in `data`. This means if you can either give it a column directly, or if you've specified a data set in the `data` argument, you can just provide a column name. The parameter `alpha` accepts a `float`. A float is a decimal number.

5. This tells you what the command gives you. In this case, the command returns a set of axes with the plot on them. We can see the axes with `plt.show()`.

6. This section is fairly unique to the well-documented packages. Maybe the command you're looking at isn't quite what you need. Here, you will see similar but functionally different commands that may better suit your needs.

7. Finally examples. It is always nice to see how things work in practice. Again, some documentation may be light on useful examples or missing this section entirely.

Most technical documents will follow a similar structure to this[22]. If we understand how to read these documents and effectively parse the information in them, then we need not be taught all the ins and outs of Python; we can expand our knowledge ourselves on our own time. The commands and packages we use most often will find their way into our memory so that we don't need to look them up, but there will always be *some* gaps in our memory or knowledge. In my experience, early students of programming seem to think looking things up in help files is a sign of weakness or somehow makes you lesser. I can assure you it does not and, much like pooping, everyone does it. The important part is doing it right[23], help files and technical documents will simultaneously give you the answers you're looking for and will help you become more knowledgeable in general. There is a habit among beginning programmers to go to message boards, forums or friends and just copy and paste pieces of code other people have written. This *may* give you the answer you're looking for[24], but you won't learn, and if you don't learn, you can never do more than just copy and paste. It's a sad hole to fall down.

---

[22]Sometimes you might have to squint to see it.

[23]The simile starts to break down here.

[24]It may also make things worse.

8.6 BAD PLOTS AND WHY THEY'RE BAD

In figure 8.9 is a plot of hospitalizations due to COVID-19 for January 2022 from the Ontario government[22]. You are seeing the whole section under that heading. There is no caption, no explanation that I have intentionally left out[25]. This plot is definitely telling a story, but is it telling an *honest* or even a *complete* story?

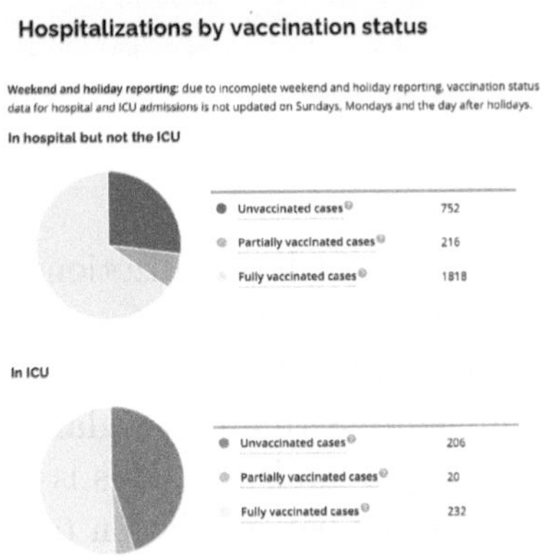

Figure 8.9: These pie graphs of COVID-19-related hospitalizations make vaccination look ineffective, or worse, risky. This is because the pie graph hides the populations from which these numbers are being drawn. The vaccinated population is roughly five times larger than the unvaccinated population!

From looking at this plot, we might come to the conclusion that a COVID-19 vaccine makes you *more* likely to end up in hospital and leaves you just as likely as not to end up in an ICU. We might come to this conclusion because that's what the graphic is telling us. The graph has neglected to give you information about the total vaccinated population vs. the total unvaccinated population. The vaccinated population is far, far greater than the unvaccinated population, and this changes our context. The pie graph format acts as an equalizer between the categories *vaccinated* and *unvaccinated*. We may be able to reason that they are not equal, but the graph does not give us this information even though it's incredibly important.

Here is a figure (figure 8.11 that incorporates the percentage of the population that is vaccinated and the percentage of the population that is

[25]That would be fairly hypocritical of me; then again, in the words of Ad-Rock, "I'd rather be a hypocrite than be the same person forever."

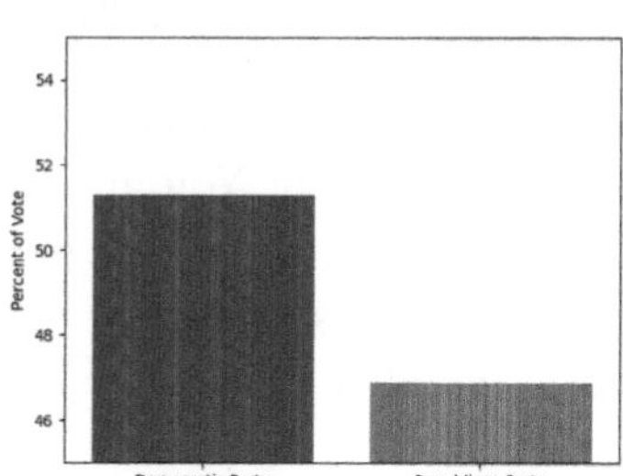

Figure 8.10: A very bar graph showing an exaggerated difference between votes in the 2020 US election because the y-axis doesn't start at 0.

not vaccinated, along with current hospitalizations and ICU occupancy.) This picture tells a very different story.

There are other, more obvious ways that data is sometimes manipulated. Looking at figure 8.10 a quick glance would suggest that the Democratic Party won by a landslide. This is because we are a comparative species. We look at one bar in relation to the other, and it may take a moment to notice that the axis isn't starting at 0 as we expect. If we redraw the figure with proper axes, we see that the U.S. election was far closer than let on by the first graph.

In figure 8.11, we see another pie graph that completely misrepresents the data within. First, the two numbers do not make up 100%, and while 46% is indeed bigger than 41%, it is *not* the majority. We haven't even

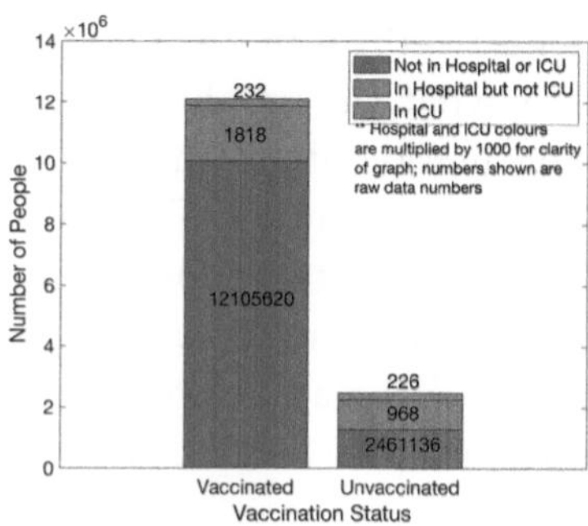

Figure 8.11: By redrawing the data, we are able to catch more of the data and give a much more realistic and honest view of the situation. While the values of hospitalization are comparable regardless of vaccination status, this graph shows that the probability of being unvaccinated and hospitalized is much higher.

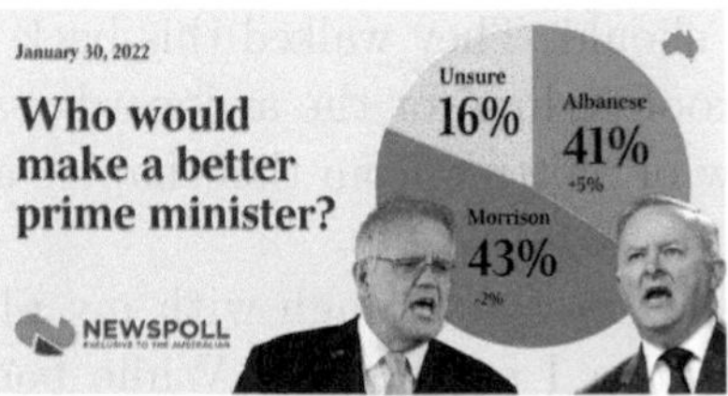

Figure 8.12: A very bad pie graph made to favour Morrison

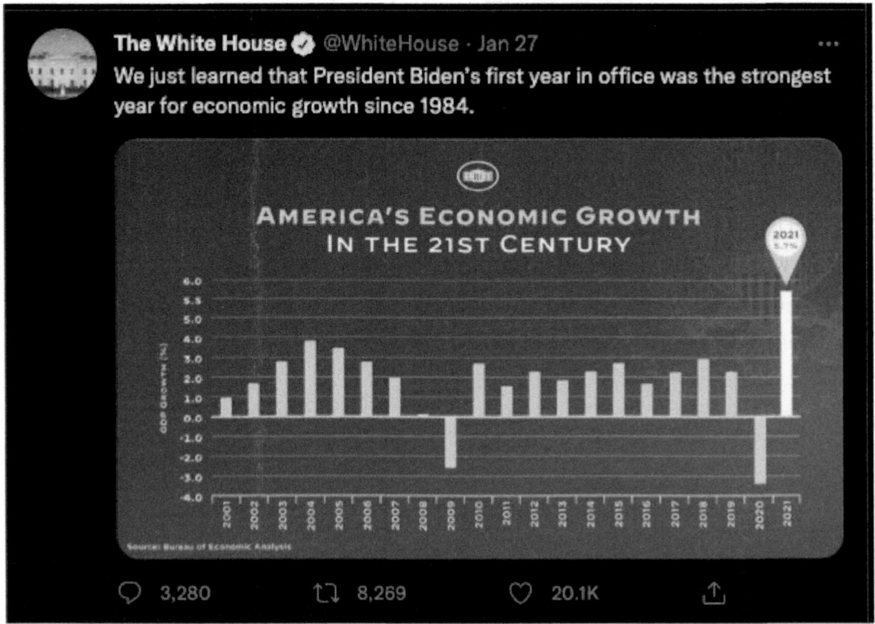

Figure 8.13: Pay close attention to the y-axis in this plot. This odd scaling makes the 2021 *bar appear larger than it should.*

talked about the ± percentages under the big numbers. Do these get added/subtracted from the numbers shown or are they already added/-subtracted from some unshown numbers? Everything in this figure is geared towards making it look like the majority of Australians prefer Morrison when in fact they are very evenly split[26].

The examples I've given may seem politicized, but data manipulation and bad visualization is something *everyone* is capable of, and no one is immune from. In figure 8.13 is an actual tweet from the White House. The y-axis is not scaled properly and leads to the bar in focus to appear

[26] And if the small numbers are to be added and subtracted, Albanese is preferred!

much larger than it should. They walked this back in a later tweet, and blamed a lack of proofreading for the awkward y-axis. Riddle me this: from what you know of plotting from this chapter alone, does this seem like a natural thing to happen?

Often times, I am not fast enough with my phone to capture bad plots on television when I see them[27]. While bad data visualization happens everywhere, anecdotally, it's particularly notorious on television where the image is but a flash in front of you and you are not given the opportunity to scrutinize it. This is why it's not enough to be able to spot false representations of data, but we need to learn to stop doing it.

The main idea here is that pictures matter; words matter; and the things you do, no matter how small, matter. These graphs were made by people, probably with good intentions. The graphs like the ones in figure 8.9 are used as evidence that vaccines kill people, or that they're ineffective. Inaccurate representations of data cause all kinds of problems in the world, and in our current world where just about everything down to *the colour of a man's suit*[14] can and will be politicized, it's more important than ever to be honest, accurate and complete.

[27]If I were a part of a data set on reaction times, I'd probably be an outlier, and not in a good way.

CHAPTER 9

Probability

I THINK WE ALL HAVE SOME SENSE OF PROBABILITY, BASED ON LIVED EXPERIENCE. We all make probability judgements all the time. Almost every decision we make occurs after some[1] analysis of probabilities. We don't always come to the right conclusion, but that's ok. We have to take a step back from the very applied, exploratory data analysis we have been doing to talk about the underpinnings of statistics: probability theory.

9.1 ETHICAL AND MORAL CONSIDERATIONS: A VERY SPECIAL EPISODE

Why? It's important because our intuitive understanding of probability can often be used against us. On the surface, an argument of *the probability you survive the day is 50%: you either do or you don't* seems to hold water if we don't know any better, but it ignores just about everything we know about the world around us. Similarly, probabilities can be used to skew us into thinking a certain way. We could say the prevalence of a disease, for example, increases from 0.1% to 0.12% in a population, or we could say the prevalence increased by 20%. One sounds more substantial than the other because one is missing information: increased 20% *from what?* Statistics and probability are also sometimes used to undermine societal efforts or to push malicious agendas.

As an example, consider the population in figure 9.1 of sheep and pigs. If a very hungry wolf were to come around for dinner and randomly pick an individual to eat, they will likely choose a pig. If another wolf, who was initially just peckish, but upon seeing her friend eat became

[1] Conscious or unconscious

DOI: 10.1201/9781003265405-9

Figure 9.1: A population of pigs and sheep, and a very hungry and poorly disguised wolf.

quite hungry, also picked a meal at random, it would likely also be a pig. If this happens several more times, and say that the 10th wolf gets a sheep, one might conclude that wolves prefer pork 90% of the time or that it is 90% safer to be a sheep than a pig! Of course this is a misuse of analysis because it ignores that the pigs outnumbered the sheep 20 to 1 in the beginning[2]. This may seem like a silly example, but it has its roots in reality[3].

The best way to combat these misuses and to be able to spot faulty uses of statistics is to understand where they come from: how probability comes to be, what it means, and how to interpret it.

9.2 COUNTING

In essence, a probability is the number of times we expect to see a particular outcome, N_p, out of all possible outcomes, N_t. When we divide the number of times a particular outcome can occur by the number of total outcomes, N_p/N_t, we get a probability, p. This means that if we run an event over and over again, say, K times, *at random*, we should expect to see our particular outcome pK times. While not important right now, this interpretation of probability is referred to as a **frequentist/objectivist** interpretation of probability[4].

In order to compute such a thing, we need to know how to count. Yes, I know we know how to count, but we don't necessarily know how

[2]You can count that there are 40 pigs and two sheep in the initial population.

[3]Similar arguments are often used to dismiss or contradict the efficacy of vaccines.[3]

[4]The opposing interpretation is the **Bayesian/subjectivist** interpretation, which isn't as straightforward.

to count *efficiently*. For instance, if I asked you to count all possible eight-character passwords that can use letters (upper or lower case) and numbers, how many would there be? Yes, we can start typing all possible passwords starting with *aaaaaaaa* and then *aaaaaaab*, *aaaaaaac*, etc., but this is going to take forever.

We can instead ask, *how many possibilities exist for each position in our password?* We have eight positions,

$$\_\ \_\ \_\ \_\ \_\ \_\ \_\ \_$$

and in the first position, we could put a capital letter (there are 26 of these), a lower case letter (there are also 26 of these), or a number (there are 10 of these, 0–9). So there are a total of 62 possibilities for the first entry in our password.

$$\underline{62}\,\_\,\_\,\_\,\_\,\_\,\_\,\_$$

Similarly, we can repeat characters in our password so there are 62 possibilities in *any* of the spots in our password.

$$\underline{62}\,\underline{62}\,\underline{62}\,\underline{62}\,\underline{62}\,\underline{62}\,\underline{62}\,\underline{62}$$

This means that there are 62^8 possible passwords using just letters and numbers. Not all of these are good passwords[5], but they are passwords none the less.

WHAT IF WE WERE ONLY ALLOWED TO USE EACH CHARACTER ONCE? What changes? In such a case, we still have 62 possible characters for our first spot,

$$\underline{62}\,\_\,\_\,\_\,\_\,\_\,\_\,\_$$

but once we make a selection, we cannot reuse that character in the second spot! So, we only have 61 choices for spot 2.

$$\underline{62}\,\underline{61}\,\_\,\_\,\_\,\_\,\_\,\_$$

Then, we only have 60 choices for spot 3 (since we've already used two characters to fill the first two spots). If we continue like this, we get the following pattern for options in spots:

$$\underline{62}\,\underline{61}\,\underline{60}\,\underline{59}\,\underline{58}\,\underline{57}\,\underline{56}\,\underline{55}$$

[5] The password *password* is one of these possibilities, and it is a notoriously bad password.

This pattern has a special name: it is a **permutation without replacement**. We will discuss it a little bit more in a moment, but first let's introduce a special mathematical operator you may or may not have seen before.

The Factorial

When we write $x!$ in mathematics, it doesn't mean we're excited[6]. The ! is a special symbol that means to multiply all **natural numbers**[7] up to and including x. For example,

$$\begin{aligned} 1! &= 1 \\ 2! &= 2 \cdot 1 = 2 \\ 5! &= 5 \cdot 4 \cdot 3 \cdot 2 \cdot 1 = 120 \\ 10! &= 10 \cdot 9 \cdot 8 \cdot 7 \cdot 6 \cdot 5 \cdot 4 \cdot 3 \cdot 2 \cdot 1 = 3628800 \end{aligned}$$

If you have read Chapter 1 of this book, you will know that different classes of functions grow at different rates. The factorial grows *fast*; faster than polynomials and faster than exponentials.

The factorial comes with a special case for 0, in that

$$0! = 1.$$

Notice that the factorial **cannot** be applied to negative numbers[8].

Looking back at our password-with-no-repeating-characters example, you can see hints of the factorial in there. We'll expand on this in the next section.

9.3 PERMUTATIONS

We can see that a permutation without replacement looks like a factorial, but it's missing a chunk. In fact, our example above looks like 62!, but it's missing the numbers 54 and below. How do we get rid of numbers in a factorial? We divide them out!

$$\begin{aligned} &\frac{62 \cdot 61 \cdot 60 \cdot 59 \cdot 58 \cdot 57 \cdot 56 \cdot 55 \cdot \cancel{54} \cdot \cancel{53} \cdot \cancel{52} \cdots \cancel{3} \cdot \cancel{2} \cdot \cancel{1}}{\cancel{54} \cdot \cancel{53} \cdot \cancel{52} \cdots \cancel{3} \cdot \cancel{2} \cdot \cancel{1}} \\ &= \frac{62!}{54!} \end{aligned}$$

[6] Much to my dismay

[7] a.k.a. the positive integers; a.k.a. the whole numbers; a.k.a. the counting numbers.

[8] Since we would descent to $-\infty$, and negative factorial is $-\infty$ and thus undefined

The 54 isn't random either; it's the number of things we have to choose from (62) minus the number of spots we have to fill (8). We call this a permutation without replacement.

Definition 9.3.1. A **permutation without replacement**, or just **permutation**, is the number of ways in which you can arrange a collection of n things into a group of size k **if the order of the things matters**. We write this as

$$_nP_k = \frac{n!}{(n-k)!}$$

For instance, in a password, the ordering of the characters definitely matters. The password `123asf` is different from `asf123`. In this case, it is proper to use permutations when counting.

Another way to think of this is that permutation[9] is counting when *all the individual things we are counting are completely distinct.* In the case of characters in passwords, `a` is distinct from `x` and both are distinct from `2`; therefore, `2a` is different than `a2`.

How many ways are there to pick four of the animals in figure 9.2? Yes, the animals are distinct, but whether the *order matters* depends on *why* we're picking them to begin with. Are we throwing them into a stew? Then order probably doesn't matter. Are we ranking them best to worst? Then ordering probably matters.

9.4 COMBINATIONS

WHAT OF THE CASE WHEN ORDERING DOESN'T MATTER? We go about this by first computing the number of ways we can arrange objects when the ordering *does* matter, and then grouping all the scenarios that have the same elements, but in a different order, and we divide that out. Confused? Probably. Let's continue by an example.

For instance, if we are indeed making a stew using four of the 12 animals shown in figure 9.2, then picking bat, cow, rabbit, and pig is the same as picking rabbit, cow, pig and bat; the stew is going to turn out

[9]Without replacement

Figure 9.2: We're making a stew of only the finest ingredients.

the same regardless. In this case, we start by calculating the number of ordered groups there are[10].

$$_nP_k = \frac{12!}{(12-4)!} = 11880$$

Now, some of these will have the same animals, just in a different order. How do we find out how many have the same order?

Again, we will continue by trying to place objects into spots. Let's say we had our four animals chosen, how many ways are there to arrange them in four positions?

$\_\ \_\ \_\ \_$

In the first spot, we can put any one of our four animals.

$$\underline{4}\_\_\_$$

Now there are three animals left that could be put in the second spot:

$$\underline{4}\,\underline{3}\_\_$$

[10] i.e. permutations

If we continue, we will see that for any four animals chosen, there are 4! ways of arranging them. Therefore, if we take our permutations and divide by 4!, we get

$$\frac{11880}{4!} = 495$$

So there are 495 ways to choose animals if we don't care about the order. This type of counting is called a **combination**.

Definition 9.4.1. A **combination** is a permutation where the order of things does **not** matter. We also assume that there is **no replacement**. Mathematically, when we want to choose k things from n things and the ordering doesn't matter, this is written as

$${}_nC_k = \frac{n!}{(n-k)!k!} = \frac{{}_nP_k}{k!}$$

The $k!$ accounts for the possible orderings that are being counted multiple times by the permutation.

Back to our example, if we wanted to choose a team[11] of four animals from a group of 12 animals, there are

$${}_{12}C_4 = \frac{12!}{(12-4)!4!} = 495$$

ways to do this, as we saw above.

What if we wanted to know the number of stews that contain alligator? In that case, one of our spots is taken up by the alligator. That means there are three spots left on the team and 11 animals left to choose from[12]. There are

$${}_11C_3 = \frac{11!}{(11-3)!3!} = 165$$

stews that use alligator as an ingredient.

We can go a little bit further so that we can segue nicely into the next section. Remember, back at the beginning we defined a probability as the

[11] I'm using the word team very loosely (and euphemistically) here.

[12] Remember, without replacement.

number of desired outcomes divided by the number of total outcomes. We can use our information here to calculate a probability. We know that the total number of four-animal stews is 495; we know that 165 of these contain alligator as an ingredient. So, we can ask the question, *if someone served us a stew at random, what is the probability that it contains alligator as an ingredient?* The answer is

$$P_{alligator} = \frac{165}{495} = \frac{1}{3}$$

One-third of the stew recipes contain alligator[13].

This probability may seem weirdly high and unintuitive. The alligator is one of *ten* possible ingredients; how can it possibly show up 33% of the time? This is because the first statement above[14] is more obvious, but less important than the fact that we are *required* to use exactly four animals in our stew. We'll expand upon this particular example in the exercises to develop a better intuition.

9.5 COMBINATIONS WITH REPLACEMENT

We've already seen that permutations with replacement are extremely simple. If we have n items and we want to pick r of them with replacement where the order matters, then the total number of possibilities is

$${}_nPR_r = n^r$$

where PR means permutation with replacement.

What if the order doesn't matter, but we are allowed to replace items before picking again? For example, if we are making a pizza, with three toppings, out of a possible 20 toppings, we could pick pepperoni, mushroom, and bacon[15], or we very well could pick pepperoni, mushroom, and more pepperoni. Just because we have used pepperoni once does not preclude us from using it again. There are four possible scenarios that we need to deal with:

1. All three toppings are unique – $\_\_\_$
2. The first and second topping are the same, but different from the third – $\_\_\_$

[13]This is bad news if you're allergic to alligator.

[14]About the alligator being one-tenth of the ingredient list

[15]A classic Canadian pizza

3. The second and third topping are the same, but different from the first – ...

4. All three toppings are the same – ...

- The first scenario is equivalent to having no replacement: ${}_{20}C_3$.
- The second scenario is the equivalent of having 20 choices and filling two spots[16]: ${}_{20}C_2$.
- The second scenario is the equivalent of having 20 choices and filling two spots: ${}_{20}C_2$.
- The last scenario is the equivalent of having 20 choices and filling one spot: ${}_{20}C_1$.

These scenarios are all mutually exclusive as well, so we can add them up to get

$$ {}_{20}C_3 + {}_{20}C_2 + {}_{20}C_2 + {}_{20}C_1 = 1540 $$

There are 1540 different pizzas with three toppings if we can double up or even triple up on certain toppings.

Now if we wanted four toppings, there are eight different scenarios:

1. ....
2. ....
3. ....
4. ....
5. ....
6. ....
7. ....
8. ....

[16] Since the second spot is completely determined by what you put in the first spot

which gives us

$$ {}_{20}C_4 + {}_{20}C_3 + {}_{20}C_3 + {}_{20}C_3 + {}_{20}C_2 + {}_{20}C_2 + {}_{20}C_2 + {}_{20}C_1 = 8855 $$

ways to make a four-topping pizza with replacement.

The number of times we repeat the inner combinations in the sums is also not random. It is related to the spaces *between* the choices; if we put dividers between our unique elements, we see there are three spaces and we either put dividers in 3, 2, 1, or 0 of the spaces.

1. . — . — . — .
2. . — . . .
3. . . — . .
4. . . . — .
5. . . — . — .
6. . — . — . .
7. . — . . — .
8. . . . .

So the number of times we repeat ${}_{20}C_2$ is exactly the same as the number of ways we can place one divider in three spaces[17]: ${}_3C_1$.

In general, to choose k things out of n possibilities with replacement, one way to write it is

$$ {}_nCR_k = \sum_{i=0}^{k-1} {}_{k-1}C_i \cdot {}_{20}C_i $$

where CR means combination with replacement.

A much simpler formula is

$$ {}_nCR_k = {}_{n+k-1}C_k $$

[17]In other words, I have three spots and I choose one of them to place my divider in.

9.6 PROBABILITY

With some intuitive idea about how we can count our way to probabilities, we can now start using the language of probability theory. As we stated at the beginning of this chapter, probability gives us some idea about how frequently a certain outcome happens when compared against all possible outcomes.

The set of all possible outcomes is often called the **sample space**, and we denote it with the symbol Ω. A sample space should satisfy two qualities:

- It should be **exhaustive**. It accounts for *all possible outcomes.*
  - If we are rolling a standard die, Ω is the set of numbers $\{1, 2, 3, 4, 5, 6\}$. If we only use the set $\{1, 2, 4, 5, 6\}$, then this cannot be our sample space, as it is not exhaustive[18].
- It should be **mutually exclusive**. One outcome occurring necessarily means all others did *not* occur.
  - If we flip a coin and the result is heads, then the coin necessarily did **not** land tails. Heads and tails are mutually exclusive.

From now on, we will call outcomes, **events**. We assign probabilities to events. The sample space, Ω, is the space of all events.

Events can be very general. An event can range from rolling a specific number on a dice, to getting sick after an encounter with a sick person, to making alligator stew. We do our best to *quantify* events so that we can use the language of mathematics to describe them and their associated probabilities.

Probability Notation

We use **function notation** from mathematics[19] to describe probabilities. Just like in math, we will use **variables** to denote **events**. When we want the probability of event A, we will write $P(A)$, which is read as *probability of event A occurring.* We can describe an event, A, in words, or we can[20] use *set notation* from mathematics. The framework for set notation looks like

$$A = \{\text{values involved} \mid \text{condition that needs to be met}\}$$

[18] Unless we have a five-sided die

[19] See Chapter 1

[20] And often do

We will see this in action in some examples below.

Some Examples

ROLLING A SIX-SIDED DIE, TWICE IN A ROW. This is such a common example in elementary probability theory, that it's almost a trope[21].

What are the possible outcomes? If we use our counting arguments from before, we know that there are 36 possible outcomes in this case.

What is our sample space, Ω? It is all possible permutations of the numbers 1 through 6 as a pair *with replacement*. We can write Ω as

$$\Omega = \{(d_1, d_2) | 1 \leq d_1, d_2 \leq 6\}$$

This says, essentially, that we are picking a pair of numbers, and those numbers must be between 1 and 6 (like on a six-sided die).

We can visualize this in two ways. The first is as a **tree diagram**; there are six possible choices for the first roll, and from each of those, six possible choices for the second roll.

Each individual event in this case is given by a branch that starts at the leftmost point and ends with a red dot. We can see that each pair of rolls has a $1/36$ chance of happening, but those aren't the only events we can ask about!

We could ask questions like the following: *what is the probability of the sum of our rolls being 6?* In this case, our event, A, is the two values adding up to 6. Written mathematically, this would look like

$$A = \{(d_1, d_2) | d_1 + d_2 = 6\}$$

There are a few branches that allow for this event. We can see from the diagram that $5/36$ events allow for a sum of 6. It stands to reason then that the probability of rolling to dice and having the values sum to 6 is $P(A) = 5/36$.

The other way to visualize this is as a grid. The only reason I bring this up is because for continuous variables (as opposed to discrete, like a die) it makes more sense than a tree. We can put die 1 on the x-axis and die 2 on the y-axis. And then we can fill in the scenarios that correspond to when we have a sum of 6. Again, we see that there is a $5/36$ chance of the sum being 6.

[21] And I'm sorry for that.

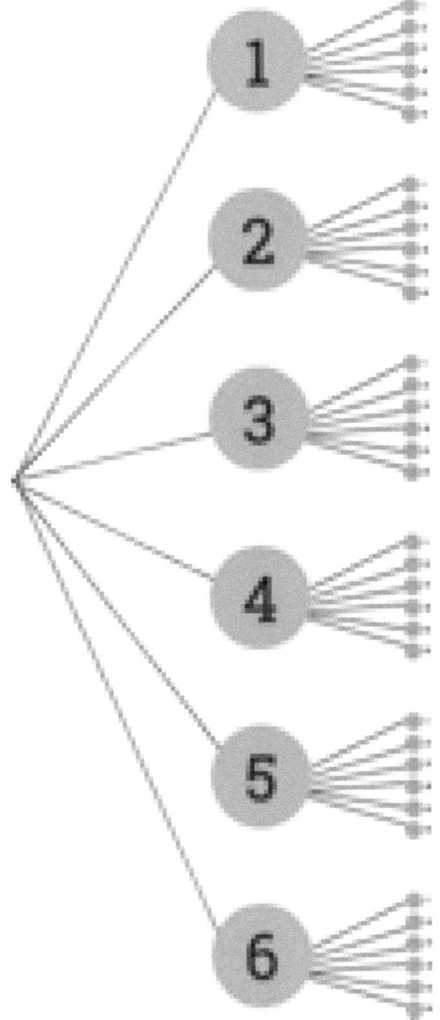

Figure 9.3: Figure showing all possible outcomes of rolling a die two times sequentially. The blue represents the first roll, and the red represents the second roll. We see that there are 36 total possibilities.

SEX OF HONEY BEES AND HUMANS[22]

All the traits that make us human are encoded in our genetic code. It takes two humans to make another human. The assigned sex of a baby human is determined by what they inherit from their parents. Your genetic code comes in pairs; one piece of that pair comes from one parent,

[22]Don't worry, we're not going to get weird with it.

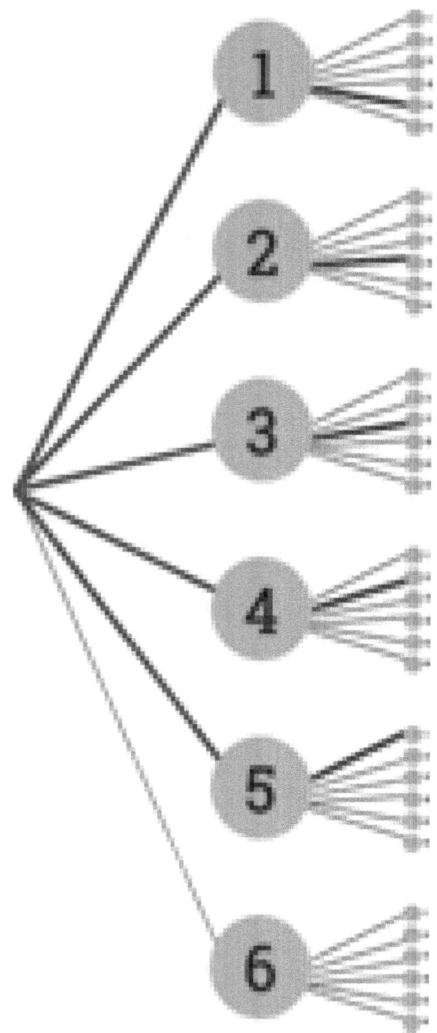

Figure 9.4: Figure showing all possible outcomes of rolling a die two times sequentially. The blue represents the first roll, and the red represents the second roll. We see that there are 36 total possibilities. The thick branches represent all branches that lead to a sum of 6.

and the other piece comes from the second parent. In order to create a viable baby[23], one partner needs to have XX as their sex chromosomes[24] and the other needs to have XY as their sex chromosome. The gametes[25]

[23] As of the writing of this, who knows what medical marvels the future holds!

[24] Chromosomes are bundles of genetic code.

[25] a.k.a. the sperm and the egg

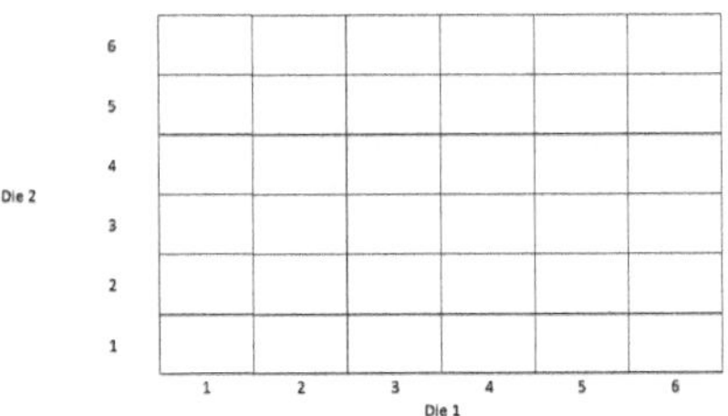

Figure 9.5: Figure showing all possible outcomes of rolling a die two times sequentially. This is another way of representing the tree diagram. Here we see we still have 36 *possibilities.*

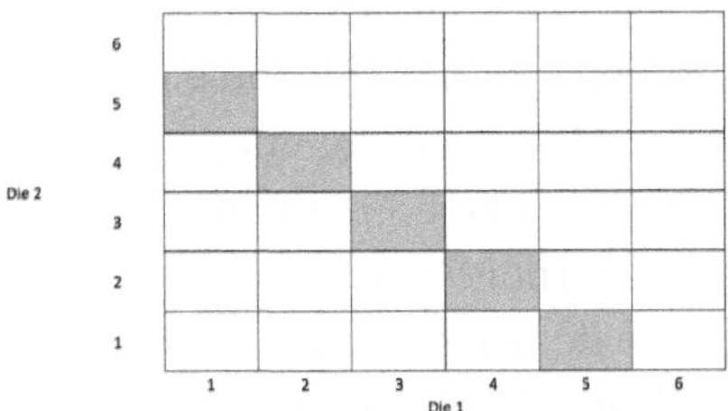

Figure 9.6: Figure showing all possible outcomes of rolling a die two times sequentially. This is another way of representing the tree diagram. Here we see we still have 36 *possibilities. The blue shaded blocks are those events where the sum of the two dice rolls is six.*

each carry one of the sex chromosomes of the parent. These gametes then combine to produce the sex chromosome pairs of the baby.

Our sample space of sexes of the baby in this case is all the possibilities of combinations. In this case, we want to pick one of the two sex chromosomes from the XX partner **and** pick one of the two sex chromosomes from the XY partner. The total number of possibilities is then

$$_2P_1 \cdot {}_2P_1 = 4$$

We can even list the four possibilities in our sample space.

$$\Omega = \{XX, XY, XX, XY\}$$

Why are they doubled up? We have two XX pairs and two XY pairs! This comes from the fact that we haven't distinguished between the two X chromosomes of the XX partner.

If the gametes exist in equal proportions in the XY partner[26], then we can determine the probability of a XX baby being produced by counting all the events *without* a Y and dividing it by the number of total events. In this case, if we call our event[27]. A, then the probability is

$$P(A) = \frac{2}{4} = 0.5$$

There is equal probability of producing an XX baby or an XY baby.

Bees reproduce slightly differently. Bees have many, many sex chromosomes. For simplicity, let's say there are four and let's call them A, B, C, and D.

Bees differ from humans[28] in that their two different sexes[29] are determined by the *number* of sex chromosomes they have: males have one; females have two. Moreover, females are only female if their sex chromosomes are *different*. The consequence of this is that a queen bee can produce male heirs without mating at all!

When producing female heirs, she mates with multiple male bees and stores their gametes. They then mix randomly with her own gametes to produce baby bees.

This is a lot. It's a lot less intuitive than human mating[30]. We can determine the probabilities with a tree diagram. Let's say our queen bee has the sex chromosomes AB and she mates with four males with sex chromosomes A, B, C, and C. We can ask the question, *what is the probability that the queen produces a viable*[31] *female offspring?*

The first thing we must do is determine our sample space Ω. This is all possible combinations of A or B from the queen, with A, B, C, or C from her male partners. This gives us

$$\Omega = \{AA, AB, AC, AC, BA, BB, BC, BC\}$$

We can double check that we indeed got everything by asking it, how many events should we have? From the queen bee, we pick one of the two sex chromosomes for Q choices

$$Q = {}_2P_1 = 2$$

[26]In other words, an X sperm occurs with the same probability as a Y sperm.

[27]i.e. XX baby

[28]In a lot of ways, but also in this way in particular

[29]Males and females, for simplicity

[30]But strangely, arguably, just as ritualistic

[31]Viable meaning two different sex chromosomes

and then we require ${}_1P_1 = 1$ choice from the first male **or** the second male **or** the third male **or** the fourth male. So from the males, we get D choices.

$$D = 1 + 1 + 1 + 1 = 4$$

Multiplying them together, we get a total number of possibilities as

$$QD = 4 \cdot 2 = 8$$

So we at least have some evidence that we got all the possibilities[32].

Since viable females are defined as those bees with two *different* sex chromosomes, we can count from Ω all of those possibilities to get the number of outcomes that satisfy our event condition. Calling the event F, we get that

$$F = \{AB, AC, AC, BA, BC, BC\}$$

Dividing the number of elements of event F by the size of the sample space Ω, we get

$$P(F) = \frac{6}{8} = 0.75$$

We see that we have a 75% chance of producing a viable female.

This esoteric[33] mating pattern also means that not all female bees are full sisters. Most, in fact, are half-sisters[34]. Again, using our queen bee, let's name her Demeter, who had four male partners, Zeus, Poseidon, Iasion, and Karmanor; we can determine the probability that two baby bees are full sisters.

We know that all the bees will share the same mother; so full sisterhood is determined strictly by whether two bees share a father. We can thus represent our sample space as a pair

$$\Omega = \{(f_1, f_2)\}$$

where f_1 is the father of the first daughter, and f_2 is the father of the second daughter. In this case, there are 64 possibilities[35].

[32] I can assure you that we did.
[33] By human standards anyway
[34] As they only share a mother – the queen
[35] How to determine this number and listing all the possibilities is left to the exercises.

If we define event S as

$$S = \{(f_1, f_2) \,|\, f_1 = f_2\}$$

we can determine how many possibilities are in this set using our counting rules.

Here we have *two* spots to fill.

$$\underline{\ }\ \underline{\ }$$

Our first baby bee can have any of the four fathers,

$$\underline{4}\ \underline{\ }$$

but once we determine the first father, there is only one possibility for the second father that results in a full sister, which gives us

$$\underline{4}\,\underline{1}.$$

Multiplying this together, we see that there are four possibilities in our event set S. Therefore the probability of two baby bees being full sisters is

$$P(S) = \frac{4}{64} = 0.0625$$

The vast majority of bees are half sisters!

Note that the more knowledge we have, the more complicated things get. You may have connected some dots that are realized that if Zeus has sex chromosome A, Poseidon has sex chromosome B, and both Iasion and Karmanor have sex chromosome C; then not all full sisters are sisters at all since some combinations produce non-viable females!

We should probably have some language and some understanding of how probabilities mix and match so that we can explore cases like the above.

9.7 PROPERTIES OF PROBABILITIES

Probabilities need to have certain characteristics to even *be* a probability. So, given an event A, the notation P is a probability if

- $0 \leq P(A) \leq 1$

- A probability *must* be between 0 and 1. Zero meaning something is impossible; one meaning it is an absolute certainty.

- $P(\emptyset) = 0$
  - The probability of *nothing* happening is zero[36]. If I ask you to roll a dice, you do not get to "choose" not to roll the dice.

- $P(A or B) = P(A) + P(B)$ *if and only if A and B are mutually exclusive.* This means that A and B cannot happen together. In other words, if A happens, then B becomes impossible.
  - If this condition is not met, then this last statement isn't true. Let's look at this third property through an example.

An example

Let's say we want to calculate the probability of the sum of our rolls being 6 *or* the first dice is a 6. Written in our mathematical notation, our event is

$$A = \{(d_1, d_2)|d_1 + d_2 = 6\}$$
$$B = \{(d_1, d_2)|d_1 = 6\}$$

What does this look like? In this case, there are two different conditions we could satisfy to belong to the event space A or B, so let's use two different colours. We see that $P(A or B) = P(A) + P(B) = {}^{11}/_{36}$. This works because the two events have no overlap. If B happens, it is *impossible* for A to happen[37].

Let's contrast this with the probability that the sum of the two dice is 6 or the first dice shows a 5. In this case,

$$A = \{(d_1, d_2)|d_1 + d_2 = 6\}$$
$$B = \{(d_1, d_2)|d_1 = 5\}$$

In this case, we get the following diagram. Here we see that ten blocks are coloured, so $P(A or B) = {}^{10}/_{36}$. If we instead try to use the above rule, we would get $P(A) + P(B) = {}^{5}/_{36} + {}^{6}/_{36} = {}^{11}/_{36}$, which is not correct.

---

[36] The symbol $\emptyset$ is a "special zero" that means *no event.*

[37] It only takes a second to figure this out if you're confused. Remember, there is no zero on a die.

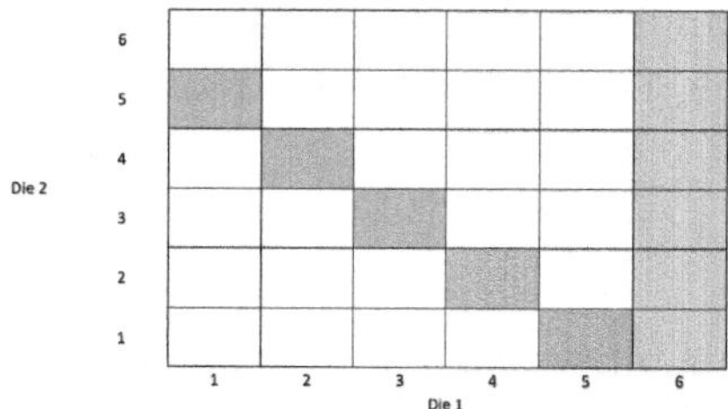

Figure 9.7: Here is a diagram of our sample space, Ω, and our event space, $A = \{(d_1, d_2)|d_1 + d_2 = 6\}$ or $B = \{(d_1, d_2)|d_1 = 6\}$. We see that $P(A or B) = 11/36$, which just so happens to be the probability of the sum being 6 $(P(A) = 5/36)$ and the first die showing six $(P(B) = 6/36)$. This works because the two events do not overlap.

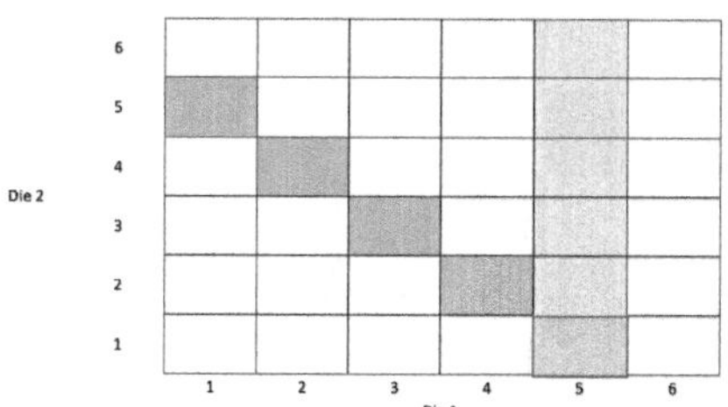

Figure 9.8: Here is a diagram of our sample space, Ω, and our event space, $A = \{(d_1, d_2)|d_1 + d_2 = 6\}$ or $B = \{(d_1, d_2)|d_1 = 5\}$. We see that $P(A or B) = 10/36$ since we have ten coloured blocks, but $P(A) + P(B) = 11/36$. What happens is we have double counted the area that is common between the two events in the second case.

What's happening is that we are counting the block at $(5, 1)$ *twice* if we assume A and B are mutually exclusive. This double counting is why the rule doesn't work when there is overlap of events.

9.8 MORE NOTATION

These ideas of mutually exclusive and overlap can all be written mathematically using the notation from *set theory*[38].

In the space of all events, Ω, the situations that satisfy the conditions of arbitrary events A and B take up space. We can visualize this using what's called a **Venn diagram**. A Venn diagram shows us things like

[38]Don't worry; we aren't going to do any real set theory here; we're just going to appropriate their framework.

where events overlap, where they don't and parts of our sample space that are left untouched by defining events. Our table representation of our dice roll above is an example of a Venn diagram.

We will use Venn diagrams to visualize and, hopefully, cement some notation in our heads. First, let's consider a sample space with a single event.

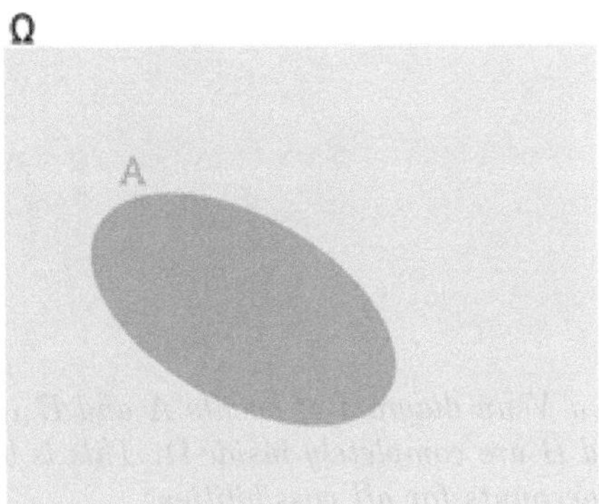

Figure 9.9: Here is a Venn diagram of an event A inside a sample space, Ω. Notice how A is completely inside Ω. This is because Ω is supposed to be exhaustive. It accounts for all possibilities.

The first thing we can define is A^C. This is called the **complement of** A. It is basically all events that are *not* in A. It's fairly easy to visualize.

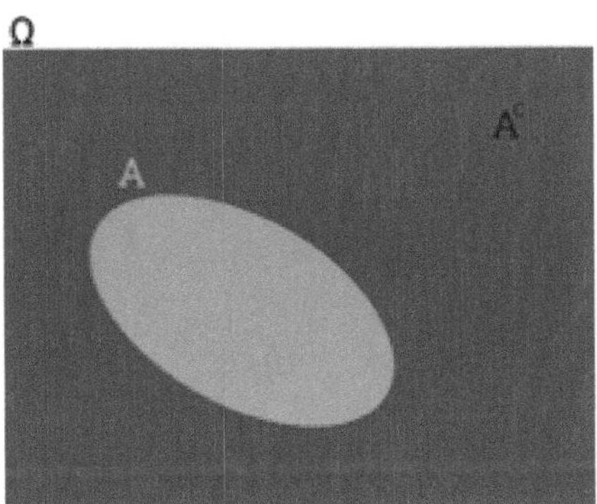

Figure 9.10: Here we see events A and A^C. Together, they make up all of Ω. This is a consequence of Ω being mutually exclusive. You either are or you are not. There is no in between. Things exist or they do not; events are in A or not in A.

Now, let's expand to see two events, A and B, within Ω. This is the most general visualization of two events.

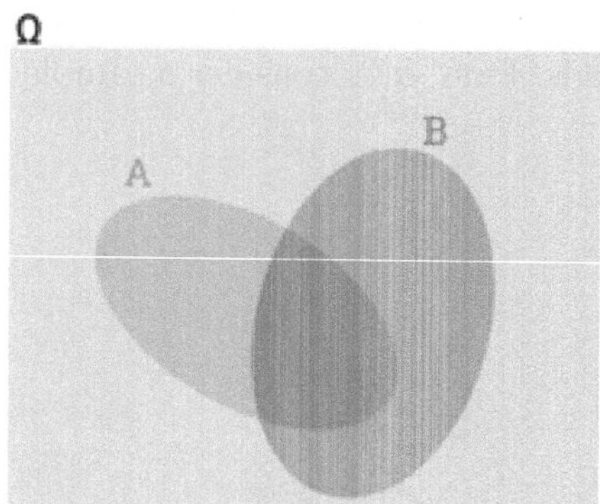

Figure 9.11: Here is a Venn diagram of events A and B inside a sample space, Ω. Notice how A and B are completely inside Ω. This is because Ω is supposed to be exhaustive. It accounts for all possibilities.

If we are looking for A **or** B to happen, this is called a **union** and has the symbol $\cup$. When we are speaking of A or B, we say $A \cup B$ and read it as A *union* B. This represents all scenarios in A or B or both.

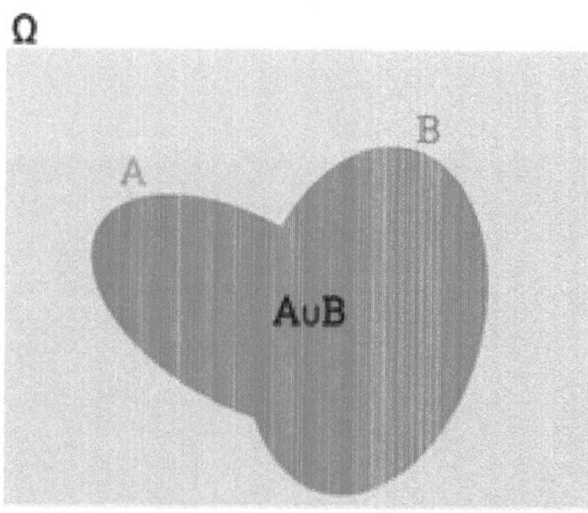

Figure 9.12: Here is a Venn diagram of events $A \cup B$. The union represents all scenarios that are in A or B (or both).

What about if we just want that overlapping bit in figure 9.11? This is called the **intersection of A and B** and is denoted by the symbol $A \cap B$. This is shown in figure 9.13. The **intersection** represents scenarios

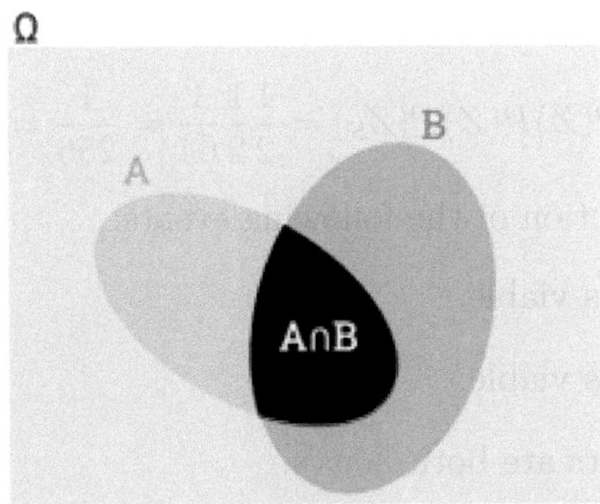

Figure 9.13: Here is a Venn diagram of events $A \cap B$. The union represents all scenarios that are in A and B.

that are in A **and** B. Two events are mutually exclusive if there is no overlap; in other words, if their intersection is empty. Figure 9.8 is an example of two events which overlap. Figure 9.7 is an example of an empty intersection (no overlap).

In our example above, we see that when we have an intersection and try to compute the total probability, we double count the intersection. A more general form of our rule that works for *any* two events is

$$P(A \cup B) = P(A) + P(B) - P(A \cap B)$$

We have to delete one copy of the intersection since it was overcounted. In our dice example, we see that $P(A \cap B) = 1/36$. Therefore, $P(A \cup B) = 5/36 + 6/36 - 1/36 = 10/36$ and the probability agrees with what we see in the picture.

LET'S GO BACK TO OUR QUEEN BEE, DEMETER, and her four suitors, Zeus, Poseidon, Iasion, and Karmanor. If Demeter has sex chromosomes A and B, Zeus has sex chromosome A, Poseidon has sex chromosome B, Iasion has sex chromosome C, and Karmanor has sex chromosome C, we can determine the probability of two baby bees being full sisters **and** indeed sisters at all[39].

The probability of Demeter and Zeus producing a one viable female is $P(Z) = 0.5$[40]. The probability of producing two viable females who

[39] i.e. both viable females

[40] There are two total possible combinations AB or AA, but only one of these is a viable female.

are both the progeny of Demeter and Zeus[41] is thus

$$P(S_Z) = P(Z)P(Z)P(Z_S) = \frac{1}{2}\frac{1}{2}\frac{1}{64} = \frac{1}{256} = 0.00390625$$

This is the intersection of the following events:

- Daughter 1 is viable.
- Daughter 2 is viable.
- The daughters are both Zeus's.

The same argument holds for Poseidon since Poseidon and Demeter both share one sex chromosome. We are just swapping some labels. Therefore,

$$P(S_{\mathcal{P}}) = P(\mathcal{P})P(\mathcal{P})P(\mathcal{P}_S) = \frac{1}{2}\frac{1}{2}\frac{1}{64} = \frac{1}{256} = 0.00390625$$

For Iasion and Karmanor, all of the offspring are viable since Demeter shares no sex chromosomes with Iasion or Karmanor. Therefore $P(I) = P(K) = 1$. The probability of both daughters being the progeny of Iasion[42] is $P(I_S) = P(K_S) = 1/64$. Since all progeny of Iasion[43] are viable, the probability of two daughters being full sisters and viable is very simple.

$$P(S_I) = P(S_K) = P(K_S) = P(I_K) = \frac{1}{64}$$

The total probability then of being viable females and full sisters is the probability of the **union** of these four events.

$$P\left(S_Z \cup S_{\mathcal{P}} \cup S_I \cup S_K\right)$$

Since these four events are **mutually exclusive**[44], we can just add the probabilities without worrying about double counting.

Thus, the probability of being viable full sisters in this case is

$$P\left(S_Z \cup S_{\mathcal{P}} \cup S_I \cup S_K\right) = P\left(S_Z\right) + P\left(S_{\mathcal{P}}\right) + P\left(S_I\right) + P\left(S_K\right) = \frac{10}{256}$$
$$= 0.0390625$$

[41] Looking back, the probability of both daughters being Zeus's daughters is $P(Z_S) = \frac{1}{64} = 0.015625$.

[42] Or Karmanor

[43] Or Karmanor

[44] If you are a daughter of Zeus, you cannot be a daughter of Poseidon.

ETHICAL AND MORAL CONSIDERATIONS, EPISODE 6

The underlying concept that I hoped to convey using the example of bee mating was, once again, **context matters**. When we initially computed the probability of two bees being full sisters, we *implicitly* assumed that all female baby bees would be viable. This was by design because I wanted you to pause and think, *wait, these two things – being sisters and being viable – are related.* If not used as a teaching tool that is trying to instil critical, independent thinking, this assumption should **not** be implicit. It should be stated, in no uncertain terms. Preferably, anytime we also state the value of said probability.

When assumptions are **not** stated explicitly, it should cause pause. It does not mean dismiss all the results outright; they could still be useful. It does not be everything is discredited. It means you should investigate further. Trust, but verify. Be critical, but open-minded.

9.9 CONDITIONAL PROBABILITY

IN THE REAL WORLD, IT'S RARE THAT WE ARE WORKING FROM A PLACE OF ZERO KNOWLEDGE. Often times, we know some things about our system, or we have added context to our problem. For instance, from our above example, we know that the probability that the sum of two dice rolls being 6 is $5/36$. This is the probability *before* we do anything. What if we roll the first die, does our probability of the two dice summing to 6 stay the same? We now *know* what the first die reads; the probability of our event will change as we gain more knowledge.

Once we see the value on the first die, a lot of our scenarios become *impossible.* So let's get them out of the way in figure 9.14. We can then ask the question again: *what is the probability of the sum of the dice*

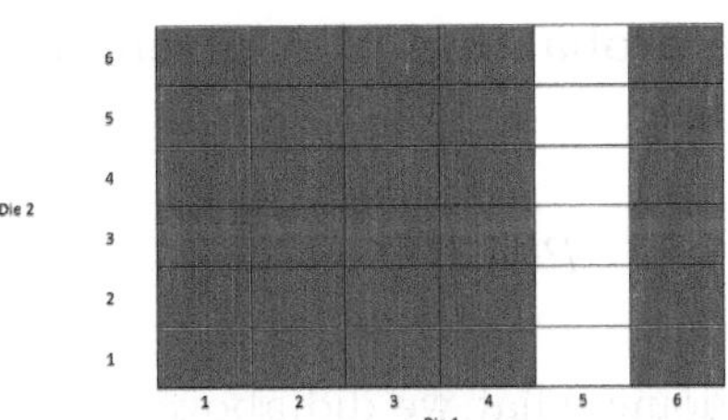

Figure 9.14: Once we know that the first die shows a 5, we can eliminate a lot of our sample space as impossible.

being 6? This time though, we are armed with information. When we ask the question now, a lot of our possibilities have already been deemed impossible; so we do the same thing we did before, but we only count entries that are *not* in the impossible area, like in figure 9.15.

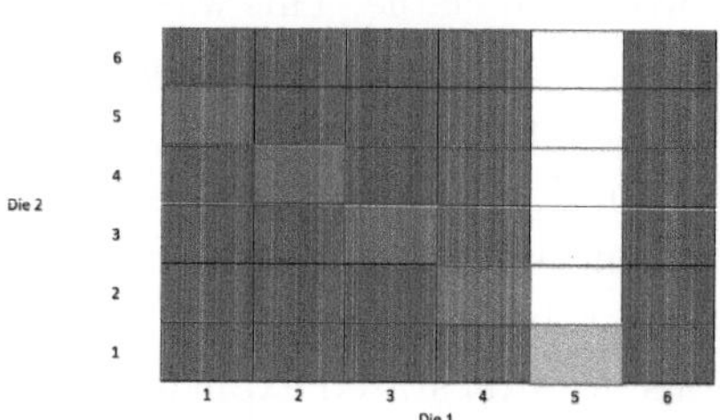

Figure 9.15: Once we know that the first die shows a 5, we can eliminate a lot of our sample space as impossible. Now when we ask about the sum being six, there is only 1/6 scenarios where this is true.

This is called a **conditional probability**, and it is exactly what it sounds like: probabilities that depend on conditions. In our example, we have two events:

$$A = \{(d_1, d_2)|d_1 + d_2 = 6\}$$
$$B = \{(d_1, d_2)|d_1 = 5\}$$

We now *know* that B happened, so we are asking, *what is the probability of A given that B has occurred?* The key words to identify conditional probabilities is **given that**. This phrase tells you what information is known.

Mathematically we write a conditional probability as

$$P(A|B).$$

This is read as **the probability of A given B**. We compute it using the following formula:

$$P(A|B) = \frac{P(A \cap B)}{P(B)}$$

Effectively, this is doing what we did above with figures 9.14 and 9.15. Our new denominator is the things left after the learning of B, and the numerator is where A and B intersect, or how much of A is left when we reduce our sample space to B.

9.10 BAYES' THEOREM

A CURIOUS THING HAPPENS WHEN WE LOOK AT TWO COMPLEMENTARY CONDITIONAL PROBABILITIES. If we want to calculate the probability of A given B, we get

$$P(A|B) = \frac{P(A \cap B)}{P(B)}$$

and if we want to calculate the probability of B given A, we get

$$P(B|A) = \frac{P(B \cap A)}{P(A)}$$

but *there is no difference between* $P(A \cap B)$ *and* $P(B \cap A)$. In fact, $P(A \cap B) = P(B \cap A)$! We can see this by drawing the Venn diagram. Which means we can rearrange one of the equations and plug it into the other one! If we rearrange the equation for $P(B|A)$, we get that $P(B \cap A) = P(B|A)P(A)$. Doing the same for the equation of $P(A|B)$, we get $P(A \cap B) = P(A|B)P(B)$. Writing these together, we get

$$P(A|B)P(B) = P(A \cap B) = P(B \cap A) = P(B|A)P(A).$$

Then removing the middle bits[45]

$$P(A|B)P(B) = P(B|A)P(A)$$
$$P(A|B) = \frac{P(B|A)P(A)}{P(B)}$$

This is called **Bayes' theorem**. It relates two conditional probabilities and is the basis of **Bayesian statistics** and the **subjectivist interpretation of probability**. Let's go through each component of Bayes' theorem to give a rough overview of the Bayesian interpretation of probability.

The Bayesian Interpretation of Probability

Let's restate Bayes' theorem, but colour code it.

$$P(A|B) = \frac{P(B|A)P(A)}{P(B)}$$

[45] In math lingo, we call this using the **transitive property**: if $a = b = c$, then $a = c$.

With the Bayesian interpretation, we add a level of abstraction to our thinking. In this case, we will interpret event B as our observations, and event A is a *belief or hypothesis* we hold as to *why* B is being observed. By looking at events this way, a probability becomes *a measure of confidence in our belief.* Contrast this with the frequentist interpretation we've been using where a probability is a frequency with which we expect a certain event if we do things an infinite number of times.

For instance, when we drop things, they fall to the ground. If we continually do this, we get a series of observations; this[46] may lead us to hypothesize about *why* things fall when we let go of them. Perhaps there is a "force" that accelerates objects at a rate, θ, towards the ground. This value, θ, quantifies our belief. If we are so strong in our belief, we can set $\theta = 2$. Then we can calculate the probability

$$P(\theta = 2|B)$$

where B is the set of observations (our data). This conditional probability gives the probability that $\theta = 2$ *given* our observations. The larger $P(\theta = 2|B)$, the more confident we can be that the true value of this force is indeed 2.

This new interpretation leads to an entire class of probability theory: Bayesian data analysis. This differs from the frequentist approach we have used so far[47]. One of the strengths of Bayesian data analysis is that we can do more with less direct data, provided we know other information about the system[48]. This makes Bayesian statistics quite a powerful tool. We won't get into Bayesian statistics here that is the subject of many books on its own[19]. The important part is knowing that it exists and that there is more than one way to interpret probabilities, and that once again: context is everything.

9.11 THE PROSECUTOR'S FALLACY

Conditional probabilities may seem almost academic in their study, but they are[49] the basis of most misunderstandings of probability and statistics. These misunderstandings can be intentional or unintentional[50]; right now I'm going to intentionally mislead you using a real-life criminal

[46] Along with our lived experience and "less tangible" observations

[47] And will continue to use in the next sections and chapters

[48] Say, from experience, wisdom, or other sources

[49] Arguably

[50] I like to believe that *most* people are good, *most* people want to do the right thing, and *most* of the time these understandings come from a misapplication of good intentions.

case. I will note that the following few examples come with a **content warning** as these are criminal proceedings[51].

Content Warning: violent crime

In 1994, Troy Brown of Carlin Nevada was charged and convicted of the rape of a young girl[21]. The key piece of evidence was a DNA match between a sample found at the crime scene and a sample taken from Troy Brown Out of interest, the rest of the evidence was mostly circumstantial at best: Troy had been drinking the night of the crime, Troy lived near the victim, etc.. The prosecution claimed that there is a 1 in $3,000,000$ chance that a random person would match the DNA profile found at the crime scene. The implication then was that there was a 1 in $3,000,000$ that Troy Brown is innocent. Makes sense, right? Let's look at this mathematically.

The probability of Troy Brown's DNA matching the crime scene DNA[52], *given that Troy Baker is innocent*, I, as presented by the prosecution in this case, is

$$\begin{aligned} P(E|I) &= \frac{1}{3000000} \\ &= 0.0000003333 \end{aligned}$$

But what is probability that Troy Brown is innocent given the evidence presented? To calculate this, we need to apply Bayes' theorem.

$$\begin{aligned} P(I|E) &= \frac{P(E|I)P(I)}{P(E)} \\ &= P(E|I)\frac{P(I)}{P(E)} \end{aligned}$$

We rewrote Bayes' theorem like this because it highlights where our conditional probabilities differ. $P(I)$ is the probability of innocence given all other evidence excluding the DNA results, and $P(E)$ is the probability that we would get a match independent of guilt or innocence[53]. In the case of a large $P(I)$ and small $P(E)$, this can lead $P(I|E)$ to be much, much larger than $P(E|I)$. In fact, any time

$$\frac{P(I)}{P(E)} > 1$$

[51]There is an example that doesn't deal with real crimes after this quick foray into true crime.

[52]i.e. the evidence, E

[53]Another way to think of this is if we chose two random people, what are the chances their DNA would match?

we are in the realm of the prosecutor's fallacy because the probability of innocence given the evidence will be *higher* than the probability of the evidence given innocence.

In the case of Troy Brown, he also had four brothers. It was noted during the trial that if one of the brothers was in fact the guilty party, the probability of seeing the evidence would be roughly 1 in 66. Again, if the other evidence is equally likely to apply to any of Troy's brothers, then the probability of innocence is $P(I) = 1/5$. This means that we can approximate

$$\begin{aligned}\frac{P(I)}{P(E)} &\approx \frac{1/5}{1/66} \\ &= 13.2\end{aligned}$$

This means the probability of innocence given the evidence is 13 times higher than the probability of observing the evidence given innocence.

Now, you may say *well* 1 *in* 3,000,000 *and* 1 *in* 300,000 *are both pretty much a sure thing*, but I challenge you to question whether an error like this should be allowed in court. In this admittedly overly simplified analysis of a legal case that has drawn a lot of attention, you may say that we could approximate $P(E|I) \approx P(I|E)$ without affecting the outcome of a legal case, but it's important to remember that this statistic was used to convict Troy Baker of life in prison.

In this case, it may not have made much difference, but in other cases it can. We will look at one case of the complement to this called the *defender's fallacy*[54].

In 1994, Nicole Brown and her friend Ronald Goldman were murdered outside of Brown's condo in Los Angeles. O.J. Simpson[55] was charged with the murders of Simpson and Brown[20]. Due to Simpson's fame, the case was highly publicized and drew incredible media attention. His defence team was built of some of the highest profile lawyers of the time. The trial team was eventually led by Johnnie Cochran and included a team of six other lawyers, including Robert Kardashian[56].

One of the pieces[57] of evidence that the prosecution presented to the court was that Simpson had a history of domestic abuse against Brown when they were married. The defence dismissed this as irrelevant stating that only 1 in 2500 cases of domestic violence result in murder; there is

[54] Mathematically, it's the same.

[55] Famous football player, actor, and Simpson's ex-husband

[56] The father of Kim, Khloé, and Kourtney

[57] There were many, many very convincing pieces.

only a 0.04% chance! Is this an honest argument? Let's do some math and find out!

The probability given is the probability of Brown being the victim of domestic violence and murder, D, given the perpetrator being the spouse[58], G, is

$$P(D|G) = 0.0004$$

But we have more information than just *Nicole Brown was a victim of domestic violence*; we also know she was in fact murdered. So the question we should ask is, *what is the probability that the victim of domestic violence and murder was murdered by her spouse?* We can frame this as a conditional probability; what is the probability that the spouse is the guilty party given that the victim was abused *and* murdered?

$$P(G|D) = \frac{P(D|G)P(G)}{P(D)}$$

We now have another situation as above. The probability $P(G)$ is the probability of guilt given all other evidence[59], and $P(D)$ is the probability seeing this evidence[60] regardless of guilt or innocence. In this instance, we can appeal to other data from other cases to estimate $P(D) = 0.00045$. This given the probability of Simpson's guilt *given that Brown was the victim of domestic violence and murder* to be

$$P(G|D) = 0.889P(G)$$

The contribution to guilt is 0.889 or roughly 8 in 9, not 1 in 2500 like the defence claimed. Given the frequency with which the spouse is implicated in domestic violence that results in murder, this piece of evidence is highly compelling, not irrelevant as the defence claimed. In other words, while domestic violence rarely ends in murder; when it *does* end in murder[61], it is more often than not committed by the spouse.

An example without a content warning

In 1968, Malcolm Collins and his wife Janet Louise Collins were convicted of committing a robbery[5]. The key piece of evidence against them was two witnesses who claimed the following:

---

[58] i.e. Simpson being guilty

[59] There was a lot in this case.

[60] A victim of domestic violence also being the victim or murder

[61] In this case, we know it did.

- They saw a blonde, white woman with a pony tail run from the scene into a yellow car.
- The yellow car was being driven by a Black man with a beard and moustache.

The prosecutor then claimed the following using "estimates" for the probabilities involved[62].

- The probability of a being a Black man with a beard: $p_B = 1/10$
- The probability of having a moustache: $p_M = 1/4$
- The probability of being a woman with blonde hair[63]: $p_W = 1/3$
- The probability of wearing your hair in a ponytail: $p_T = 1/10$
- The probability of being an interracial couple in a car: $p_I = 1/1000$
- The probability of having a yellow car: $p_Y = 1/10$

The prosecutor then argued that all of these things had to happen together. What is the probability of two random people satisfying all these conditions? In other words, if we take the events to be the subscripts in the above probabilities, what is probability of $B \cap M \cap W \cap T \cap I\ capY$? The prosecution said we just need to multiply these probabilities together so that

$$P(B \cap M \cap W \cap T \cap I\ capY) = p_B p_M p_W p_T p_I p_Y = \frac{1}{12000000}$$

the chances are exceedingly rare of matching all these conditions, the prosecution argued, and therefore it is exceedingly rare that the Collins couple were innocent!

There are two glaring issues with this argument. The first is that these probabilities are **not** independent. We cannot just multiply them together. The probability of a man having a beard may be 0.1, and the probability of a random man having a moustache may indeed be 0.25, but what is the probability of a man having a moustache *given that he has a beard*? Surely, it is much closer to 1 than 0.25. The correct approach

[62] It seems, in hindsight, that many of these estimates were just full-on fabrications.

[63] Presumably, the prosecutor in this case thought there were three hair colours and they were equally distributed among white people. This in itself is suspect.

would be to use the conditional probability definition to calculate the probability of $B \cap M$.

$$P(B \cap M) = P(M|B)P(B)$$
$$P(B \cap M) = P(M|B)p_B$$

The same argument, perhaps more obviously, could be used to nullify the rarest of the probabilities put forth by the prosecution: if you have observed a Black man and a white woman together, that couple is by default inter-racial. While $P(I)$ might indeed be 1 in 1000, $P(I|B \cap W) = 1$. Multiplying by $1/1000$ is the most egregious example here and alone makes the probability 1000 times less likely than it should be. A probability of $1/12000$ is certainly far less impressive than $1/12000000$[11].

Even if all of these probabilities were correct, and even if these traits were all independent, the prosecutor still fell into the trap of equating the fact that $P(E|I)$ is the same as $P(I|E)$; that is, the probability of observing the evidence given innocence is the same as the probability of innocence given the evidence. In this case, the probability of observing the evidence at all independent of guilt or innocence, $P(E)$, is very small but considering this eyewitness testimony seems to have been the only evidence, the probability of innocence without the testimony, $P(I)$, was quite high. This means that $P(I)/P(E)$ would be a very big number and thus

$$P(I|E) = P(E|I)\frac{P(I)}{P(E)}$$

will be very different.

In this case, it is important to note that the probability of $\frac{1}{12000000}$ was provided by an expert witness who was also a professor of mathematics[11]. Everyone is susceptible to making these mistakes, of failing to consider dependence, or ignoring conditional probabilities. It's important to **always** think critically when doing statistics.

9.12 THE LAW OF TOTAL PROBABILITY

Canada is broken up into ten provinces and three territories. Population estimates for these regions are given in table 9.1. If we wanted to determine the population of Canada from this information, how might we proceed? Hopefully it's pretty clear that we should add up the numbers to get the total.

Table 9.1: Populations for the ten provinces and three territories of Canada.

Province	Population
NL	520194
PE	161402
NS	981552
NB	783257
QC	8579476
ON	14740704
MB	1380447
SK	1178164
AB	4424557
BC	5156587
YT	42300
NT	45265
NU	39109

Probabilities work the same way. If we have a set of events that together make up our entire sample space

$$\Omega = B_1 \cup B_2 \cup \cdots B_N$$

and only know probabilities of event A given event B_i for each $1 \leq i \leq N$, we could find $P(A)$ so long as we also know the probabilities for each of the B_i events. This is called the law of total probability.

Definition 9.12.1. The **law of total probability** states that if there is a set of events B_n that together make up our entire sample space Ω, then the probability of event A occurring can be computed as

$$P(A) = \sum_{n=1}^{N} P(A|B_n)P(B_n)$$

Using Bayes' theorem, this can be rewritten as

$$P(A) = \sum_{n=1}^{N} P(A \cap B_n)$$

The law of total probability is extremely useful because often times it is difficult to measure probabilities that *aren't* conditional. As we saw with the examples given of the prosecutor's fallacy, stated probabilities often have some context behind them and some information is often already known; to then get the raw probability requires the use of the law of total probability.

As an example, let's revisit the prosecutor's fallacy. Say a piece of evidence has a probability of 0.999 of showing guilt. This gives us a probability.

$$P(E|G) = 0.99999$$

The probability of seeing this evidence given guilt is quite high, but it's not the probability we are interested in. We are interested in $P(G|E)$, the probability of guilt given we see the evidence[64]. We use Bayes' theorem to show

$$P(G|E) = \frac{P(E|G)P(G)}{P(E)}$$

In this case, $P(E)$ and $P(G)$ may be difficult to quantify. The probability $P(E)$ is the probability of seeing the evidence at all. Let's say it is in fact DNA evidence; then $P(E)$ is the probability of finding DNA at a crime scene. This becomes more interpretable if we can rewrite it as a total probability.

$$P(E) = P(E|G)P(G) + P(E|G^C)P(G^C)$$

This tells us that the probability of finding that DNA evidence at the crime scene is the sum of finding the evidence given the defendant is guilty plus the probability of finding the evidence given the defendant is *not* guilty. We know the first. Since the defendant is either guilty or not guilty, we can rewrite $P(G^C) = 1 - P(G)$. While this may still look confusing, these probabilities are much easier to compute.

The probability $P(E|G)$ is the true positive rate for a DNA test. It's the probability that the DNA test did its job and does indeed implicate the actual criminal. The probability $P(E|G^C)$ on the other hand is the false positive rate; it's the probability that you find a DNA match with an innocent person.

[64]Since we know we are seeing the evidence, presumably it is in front of us and its existence a sure thing

Pause. Take a beat. It is very, very likely here that you will assume that $P(E|G^C) = 1 - P(E|G)$ that the true positive rate and the false positive rate must together add to 1. This is a very easy mistake to make, but it is not true, not in the slightest[65].

It might help to think about it in a slightly different context than DNA evidence. Consider a very hungry dog who just loves beef. If there is beef, he will always sniff it out and eat it. The true positive rate for this puppy is one; they are always able to find beef if it is in front of them. On the other hand, it is indeed just a puppy; it will eat anything. It also thinks that chicken, pork, and lamb are beef. The false positive rate is *also* one because it is identifying *everything* as beef.

Back to our example with our crime scene DNA, we know the true positive rate: 0.99999. If it is indeed a match, we will almost definitely pick it up. The false positive rate for a DNA test can be as high as 0.01. Plugging these into our example, we get

$$P(G|E) = \frac{0.99999P(G)}{0.99999P(G) + 0.01(1 - P(G))}$$

We see that the usefulness of our DNA evidence is highly dependent on the probability of guilt determined through other means. If this crime occurred in New York City and there was no other evidence pointing to the suspect, we could argue that they are as likely as anyone else in the city to have committed the crime. If $P(G) = 1/8415000$, then

$$P(G|E) = 0.00001$$

There is less than tenth of a percent chance the defendant is guilty. This is a far cry from the 0.99999 touted by our pretend prosecution!

[65]The exercises will shed some light on this issue.

CHAPTER 10

Probability Distributions

We saw in Chapter 7 how we can use the empirical rule as a quick test to see if our data follows a normal distribution. We also saw in Chapter 8 how we can build histograms to visualize the distribution of a particular measurement.

In almost all cases, data we collect is a sample, but often what we want to do is say something about the population. If our sample is taken properly[1], we can treat the sample as an approximation of our population.

We have specific mathematical formulae to express idealized distributions. These formulae are well studied, and a lot of work has been put into understanding which applications each is best suited for, and how to best interpret the probabilities that are the output of the mathematical expressions. For instance, one of the most desirable properties of a normal distribution[2] is that the mean and standard deviation are independent of one another. In other words, changing the mean[3] does not change the standard deviation[4]. This property allows us to draw many, many conclusions about data that follows a normal distribution that we, generally, can't achieve for other distributions, and leads to many of the test and tools used in hypothesis testing[5].

We use our sample data as an approximation to the underlying mathematical distribution, which is used to approximate the way the population behaves. In this chapter, we will discuss some common distributions, their most common interpretations, and how we might use them to answer questions about data sets.

[1] A random sample with a proper representation of our population

[2] Also known as the bell curve

[3] Sliding the curve around the x-axis

[4] As a percentage of the mean

[5] See Chapter 11

DOI: 10.1201/9781003265405-10

Table 10.1: Table of exam letter grades.

F	A+	D-	C	B-	D+
A-	A+	D	C	F	F
F	D-	C-	B-	F	F
F	F	F	C+	F	C
D-	F	F	D-	C-	B-
B-	F	F	F	A+	F
D-	B-	C	B	F	B-
F	A-	B-	F	D+	A-
F	C	F	F	C	F
D-	F	C	A	A	F
D+	A-	F	F	F	F
B	C	F	D-	F	F
F	F	B	A-	B+	D-
F	C	A	B-	A-	F
B-	B	F	F	C-	D-

10.1 DISCRETE PROBABILITY DISTRIBUTIONS

Table 10.1 shows some sample grade data from two final exams all mixed up together and anonymized. We can calculate some probabilities from this data, like *what is the probability that a student chosen at random from this list got a B?*

$$P(x = B) = \frac{4}{90} \approx 0.044$$

Or we can calculate the probability that a student chosen at random failed[6]

$$P(x = F) = \frac{37}{90} \approx 0.411$$

We could do this for every possible letter grade in our sample set, $\Omega = \{F, D-, D, D+, C-, C, C+, B-, B, B+, A-, A, A+\}$, and plot the inputs versus the outputs. Do we know of a plot that shows us the count of a set of discrete data? Of course we do! It's our friend the histogram!

In order to use this data in Python, we would need it to be in a verb+csv+ file, and written as a single column with a header. Say the data was stored in `s_grades.csv` and the column was called `Letter Grade`.

Using seaborn, we can easily tell our histogram to plot the probabilities of a categorical or discrete variable on the y-axis instead of the count:

[6] These were notoriously difficult exams.

```
student_grades = pd.read_csv('s_grades.csv')
sns.histplot(data=student_grades,
             x='Letter Grade', stat='probability')
```

This will rescale the y-axis to give us probabilities.

This is an example of a **discrete probability distribution** or otherwise known as a **probability mass function**. Notice that if we sum up the *heights* of all the bars, we get a value of 1. This makes sense because each student must receive a letter grade. The probability of a student receiving an F or D- or D or D+ or any other grade is a certainty. Mathematically we can write

$$\begin{aligned} P(F \cup D- \cup D \cup D+ \cup \cdots \cup A- \cup A \cup A+) &= P(x = F) + P(x = D-) \\ &\quad + \cdots + P(x = A+) \\ &= 1. \end{aligned}$$

In fact, this is the main property of a probability mass function.

The stat `probability` in Python ensures that the sum of all the bars in our histogram add to 1. In other words, if we pick a random data point within our data set it definitely falls into one of the bins in our data set[7].

Definition 10.1.1. A **probability mass function** is a special type of function that takes *discrete* inputs and returns a number between 0 and 1 for each input (i.e. a probability). In order for a function to be a probability mass function, the following must hold *in addition to* the input being discrete:

- The sum of all probabilities for all possible inputs must be 1.
- The probability for each input must be between 0 and 1.

A probability mass function describes probabilities for **categorical** data or **discrete numerical** data. The letter grades above are one example, but these letter grades come from numerical grades, given in table below. Is this also discrete? By the nature of how things are graded, yes it is discrete. There were 100 marks to be earned on each exam, and

[7]This is a tautology: if we pick a data point from our data set, there is a 100% chance that the data came from the data set from which we just picked it.

Table 10.2: Table of numerical exam grades

30.0	91.5	55.0	63.0	70.5	57.0
81.5	96.0	55.5	65.0	49.0	3.0
9.0	54.0	62.0	71.0	52.0	26.0
48.5	24.0	25.0	68.0	12.0	63.5
51.0	41.0	32.0	51.0	60.0	71.5
71.5	34.5	49.5	14.5	90.0	43.0
52.0	71.5	67.0	75.0	22.0	70.5
35.0	83.0	71.5	30.0	59.0	83.0
31.0	67.0	43.0	37.0	65.0	38.0
51.0	34.0	67.5	89.0	88.0	14.0
58.0	82.0	48.0	46.5	8.0	22.0
73.0	64.5	46.0	52.0	36.0	47.0
34.0	46.0	75.0	83.5	79.5	50.0
31.0	67.0	85.0	70.0	81.0	40.5
72.0	73.0	21.0	36.0	60.0	51.5

I gave out half marks so it would be impossible for a student to receive a grade of, say, 72.25 on the exam. All numbers are not possible, so it is discrete but as we will see later, it's often desirable to *approximate* numbers like this as continuous.

10.2 THE BINOMIAL DISTRIBUTION

The most common discrete probability distribution is the **binomial distribution**. The typical interpretation is that given the probability of an event A happening once, denoted by p, the binomial distribution is the probability that in n events, called trials, A happens k times[8], provided each event is independent. This last clause is important: it means that the results of the second trial do not depend on the results of the first trial. The event has no memory of what came before and no foresight of what will happen after.

For instance, we know that if we flip a coin there is a $1/2$ chance that it lands heads. If event A is our coin landing heads, we can use the binomial distribution to answer the question, *what is the probability of seeing eight heads if we flip a coin ten times?*

[8] And as a result doesn't happen $n - k$ times

Does this satisfy our independence condition? Yes! A coin has no memory of the results of previous flips. A coin being flipped heads 100 times in a row does not guarantee or increase the chance of the 101st flip being tails. The coin does not care what happened before and does not care what will happen after. The coin lives in the moment; the coin is very zen.

Belief in the fact that the coin *does* have some kind of memory, or equivalently that the universe has some grand design that compels the 101st flip to be tails after 100 heads, is known as the gambler's fallacy. It is a trap to think that independent events *must* act according to past events. However improbable it may be to observe 100 heads in a row, the next flip always has a probability of $1/2$.

Definition 10.2.1. The **binomial distribution** is a discrete probability distribution[9] defined as

$$B(k; n, p) = \frac{n!}{(n-k)!k!} p^k (1-p)^{n-k}.$$

We can read this as the union of n events. If we list our events as A or $\texttt{not}A$, we can write the binomial distribution for k out of n events being A as

$$\underbrace{A \textbf{ and } A \textbf{ and } A \textbf{ and } A \cdots \textbf{and} A}_{k\,\texttt{times}} \textbf{ and } \underbrace{\texttt{not}A \textbf{ and } \texttt{not}A \textbf{ and} \cdots \texttt{not}A}_{n-k\,\texttt{times}}$$

Using this list, we can see that probability p happens k times and probability $1-p$ happens $n-k$ times. We know that when dealing with independent events, we can multiply probabilities when we see the word **and**. This is how we get the factors p^k and $(1-p)^{n-k}$. Where does the factor in front come from?

Notice that the binomial distribution tells us nothing about the *order* of the k events A. In 500 coin flips, the binomial distribution evaluated at k tells us how many times we can expect to see k heads in those 500 flips. It does *not* tell us whether those events happen in a row, if they are the first k flips or the last k flips, if they happen randomly throughout the 500 flips, or if $k/2$ happens at the beginning and the other

[9] i.e. a probability mass function

$k/2$ happens at the end. The order *doesn't matter*. We multiply by all the possible combinations of how we can arrange 100 heads in 500 coin flips. This factor is exactly ${}_{500}C_{100}$. Another way to think of this is we have 500 positions, and we need to select 100 of them to label as heads. Generalizing from 100 heads in 500 flips, we can say that we have ${}_nC_k$ ways of arranging k heads in n positions.

The binomial distribution is a function of k. The **domain** of this function is $k \in [0, n]$. If we plot all values for the binomial distribution, we get a distribution that looks like the one in figure 10.1, which gives the probability of seeing k heads in 500 coin flips.

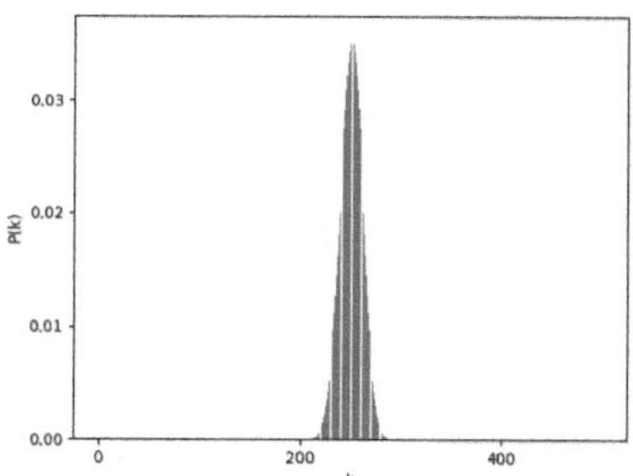

Figure 10.1: The binomial distribution for the number of heads, k, in 500 *fair coin flips*

An example that isn't coins (or dice)

Creutzfeldt-Jakob disease (CJD) is a brain disorder that is caused by prions[10]. It is interesting as a disease because it can occur spontaneously in a human. The probability of developing CJD spontaneously in a human per year is estimated at 0.000015% [12]. It is usually fatal within one year of symptom onset. It can also be transmitted during surgery if there is equipment contaminated with CJD-associated prions.

Let's look at a population of 140,000 people and say that we see two cases of CJD in this population appear at roughly the same time. What is the probability that these two cases that appeared in the same year are unrelated?

[10]Misfolded proteins that encourage other proteins to also misfold; in case the *disease* part of the name didn't make it clear: misfolded proteins are not good for the brain.

We are looking for a crude probability of event A, developing spontaneous CJD, happening twice in 140,000 trials. We can use the binomial distribution with $k = 2$, $n = 140000$, and $p = 0.00000015$.

$$\begin{aligned} B(2; 140000, 0.00000015) &= \frac{140000!}{139998!2!} 0.00000015^2 (1 - 0.00000015)^{139998} \\ &= 0.00022 \end{aligned}$$

What we can gather from this is that it's pretty rare to see two cases in such a small population[11] in one year. That's not to say it *can't* be the case just suggests that perhaps we should look at the possibility that some surgical equipment[12] might be contaminated.

In fact, with this probability we can ask *another* question:

> What is the probability that in 50 years, we see two cases in one year once?

In this case, $p = 0.00022$, $n = 50$, and $k = 1$ and the probability is

$$B(1; 50, 0.00022) = \frac{50!}{49!}(0.00022)^1(1 - 0.00022)^{49} \approx 0.011.$$

This is still really small! Again, this doesn't mean it couldn't have happened, just that we may want to look into other causes before dismissing the two cases as spontaneous.

Usually when speaking of the interpretations of the binomial distribution, you will hear the term k success in n trials. The word success here is used as loosely as possible. The binomial distribution gives estimates of probabilities when there is at most two outcomes. The outcome that is desirable in our current context is called the success[13]. Sometimes the word success isn't the best choice, as in the above example. It's rather callous to call a degenerative neurological disease a success. The key is that the events must be independent, and there must be two outcomes[14] A and A^C which exhaust the entirety of our sample space.

A note on notation

If for any reason you skipped Chapter 1 of this book[15], you may be wondering why there is a semicolon in $B(k; n, p)$. This is to separate the

[11] Relatively small

[12] Say, scalpels used in cornea surgery

[13] I think of this as *we successfully observed what we wanted.*

[14] Something either happens or it doesn't.

[15] I suggest you go back and read it; it's fun.

variable of the distribution, k, which we don't know and we can plug in from the *parameters*. The numbers n and p are considered parameters because we assume that they are fixed or at least knowable given the context of what we are applying the binomial distribution to.

Depending on the other books you may read, sometimes the variable is *implied*. Once we become familiar and comfortable with the binomial distribution, we will know that it takes an integer number from 0 to n as its input variable. As such, you may sometimes see the binomial distribution[16] written with *just* the parameters as inputs. You may see the binomial distribution defined as

$$B(n,p) = \frac{n!}{(n-k)!k!}p^k(1-p)^{n-k}.$$

This doesn't mean k is fixed, but means that the implication is that k is varied. We will *never* write distributions in this way because it is less clear, and since we are presumably learning this for the first time, we should be as clear as possible.

10.3 TRINOMIAL DISTRIBUTION

Using what we know about the binomial distribution and how it was constructed, we can extend the idea to account for *two* distinct, mutually exclusive[17] events, A and B, happening or neither happening. Let's say we have event A that occurs with probability p_A and event B that occurs with probability p_B. What is the probability that A happens i times and B happens j times in n trials?

Writing this in words, a single set of trails might look like

$$\underbrace{A \textbf{ and } A \cdots \textbf{and } A}_{i\,\texttt{times}} \textbf{ and } \underbrace{B \textbf{ and } B \cdots \textbf{ and } B}_{j\,\texttt{times}} \textbf{ and } \underbrace{\texttt{not } (A \text{ or } B) \textbf{ and}}_{n-i-j\,\texttt{times}} \cdots$$

Since, like the binomial distribution, we assume the individual trials are all independent, we can just the individual probabilities together. This means that the probability of A happening i times is p_A^i, the probability of B happening j times is p_B^j, and the probability that neither happens $n-i-j$ times is $(1-p_A-p_B)^{n-i-j}$, and all three of these things happening has probability $p_A^i p_B^j (1-p_A-p_B)^{n-i-j}$. Once again, we have to acknowledge that we are not asking these events to

[16] Or any probability distribution for that matter

[17] In each trial, A can happen or B can happen, but A and B can't happen at the same time.

happen in any particular order; therefore we have to account for all possible orderings. In this case, we need to choose i spots out of our n and fill them with event A; this is exactly ${}_nC_i$. Then, from the remaining $n-i$ positions, we choose j of them to fill with event B, which gives us ${}_{n-i}C_j$. Since both of these need to happen, ${}_nC_i$ **and** ${}_{n-i}C_j$, we multiply the two numbers. This gives a coefficient of

$$\begin{aligned} {}_nC_{i\,n-i}C_j &= \frac{n!}{(n-i)!i!}\frac{(n-i)!}{(n-i-j)!j!} \\ &= \frac{n!}{i!j!(n-i-j)!}. \end{aligned}$$

Using this we can define the **trinomial distribution** as

$$T(i,j;n,p_A,p_B) = \frac{n!}{i!j!(n-i-j)!}p_A^i p_B^j \left(1-p_A-p_B\right)^{n-i-j}.$$

IN CHESS, THERE ARE THREE OUTCOMES TO A GAME: white wins, black wins, or there is a draw. These probabilities are not equal! According to chessgames.com, a website that keeps track of all reported chess games, white has an advantage because they historically play first. According to the database, white wins roughly $p_W = 37\%$ of the time and black wins $p_B = 28\%$ of the time. The rest of the games result in a draw.

We might ask the question, *if two players play five games, without switching colours, what is the probability that they each win two games?*

In this case, $i=2$, $j=2$, and the probabilities are as above. The probability of each winning two games is given by

$$\begin{aligned} T(2,2;5,0.37,0.28) &= \frac{5!}{2!2!1!}\left(0.37\right)^2\left(0.28\right)^2\left(1-0.37-0.28\right)^1 \\ &\approx 0.11. \end{aligned}$$

There's 11% chance that white and black each win two games and draw the last one. What are the chances that white wins two games, black wins one game, and they draw twice?

$$\begin{aligned} T(2,1;5,0.37,0.28) &= \frac{5!}{2!1!2!}\left(0.37\right)^2\left(0.28\right)^1\left(1-0.37-0.28\right)^2 \\ &\approx 0.14 \end{aligned}$$

White clearly has an advantage in a multigame scenario as well!

10.4 CUMULATIVE PROBABILITY DISTRIBUTIONS

So far, we have looked at individual scenarios. What is the probability of seeing 100 heads in 500 coin flips? What is the probability black wins exactly one game and white wins exactly two games out of five?

These are not the only questions we can ask when talking about probability.

Roulette: Why the house always wins

One of the ways to play roulette is to bet on a colour: red or black. The issue comes when you look at a roulette wheel, like the one I painstakingly drew in figure 10.2, and notice that not all the possible spaces are red or black. Two of them are green!

Figure 10.2: This is an American roulette wheel; a ball will travel around this wheel and land randomly on a number. Each number is also assigned a colour, which we can bet on.

We will keep things simple and work with a roulette wheel with only 1 green spot. There are 13 red spots, 13 black spots, and 1[18] green spot. The probability of the ball landing on a red space is $p_R = 18/37$, the probability of landing on a black space is $p_B = 18/37$, and the probability of the ball landing on the green space is $p_G = 1/37$. If you bet on red or black and win, you double your money[19].

Let's say you play 50 games and bet the same amount of money every time and always bet on red. In order to break even[20], you would need to win half your games. This is simply an application of the binomial distribution. The probability of winning is $p_W = 18/37$, and the probability

---

[18]Sometimes two: a roulette wheel with two green spots is called American Roulette; it just gives the house an even greater example.

[19]In casino talk, we would say the payout is 1:1; you bet one and if you win, you get your original bet back and get that amount from the dealer as well.

[20]End up with the same amount of money you started with after 50 games

of losing is 19/37 because of the green space. The probability of breaking even is the case where we win exactly 25 games, or

$$B\left(25; 50, \frac{18}{37}\right) = \frac{50!}{25!25!}\left(\frac{18}{37}\right)^2 5 \left(\frac{19}{37}\right)^2 5$$
$$\approx 0.11.$$

There's an 11% chance that after 50 games we have exactly as much money as we started with.

What about the probability that we have *less* money than we started with? That can happen in multiple ways. If we bet \$1 and win, we have \$2. If we then lose the next 49 games, we would be down \$48. If we win two games and then lose the next 48 games, we will be down \$47. In fact, if we lose more games than we win, we will always end up with less money than we started with[21]. We lose money if we win one game **or** two games **or** three games or $\cdots$ or 24 games. Since each spin of the roulette wheel is independent, we can add these probabilities together to get the probability that we lose money. Let's call this probability P_{loss}.

$$P_{loss} = B\left(0; 50, \frac{18}{37}\right) + B\left(1; 50, \frac{18}{37}\right) + B\left(2; 50, \frac{18}{37}\right) + \cdots$$
$$+ B\left(24; 50, \frac{18}{37}\right)$$

The probability that we lose money is then

$$P_{loss} \approx 0.52.$$

The majority of the time we will lose money[22].

THIS PROBABILITY OF LOSS IS AN EXAMPLE OF A CUMULATIVE PROBABILITY. It is the probability not of $k = x$, but of $k < x$. It is the probability that our event A occurs *up to* x times instead of *exactly* x times. When dealing with a probability mass function, the associated cumulative probability function is the sum from 0 up to x of the associated probability mass function.

[21] This should make intuitive sense; losing more than winning results in losing.

[22] It's a very small majority, but a majority none the less. Even with this small 2% advantage, a casino stands to make a lot of money over time.

Definition 10.4.1. The **cumulative probability function**, $C(x)$ of a probability mass function, $P(k)$, is given as

$$C(x) = P(k < x) = \sum_{i=0}^{x} P(i).$$

This gives the probability *up to* x instead *exactly* x.

For the binomial distribution, we define the associated cumulative probability function as

$$C_{binomial}(x) = B(k < x; n, p) = \sum_{k=0}^{x} B(k; n, p).$$

We can also use the concept of the cumulative probability on the trinomial distribution.

For our chess example, we can ask, what is the probability white wins *up to* three games? Since the trinomial distribution takes two input variables, we have to account for many more cases. We can start to list all the possibilities in a table.

White	Black	Draw	White	Black	Draw	White	Black	Draw
0	5	0	1	4	0	2	3	0
0	4	1	1	3	1	2	2	1
0	3	2	1	2	2	2	1	2
0	2	3	1	1	3	2	0	3
0	1	4	1	0	4			
0	0	5						

Each of these scenarios has a probability associated with it, and by adding all of these probabilities, we can get the probability that white wins *up to* three games. We can write this in summation notation as

$$C = \sum_{i=0}^{2} \sum_{j=0}^{5-i} T\left(i, j; 5, p_W, p_B\right).$$

WITH CUMULATIVE PROBABILITY DISTRIBUTIONS, IT OFTEN HELPS TO REMEMBER THAT THE TOTAL PROBABILITIES MUST SUM TO ONE; there is a 100% chance that *something* happens. This can help us when things seemingly get unwieldy. For instance, what is the probability that black wins at least one game?

White	Black	Draw
0	5	0
0	4	1
0	3	2
0	2	3
0	1	4
1	4	0
1	3	1
1	2	2
1	1	4
2	3	0
2	2	1
2	1	2
3	2	0
3	1	1
4	1	0

We can make a table of all the possibilities: then we can compute the individual probabilities of all of these and add them up.

Alternatively, we can recognize that black winning at least one game is the complement of black winning no games at all. In that case, there are far fewer scenarios we need to account for:

White	Black	Draw
5	0	0
4	0	1
3	0	2
2	0	3
1	0	4
0	0	5

The probability black wins 0 games is thus given as

$$
\begin{aligned}
C_{B=0} &= \sum_{i=0}^{5} T\left(i, 0; 5, p_W, p_B\right) \\
C_{B=0} &= \frac{5!}{0!0!5!} 0.37^0 0.28^0 \left(1 - 0.37 - 0.28\right)^5 \\
&\quad + \frac{5!}{1!0!4!} 0.37^1 0.28^0 \left(1 - 0.37 - 0.28\right)^4 \\
&\quad + \frac{5!}{2!0!3!} 0.37^2 0.28^0 \left(1 - 0.37 - 0.28\right)^3 \\
&\quad + \frac{5!}{3!0!2!} 0.37^3 0.28^0 \left(1 - 0.37 - 0.28\right)^2 \\
&\quad + \frac{5!}{4!0!1!} 0.37^4 0.28^0 \left(1 - 0.37 - 0.28\right)^1 \\
&\quad + \frac{5!}{5!0!0!} 0.37^5 0.28^0 \left(1 - 0.37 - 0.28\right)^0 \\
&\approx 0.193.
\end{aligned}
$$

The probability that black wins at least one game is the same as the probability that black *doesn't* win no games[23]. Therefore,

$$
C_{B\geq 1} = 1 - C_{B=0} \qquad = 0.807.
$$

The 1 represents the probability that something, anything happens; there's a 100% chance that an outcome will occur. The subtraction precedes the probability of events that *don't* happen. It is our mathematical equivalent of the word *not*, and $C_{B=0}$ is the probability that black wins no games.

10.5 CONTINUOUS PROBABILITY

Even though a lot of data that we deal will be discrete, usually due to measurement limitations, as stated before, it's often useful to approximate otherwise discrete data as continuous. With the test data above, we can approximate the data as continuous and this gives us a whole new set of tools that we can use; particularly, continuous probability distributions.

Much like their discrete counterparts, continuous probability distributions describe how different outcomes are related to each other, but there is one significant conceptual difference between discrete and continuous probability distributions.

[23]The language is kind of awkward for the sake of consistency and, weirdly, clarity.

If you recall[24], the value given by a discrete probability mass function gives the probability that the specific outcome X happens *exactly*. When we looked at the binomial distribution for roulette, the probability that we win exactly one game out of two played games is given by

$$B\left(1; 2, \frac{18}{37}\right) = \frac{2!}{1!1!}\left(\frac{18}{37}\right)^1\left(\frac{19}{37}\right)^1$$
$$\approx 0.4996.$$

This is quantifiable.

When it comes to continuous numbers, things get less intuitive because we must introduce concepts of infinity. Infinity can often lead to strange behaviours.

Think, for instance, the ratio of the length of your index finger[25] and your ring finger[26]. You may never have thought about it, but there's no reason for these two fingers to be the same length, and, as it turns out, they're not. Generally, your ring finger is slightly longer than your index finger. How much shorter your index finger is called the 2D:4D ratio. This number, for all intents and purposes, exists on a continuum.

Let's say we pick a number, $0.944412921093218900000\cdots$, where the 0's carry on infinitely. We can ask, what is the probability that we find a person with *exactly* this 2D:4D ratio? You are probably, correctly, mulling around a word like *unlikely*, or *impossible*[27]. That's a very specific, rather esoteric, number I've written down.

What if we asked the same question: *what is the probability that we find someone with a specific 2D:4D ratio?* But instead we use the specific number 0.9. You may think this is more likely than the number I have written above. It somehow looks or feels nicer. Even if I write it with all the infinitely many zeros, $0.90000000000\cdots$, it still *seems* nicer. It seems more probable, somehow.

The key idea about probability and continuous numbers is to recognize that these numbers that *feel* nicer aren't any different than numbers that feel *too specific*. When we are tasked to find someone with a 2D:4D ratio of 0.9, then 0.9000000000000001 simply won't do, neither will 0.8999999999999999999999909. The number 0.9 is just as specific as $0.944412921093218900000\cdots$.

---

[24] If not, it's just a couple of pages back.

[25] The one next to your thumb

[26] The one next to your pinky finger

[27] Perhaps predicated by a *damn near*

When dealing with continuous numbers, the probability of measuring[28] any one particular number is always 0. Doesn't matter what that number is. There are infinitely many numbers in a continuous range, and they are all equally unlikely to be picked exactly.

In math terms, this means that if we have a probability function $P(X)$, where X is continuous, then the probability that $X = y$ is $P(X = y) = 0$. No matter what y is.

10.6 CONTINUOUS VS. DISCRETE PROBABILITY DISTRIBUTIONS

As we saw above a discrete probability distribution, a probability mass function, $P_M(X)$, gives the probability of event X happening and shows how the probabilities of the events are related to each other.

We also saw above that when it comes to continuous values, the probability of a continuous event Y happening is always zero. Does that mean that the probability distribution $P_D(Y) = 0$ no matter what? Of course not.

When discussing continuous events, we need to refine our definition of what it means for something to *happen*. As we've seen above, the probability that something on a continuum happening *exactly* is zero, but the probability of something happening *near* any given number is necessarily not zero.

As a quick example, I am thinking of a number between 0 and 1. Try to guess it, without looking at the answer[29]. The probability that you guessed my exact number (before looking at the answer) is 0. There are just *too many* numbers to choose from. However, if you were to defy the rules of our game and instead state *the number you have chosen is between* 0 *and* 0.5 should you still have a probability zero of being right? No, the probability that your statement[30] is correct is 50%, or 0.5. If I must pick a number between 0 and 1, there is a 50% chance that if you say my number is less than 0.5, that you are correct. The idea that you can pick the number *exactly* is nearly impossible, but the fact that you can give a range of values that might contain my number is very possible. In fact, it's so possible we can measure it.

[28] Or seeing, identifying, etc.

[29] The number is 0.325252.

[30] However rebellious

10.7 PROBABILITY DENSITY FUNCTIONS

The continuous equivalent to a probability mass function is called a **probability density function**. Unlike a probability mass function, the values of a probability density function do not give the probability of an event happening, but instead when compared to another event gives how much more or less probable event Y is than event X.

For instance, consider our picking of a number between 0 and 1. For our purposes, we can say that every value was equally likely to be picked[31]. This means that the probability density function should look as in figure 10.4. Each number has no more or less a chance than any

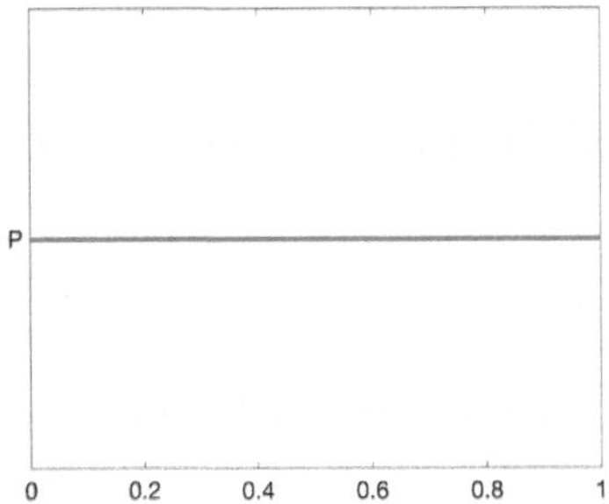

Figure 10.3: The probability of picking any number between 0 and 1 should be equal across all numbers. We don't know what the value of P should be yet, but we know every number should have the same probability.

other. When compared, the value of $P_D(X)$ is the same as $P_D(Y)$; they have the same chance of *occurring* whatever that chance may be.

We can add numbers to the y-axis for our probability density function, P_D, and when we do we get figure 10.4. We can say that the probability density function for picking a random value between 0 and 1 is

$$P_D(x) = \begin{cases} 1 & \texttt{if } 0 \leq x \leq 1 \\ 0 & \texttt{otherwise} \end{cases} .$$

Why did we pick the value of 1? Why can't we use the function, $P_D(x) = 0.3$ or $P_D(x) = 50$? This is because our probability density function must satisfy certain properties. It's these properties that force the value of 1 on the right-hand side of our equation.

[31] In reality, because of page limits, atoms in the universe, and a whole host of psychological phenomena, not all numbers are truly equally probable.

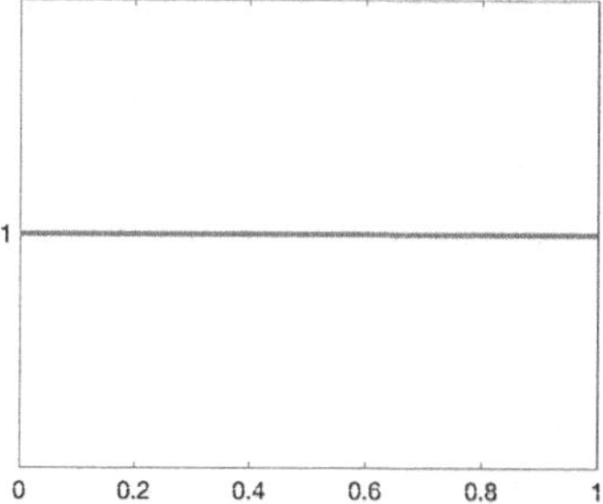

Figure 10.4: The probability of picking any number between 0 *and* 1 *should be equal across all numbers. The value of* P *we must pick is* 1 *so that our probability density function satisfies some conditions.*

The **properties of the probability density function**, $P_D(x)$, are the following:

- $0 \leq P(x) \leq 1$ for any x
- The probability that *something* happens is 1.

The second property is one we've seen before. For a discrete probability mass function, if we have three possible events, X, Y, and Z, then the probability that X or Y or Z happens is $P_M(X) + P_M(Y) + P_M(Z) = 1$ because something *must* happen.

With continuous events, we run into trouble because we have to now add up infinitely many things. If you've read Chapter 6, then you might have an idea of how we add up infinitely many things; if not, I suggest you maybe take a quick look back.

The accumulation of infinitely many small changes, as defined in Chapter 6, is called an **integral**. It is the **area under the probability density function**.

For our random number between 0 and 1 distribution, we had to choose 1 because the probability that we choose a number between 0 and 1 must be 1. Therefore, the area under our rectangle must be 1.

If we wanted to pick a number between 0 and 2 our probability density function will have to change. We will still have a rectangle[32], but the

[32] Assuming that all numbers are equally probable

width of our rectangle is now 2, and we require an area of 1. What should our length, l, be?

$$2l = 1$$
$$l = \frac{1}{2}$$

So if we want to write a probability density function for a number between 0 and 2 it would be

$$P_D(x) = \begin{cases} 1/2 & \texttt{if } 0 \leq x \leq 2 \\ 0 & \texttt{otherwise} \end{cases}$$

This ensures that the area under $P_D(x)$ is 1, as it should be.

With a discrete probability mass function, $P_M(X)$, we would write this condition mathematically as

$$\sum_{i=0}^{n} P(X_i) = 1.$$

When we are dealing with a continuous probability density function, $P_D(x)$, we use the continuous version of the sum called the integral.

$$\int_{-\infty}^{\infty} P_D(x)\mathrm{d}x = 1$$

This says that the *area* under the probability density function must be one: something *must* happen.

This same notion of area under the curve can be used to talk about the probability of ranges of values. We could intuit that if we said *the number that you picked between* 0 *and* 1 *is less than 0.5*, we would be correct 50% of the time. Why is this?

We can think of this statement as an addition of infinitely many individual probabilities. When we say *the number you picked is less than* 0.5, what we are actually saying is *the number you picked is* 0 **or** $0.0000000\cdots1$ **or** 0.00000002 **or**....

On the surface, this looks like we're adding up a bunch of zeroes, because the probability of picking any individual number is 0; but this is one of those times where infinity makes things weird. Adding zero to zero infinitely many times[33] isn't zero. Infinity makes things weird. If this is

[33] As we have to do, because there are *infinitely many* numbers between 0 and 0.5

too far for you, if you are ready to shut this book and say *if you're going to tell me that a lot of nothing eventually ends up being something, then I quit.* I don't blame you. There's a lot hiding under the hood here.

The truth is the probability of picking any individual number out of a continuum is in fact zero *in the limit.* All of this talk of continuous numbers, integrals, only makes sense in terms of limits. The probability is zero *in the limit.* It's so imperceptibly small that it is indistinguishable from 0. So it's not necessarily that we're adding up nothings and getting something. We are adding up imperceptibly small bits, an imperceptible many times. When we combine the very, very small with the very, very big, things get *wild.* This is why we have two different concepts for the discrete[34] and the continuous[35].

SO, BACK TO OUR CONUNDRUM. How do we *know* we have a 50% chance of being right when we say *the number you picked is between* 0 *and* 0.5? If there were a discrete set of numbers between 0 and 0.5, then we would just add up the individual probabilities. Since this isn't the case, we need to use integration. To get the probability that the number is between 0 and 0.5, we take the **integral of the probability density function.**

$$\int_0^{0.5} P_D(x)\mathrm{d}x = 0.5$$

This is itself 0.5 because the area created by the rectangle is exactly 0.5.

Definition 10.7.1. The **probability** of a continuous random variable with probability density function $P_D(x)$ taking on a value between a and b is given by

$$p = \int_a^b P_D(x)\mathrm{d}x.$$

The probability then of x being less than a certain value, y, i.e. the **cumulative probability function**, is

$$C(y) = \int_{-\infty}^{y} P_D(x)\mathrm{d}x.$$

[34]The concept of summation

[35]The concept of integration

10.8 THE NORMAL DISTRIBUTION

Earlier we mentioned the **empirical rule** for determining how bell shaped our data is. That bell shape has a particular mathematical expression, and when it is a probability distribution, we call it the **normal distribution**.

A bell-shaped curve has the expression

$$f(x) = \mathrm{e}^{-x^2/2}$$

shown in figure 10.5.

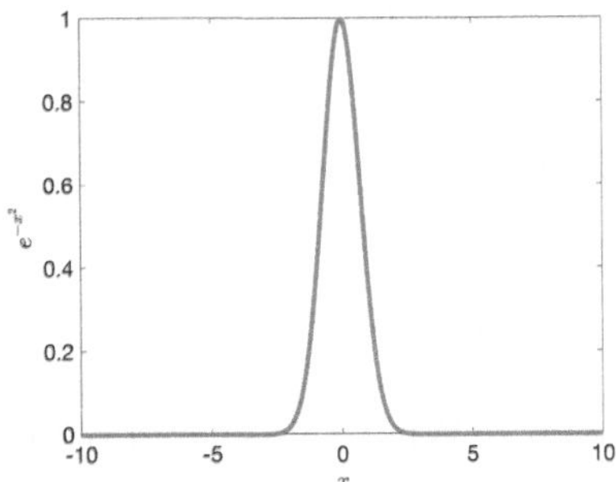

Figure 10.5: A canonical bell curve, $f(x) = \mathrm{e}^{-x^2}$.

Remember, the goal is to build a probability density function. The peak of the bell curve means that the value of x where the peak occurs should be more probable than all other values. In this case, that means $x = 0$ is more probable than any other outcome. Using $f(x)$ without modification means we could only use an associated probability density function if $x = 0$ is the most probable event. This seems needlessly restrictive, so let's fix that by defining

$$g(x) = \mathrm{e}^{(x-\mu)^2/2}.$$

Plotting this function in 10.6, we can see that adding the parameter μ allows us to move the peak. With $g(x)$, μ is the most probable outcome and μ can be whatever we want.

Now let's pick two specific values for x and compare them. For simplicity, I'll pick $x = \mu$ and $x = \mu + 1$. If we check the ratio $g(\mu)/g(\mu+1)$, we see that $x = \mu$ is $1/\mathrm{e}^{-0.5}$ more likely to occur than $x = \mu + 1$. This is true no matter what μ is. The **relative probabilities** are always the same. Again, needlessly restrictive, $x = \mu$ and $x = \mu + 1$ should have

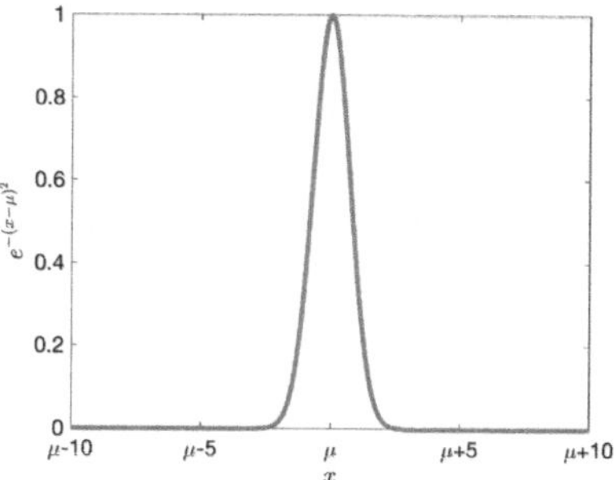

Figure 10.6: By adding a parameter, μ, to $f(x) = \mathrm{e}^{-(x-\mu)^2}$, we can change where the centre of our bell curve is.

whatever relative probability that we want. We can do this by adding a second parameter.

$$h(x) = \mathrm{e}^{-(x-\mu)^2/\sigma^2}$$

This is shown in figure 10.7. What this second parameter does is change the width of the bell.

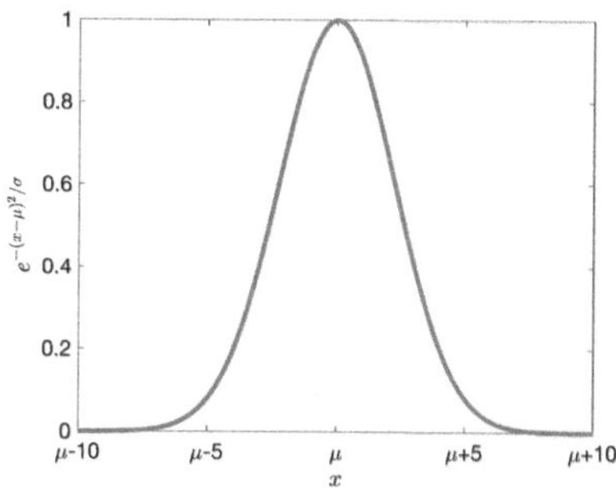

Figure 10.7: By adding a parameter, σ, to $f(x) = \mathrm{e}^{-(x-\mu)^2/\sigma}$, we can change the width of our bell curve.

If we calculate the relative probability of $x = \mu$ and $x = \mu + 1$, at the very least, we can set. If we compute this relative probability, we see that

$$\frac{h(\mu)}{h(\mu + 1} = \frac{1}{\mathrm{e}^{-1/2\sigma^2}}.$$

So we can at least have some say over the relative probabilities of points by picking sigma.

In reality, these parameters μ and σ are determined by our data. The parameter μ is the **mean** and σ is the **standard deviation**.

Is what we have a probability density function? Not quite. There is no guarantee that the area under the curve is one, but we can make it so. The total area is given by the integral from $-\infty$ to ∞. *How* this integral is done is a bit beyond the scope of our treatment, so we'll just provide the result and trust.

$$\int_{-\infty}^{\infty} \mathrm{e}^{(x-\mu)^2/2\sigma^2}\,\mathrm{d}x = \sigma\sqrt{2\pi}$$

Knowing this, if we divide our function by this number, then we get a total area of 1!

Definition 10.8.1. The **normal distribution** also known as a Gaussian distribution is a probability density function defined by

$$N(x;\mu,\sigma) = \frac{1}{\sigma\sqrt{2\pi}}\mathrm{e}^{-(x-\mu)^2/2\sigma^2}$$

where μ is the population mean and σ is the population standard deviation.

If data follows the empirical rule, we can often use the mean and variance of the data to *fit* a normal distribution to the data.

For instance, if we calculate the 2D:4D ratio for many people as seen in the data set in Appendix B, we see that the ratio is normally distributed, with mean $mu = 0.95671$ and standard deviation $\sigma = 0.028869$. Plugging these into our expression for the normal distribution, we get

$$N(x;0.95671,0.028869) = \frac{1}{0.028869\sqrt{2\pi}}\mathrm{e}^{-(x-0.95671)^2/0.00168}.$$

Why might we need this? We already have the data. A theoretical underpinning for our data can help us fill in the gaps in our data. Our sample size isn't really *that* big. We can view the data as an approximation of this probability density function, or the probability density function is the *limit* of the data if we had infinitely many samples. Having theoretical expressions that we can attribute to our data allows us to do more with less, in a sense.

For instance, we can now compute the probability of finding someone with a 2D:4D ratio between 1.04 and 1.06. We can go out and just keep taking measurements until we get a measurement in this range, or we can use what we already know to estimate the probability that any one individual we choose falls in this range.

As above, in order to compute probabilities, we need to compute an integral.

$$P(1.04 \leq y\ leq 1.06) = \int_{1.04}^{1.06} N(x; 0.95671, 0.028869)\mathrm{d}x = 0.0018$$

It's quite rare indeed!

What is normal anyway?

Why is it called the *normal* distribution? What's an un-normal distribution? It is normal in the sense that it is the norm[36]. Many, many phenomena in the natural and social sciences can be approximated by a normal distribution. We've seen that people's 2D:4D ratio is normally distributed around a value of 0.95671.

There's also what is called the **central limit theorem**. Without getting into the math, the central limit theorem states that if an outcome is the sum of many small, independent, additive processes, then the population mean of those outcomes will be normally distributed. Our 2D:4D ratio is the product of genetics, hormones, and environmental factors among other things. By the central limit theorem, we shouldn't be surprised that it is normally distributed. The obvious failure of the central limit theorem is when there is one overwhelming driving force that produces the observed outcome. The other key point is that the effects need to be *additive*. Some outcomes are the result of multiplicative effects, and we shouldn't expect the outcomes to be normally distributed in this case.

Besides being incredibly useful, it has a lot of mathematical properties. The normal distribution is as follows:

- It is symmetric around the mean μ.
- The mean, mode, and median are all the same.
- Mathematically, it is infinitely (and relatively easily) differentiable.

[36] In fact, the original name was normal distribution of residuals because errors in measurements tend to be random and symmetric.

Combining Discrete and Continuous Probabilities

Our 2D:4D data set has 285 entries, and we know that there is roughly a probability of 0.0018 of any one observation being between 1.04 and 1.06. What is the probability that we see *exactly one* measurement between 1.04 and 1.06 in our 285 measurements?

The number of measurements we take is discrete, and so we can fall back on our tools for discrete probabilities. If we take our ratio measurements, and we see a number between 1.04 and 1.06 at least once time, we yell "success!" and fire off some fireworks. What is the probability we get to see our fireworks?

We have two outcomes: we see a measurement between 1.04 and 1.06 with probability 0.0018, or we don't. We want to see this exactly once. We can use the binomial distribution to determine how many times this will happen.

$$\begin{aligned} B(1; 0.0018, 285) &= \frac{285!}{284!}(0.0018)(0.9982)^{284} \\ &\approx 0.31 \end{aligned}$$

We shouldn't expect to see such a measurement with our relatively small sample size.

We can also ask, *what should our sample size be in order to all but guarantee seeing a measurement between* 1.04 *and* 1.06 *at least once?* Of course, this comes with all the assumptions that are built-in to the tools that we're using. A reminder of some of those assumptions is as follows:

- The 2D:4D ratio is indeed normally distributed in humans.
- The measurements are independent as required by the binomial distribution.

The second bullet above highlights the importance of random sampling, or at least non-biased sampling. There is evidence that the 2D:4D ratio is highly heritable[18]; this means that it is often similar between related individuals. If we sample only a single family, our measurements are *not* necessarily independent and the binomial distribution is not valid[37]. If we only record a measurement if it's bigger than the last measurement we took, then our measurements are not independent. Random sampling ensures that we can use our tools most effectively as you will see that many of our tools work best on independent events.

[37]The 2D:4D ratio may not even be normally distributed anymore.

BACK TO OUR EXAMPLE, in this case we want to solve for n. In order to do so, we have to set a probability. Let's say we want a 99% chance of seeing a 2D:4D measurement in the range of 1.04 and 1.06 at least once; the problem we have to solve is

$$0.99 = \sum_{i=1}^{n} B(i; 0.0018, n),$$

which is a little tough.

We can think of the opposite problem. What does n have to be so that there is a probability of 0.01 that we *never* see a measurement between 1.04 and 1.06. This is a much easier problem.

$$\begin{aligned}
0.01 &= B(0; 0.0018, n) \\
0.01 &= \frac{n!}{0!n!}(0.0018)^0(0.9982)^n \\
0.01 &= 0.9982^n
\end{aligned}$$

Using our logarithm rules[38], we can solve this for n.

$$\begin{aligned}
n &= \frac{\ln(0.01)}{\ln(0.9982)} \\
n &= 2556.1246,
\end{aligned}$$

but remember that n is **discrete**. It needs to be an integer. We can take 0.1246 of a measurement, so we better round up to be sure that we have our 99% chance of seeing our desired measurement at least once. We would say that $n = 2557$ is the minimum sample size[39] we would require to all but guarantee we see a measurement between 1.04 and 1.06.

10.9 OTHER USEFUL DISTRIBUTIONS

The binomial distribution and the normal distribution are arguably the most ubiquitous distributions, but they are far from the only *useful* distributions. In this section, we will discuss some of the other distributions that show up in the wild and the properties that make them useful.

[38] Way back in Chapter 1 if you need a refresher

[39] So long as our participants are randomly sampled

The Exponential Distribution

Definition 10.9.1. The **exponential distribution** is a probability density function that is strictly decreasing[40]. It is given mathematically as

$$E(x;\lambda) = \lambda e^{\lambda x}$$

where $1/lambda$ is the mean and $0 \leq x < \infty$.

The exponential distribution has all the same properties of any probability density function: the total area is 1[41], and all values of $E(x)$ are between 0 and one.

The mean, as mentioned above, occurs at $1/\lambda$, but because the distribution is not symmetric, the median and mode are **not** equal to the mean. The median of the exponential distribution occurs at $x = \ln(2)/\lambda$, and the mode is $x = 0$. The standard deviation, strangely, is $1/\lambda$ like the mean.

Because the mode is 0, and it also happens to be the highest point in the distribution, we can say that $x = 0$ is more likely to occur than any other value. Due to the high probability of having extremely large values though, the mean is shifted to the right, to $1/\lambda$. There is a balancing of probabilities: the events happening at $x = 0$ are offset by the relatively high-probability events[42] happening for values of x much, much larger than the mode.

The exponential distribution is the under-the-hood driving force of parameters in ordinary differential equation mathematical models[43].

When we are looking at differential equations, we can interpret any of the parameters of our model to be exponentially distributed. For instance, consider the differential equation

$$\frac{dN}{dt} = bN - \mu N.$$

---

[40]i.e. the most probable outcome is 0.

[41]Keep in mind that the **domain** of the function neglects negative events – values of $x < 0$ have probability density of 0.

[42]Compared to, say, the normal distribution

[43]Discussed briefly in Chapter 5

In this equation, N is the *n*umber of people in a population, b is the birth rate, and μ is the death rate. The equation described how the population changes over time, t. These parameters are exponentially distributed, meaning most individuals are born and die immediately, but this is offset by those that live an infinitely long time or take an infinitely long time to be born. The balancing of these two probabilities puts the average birth rate at b and the average death rate at μ. We say that birth and death are exponentially distributed over time. On average a new person is born every $1/b$ time steps and a person dies, on average, every $1/\mu$ time steps[44].

This seems like it *shouldn't* work. The idea that most people are born and die at the beginning of time and then others are never born and others still never die seems unrealistic, and it is. Exponentially distributed parameters like this work surprisingly well in modelling human and animal behaviours, whether those behaviours be biological or social. Many mathematical models in biology, economics, and the social sciences assume exponentially distributed parameters to great effect.

The Gamma Distribution

The gamma distribution is a widely applicable family of probability density functions, of which the exponential distribution is a special case.

Definition 10.9.2. The **gamma distribution** is a probability density function that is defined by

$$G(x;\theta,k) = \frac{1}{\Gamma(k)\theta^k} x^{k-1} \mathrm{e}^{-x/\theta}$$

where $\theta > 0$ and $k > 0$ are called **scale and shape parameters**, respectively. The function $\Gamma(k)$ is called the **Gamma function**[45], it is an extension of the factorial function to non-integer values.

The Gamma distribution $G(x;\theta,1)$ is the exponential distribution, where $\lambda = 1/\theta$. Put more simply,

$$G(x;\theta,1) = E\left(x; 1/\theta\right).$$

[44] Even though most of these births and deaths are happening immediately, and some are effectively never happening

[45] Gamma function is not the Gamma distribution, confusing I know.

The shape parameter k can be any number greater than 0. If k is an integer, the Gamma distribution is often called the **Erlang distribution**.

If the scale parameter, $\theta = 1/2$, and $2k = \kappa$, then the gamma distribution is called the **chi-squared distribution**, with parameter κ.

The Gamma distribution is again not symmetric, so the mean, median, and mode are not equal to each other. Generally, θ and k act in tandem to move the peak of the distribution to the right[46].

The mean is given by $k\theta$. The mode is given by $(k-1)\theta$; it's just to the left of the mean. This should tell us that the right tail of the distribution is much longer than the left tail, and of course it is: on the left we're stopped by $x = 0$, whereas on the right we can go all the way to ∞.

The median in this case cannot be written in a nice, compact closed expression.

The Gamma distribution can be used in a wide variety of applications, specifically when k is an integer[47].

The Erlang distribution was first used in an analysis of call centres, to determine the probability of receiving two calls within x minutes of each other. This could be used to help inform staffing and resource requirements. The mean then gives the mean time between calls, also a useful statistic.

The Erlang distribution is also useful for determining the probability of an individual recovering[48] at time x after exposure to a disease. Generally, if there are k steps of progression of a disease[49], each lasting $1/\theta$ time steps, we can find the mean period of infectiousness using the Erlang distribution.

Let's say you lived in a musical, where sometimes you're asked a question like, what *did you get up to this summer*? and have the inescapable urge to break into song. Of course, the only difference between living in a musical and being *that* person is how many people are involved. Let's say you break into a song that is θ beats per minute. The mean of the Erlang distribution with $k = 1$ is the average amount of time you will have to wait until one person joins you in the song. The mean of the Erlang distribution with $k = 2$ is the average amount of time you will

[46]When $k = 1$, the peak is as far left as possible; when $0 < k < 1$, there is no peak at all!

[47]i.e. the Erlang distribution

[48]Or, at least, no longer being infectious

[49]Often times, if you are exposed to a disease, it takes time before you are infectious, and possibly even *more* time until you are symptomatic, and possibly after even more time, you are less infectious again.

have to wait until two people join you, and you are then invariably fully in musical territory.

The Beta Distribution

So far, the probability density functions that we have discussed have at least one tail that goes to ∞. This means that there is a small, non-zero[50] that an event at $x = 4 \times 10^{23918}$ happens. No matter what we are measuring, this number is bigger than it. It is a number much bigger than the age of the universe in seconds, the number of atoms in the universe, and the radius of the observable universe.

The beta distribution is a probability distribution with **compact support**. Compact support means that *both* the left and right end points of the domain are finite.

Definition 10.9.3. The **beta distribution** is a probability density function defined by

$$B(x; \alpha, \beta) = \frac{\Gamma(\alpha + \beta)}{\Gamma(\alpha)\Gamma(\beta)} x^{\alpha-1}(1-x)^{\beta-1}$$

where $\alpha > 0$ and $\beta > 0$ are shape parameters, and $0 \leq x \leq 1$. The function $\Gamma(x)$ is once again the Gamma function. Outside the domain $0 \leq x \leq 1$, the probability is identically zero.

As with the Gamma distribution, the median has no closed form expression. The mean of the beta distribution is given as $\alpha/(\alpha+\beta)$.

The mode of the beta distribution depends on the values of α and β.

- If $\alpha > 1$ and $\beta > 1$, the mode is $(\alpha-1)/(\alpha+\beta-2)$.
- If $\alpha = \beta = 1$, the mode is all numbers between 0 and 1.
- If $\alpha < 1$ and $\beta < 1$, there are two modes: 0 and 1.
- If $\alpha > 1$ and $\beta \leq 1$, then the mode is 1.
- If $alpha \leq 1$ and $\beta > 1$, then the mode is 0.

[50] But, for all practical purposes, it might as well be zero.

Why these are the cases is evident from the visualization of the different forms of the beta distribution.

The beta distribution should look familiar. It has a very similar form to the binomial distribution, except all the parameters are continuous. If we set $x = p$, $\alpha = k + 1$, and $\beta = n - k + 1$, where n and k are integers, we get

$$B(p; k+1, n-k+1) = \frac{\Gamma(n+2)}{\Gamma(k+1)\Gamma(n-k+1)} p^k (1-p)^{n-k}.$$

Since n and k are integers,

$$B(p; k+1, n-k+1) = \frac{(n+1)!}{n!n-k!} p^k (1-p)^{n-k}.$$

Which looks suspiciously similar to the binomial distribution!

It turns out the two are related, but are telling us slightly different things. With the binomial distribution, we assume we *know* the probability p of a single success, but we *don't know* the number of times we will succeed. With the above formulation of the beta distribution, we assume that we have already seen k successes in n independent events, but we don't necessarily know the probability of a single success.

The beta distribution in this case can be interpreted as giving us a probability about a probability. If we have seen k successes in n trials, then the beta distribution gives us the probability that an individual trial results in success with probability p. If that was a little confusing, I get it. It's using the word probability to mean two different things, which inevitably leads to problems[51].

The key comes in the *if...then* statement. We can rewrite it as a question and clarify what is happening: *given that we have seen k successes in n trials, what is the probability, p, the next trial will be a success?* This is a conditional probability. We are looking for p given some previous knowledge. The probability given by the beta distribution is thus a conditional probability.

$$P(p|k,n) = B(p; k+1, n+1)$$

is the probability that each value of p is the true value. This is a good example of the Bayesian interpretation of probability. In this case, $P(p|k,n)$

[51]The sentence *Buffalo buffalo Buffalo buffalo buffalo buffalo Buffalo buffalo* uses the same word as a noun, proper noun, and verb, which is what makes the sentence confusing.

gives the confidence we can have in each value of $0 \leq p \leq 1$ being the true value of p.

As a concrete[52] example of this, we can use the beta distribution to check if the outcomes of n roulette games came from a fair wheel. In this case, if we bet *red* every time, we can tabulate our results as a 1 for winning or 0 for losing. In this case, we know that for a fair wheel the true value is $p_{fair} = {}^{18}/_{37}$. Below is the result of 12 games.

Game	1	2	3	4	5	6	7	8	9	10	11	12
Result	0	0	0	0	0	1	0	1	1	0	0	1

Figure 10.8 shows the confidence we can have in different values of p.

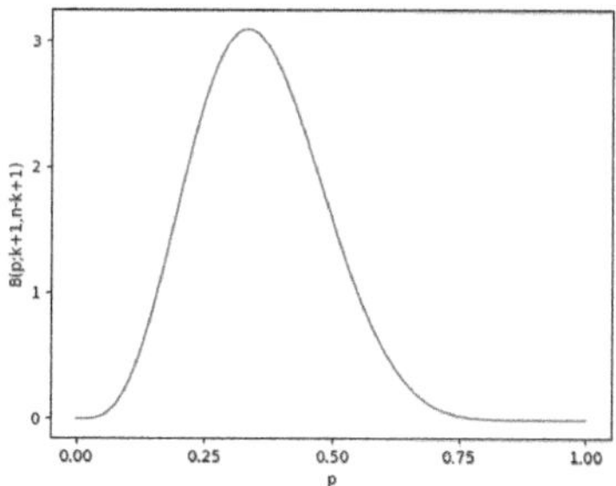

Figure 10.8: The probability density for p, the probability that a single game on our roulette table will result in a win if we bet red every time, is based on our 12 games. The most likely value for p is 0.3333; but playing more games would change this and result in a more accurate prediction.

The highest point[53] is a little to the left of the true fair value at $p = 0.33$. The mean is $p = 0.3846$, which is closer to our fair value. We have to be careful about our interpretation in this case, because of our sample size. In this case, we only have 12 games to *inform* our distribution. The more data you have, the better your estimate of p will be. If we keep playing, we may learn the true nature of this particular roulette table.

10.10 MEAN, MEDIAN, MODE, AND VARIANCE

We know how to calculate our mean, median, mode, and variance when we have data laid out in front of us; these are known as our **sample**

[52] And relatively common

[53] The value we should have the most confidence in.

statistics because they come from, well, samples[54]. We can also estimate **population statistics** from the underlying probability distributions.

When describing different distributions, we stated the mean, median, mode, and variance when we could, but we didn't say where these came from. What if we have a different distribution than one of the above? How will we find these things?

The reason I left these for the end is that we need to know some things about derivatives, sums, and integrals in order to compute some of these things. Depending on your interests, you may have skipped right past Chapter 6[55]. If you did, feel free to go back[56] or skip on ahead.

Mode

Whether we have a discrete probability mass function, or a continuous probability density function, the **mode** is always **the highest point on the distribution**. Mathematically, if we have a probability distribution, $P(x)$, the mode, m is the point such that

$$P(m) \geq P(x) \texttt{ for all } x \texttt{ in the domain.}$$

In the first half of the book, we discussed some tools for finding the highest[57] points of a function. We can use these to find the mode for any probability distribution.

Median

Just like with the sample median, the population median is the exact midpoint, where you have a probability of 0.5 of being less than or greater than the median value. Conceptually, this makes an expression for the median simple. It is the point where the cumulative probability is exactly 0.5.

For a discrete probability mass function, $P_M(x_i)$, this means we need to find an n such that

$$\sum_{i=0}^{n} P_M(x_i) = 0.5.$$

If we can find this n, then x_n is the median.

---

[54] i.e. data

[55] And 1, 2, 3, 4, and 5

[56] As usual, I'll wait

[57] And lowest

For a continuous probability density function, $P_D(x)$, we need to find a y such that

$$\int_{-\infty}^{y} P_D(x)\mathrm{d}x = 0.5.$$

Then y is the median.

Mean

The mean is given as a weighted average of the possible values. In general, this means to calculate the mean, we multiply each value by its probability and sum up all possibilities, in the case of a discrete distribution, or its relative probability and integrate over the entire domain in the case of a continuous distribution.

Mathematically, we write this as

$$\mu = \sum_{i=0}^{\infty} x_i P_M(x_i)$$

for a discrete probability mass function $P_M(x)$.

For a continuous probability density function, $P_D(x)$, the mean, m, is given by

$$\mu = \int_{-\infty}^{\infty} x P_D(x)\mathrm{d}x.$$

Variance

The variance is the average deviation from the mean, so essentially, what we want to do is compute the mean, but with the deviation. We write deviation from the mean as

$$dev = (x - \mu)^2.$$

We can then define the variance as

$$\sigma^2 = \sum_{i=0}^{\infty} (x_i - \mu)^2 P_M(x_i)$$

for a discrete probability mass function, $P_m(x)$, or

$$\sigma^2 = \int_{-\infty}^{\infty} (x - mu)^2 P_D(x)\mathrm{d}x$$

for a continuous probability density function, $P_D(x)$.

We can simplify this simply by expanding the binomial term, $(x-\mu)^2$. Since

$$(x-\mu)^2 = x^2 - 2x\mu + mu^2.$$

we can rewrite

$$\begin{aligned}\sigma^2 &= \int_{-\infty}^{\infty} \left(x^2 - 2x\mu + \mu^2\right) P_D(x)\mathrm{d}x \\ &= \int_{-\infty}^{\infty} x^2 P_D(x)\mathrm{d}x - 2\mu \int_{-\infty}^{\infty} x P_D(x)\mathrm{d}x + \mu^2 \int_{-\infty}^{\infty} P_D(x)\mathrm{d}x\end{aligned}$$

Now we can look at these terms individually. There isn't much we can do with the first term, so we'll skip it. The second term should have a piece that looks familiar. We know from above that

$$\mu = \int_{-\infty}^{\infty} x P_D(x)\mathrm{d}x.$$

So our second term is just $2\mu \cdot \mu = 2\mu^2$.

Our third term simplifies as well. Remember that for a function to even *be* a probability distribution at all, the total area under it must equal 1[58]. Since we know that

$$\int_{-\infty}^{\infty} P_D(x)\mathrm{d}x = 1,$$

we can simplify our expression for the variance as

$$\begin{aligned}\sigma^2 &= \int_{-\infty}^{\infty} (x-\mu)^2 P_D(x)\mathrm{d}x \\ &= \int_{-\infty}^{\infty} x^2 P_D(x)\mathrm{d}x - 2\mu^2 + \mu^2 \\ &= \int_{-\infty}^{\infty} x^2 P_D(x)\mathrm{d}x - \mu^2.\end{aligned}$$

10.11 SUMMING TO INFINITY

In Chapter 6, we briefly discussed some tools for dealing with integrals that either start at $-\infty$ or go to ∞, and you may have noticed these types of integrals come up quite a bit in probability theory[59].

[58] In the case of discrete probability distributions, it's just the sum.

[59] Particularly when calculating mean or variance or other summary statistics that exist but we haven't spoken about

When it comes to discrete probabilities, we also sometimes need to sum to ∞. This is only really possible in specific instances using the identity.

$$\sum_{n=0}^{\infty} a^n = \frac{1}{1-a} \quad \text{for } |a| < 1$$

Yes, you've read that right. The sum of infinitely many numbers is not necessarily infinite. We can kind of reason through why this may be. As long as $1 < a < 1$, then a^n gets smaller as n gets bigger, so even though we're adding things forever, we're adding very, very little each time. Eventually, we're adding so little each time that we never actually reach $1/(1-a)$ unless we go on adding for infinitely long.

Compare this to the case where we have

$$\sum_{n=0}^{\infty} 1^n.$$

We know that $1^n = 1$, which means that as n gets larger, 1^n does **not** get any smaller. So we're never really approaching any number as we keep adding things to our sum. For instance, after one element, our sum is 1; then adding the second element, our sum is 2; if we were being silly, we could say, "*hm, it looks like we're approaching* 3, but adding just two more elements our sum blows right past 3 onto 4!" Maybe we're approaching 5 then, but again, just adding two more numbers allows us to blow right past 5 as well. No matter what number we choose to "approach," we can always just take that number of elements in our sum, plus a few more and blow past it. There's no limit to our sum.

Using the identity

$$\sum_{n=0}^{\infty} a^n = \frac{1}{1-a}.$$

we can also figure out what the sum would be if we start at $n = N$ instead of at $n = 0$, or to write it more precisely,

$$\sum_{n=N}^{\infty} a^n.$$

To work out what this is, we should try to make it look like our identity. The easiest way to do this is to replace n with another value, call it

$\hat{n}+N$. We choose this so that when we put replace n with $\hat{n}+N$ in the expression, $n=N$, we get

$$\begin{aligned} n &= N \\ \hat{n}+N &= N \\ \hat{n} &= 0, \end{aligned}$$

and the beauty of ∞ is that

$$\begin{aligned} n &= \infty \\ \hat{n}+N &= \infty \\ \hat{n} &= \infty - N \\ \hat{n} &= \infty \end{aligned}$$

since subtracting a little bit for ∞ is still ∞.

Now our sum looks like

$$\sum_{\hat{n}=0}^{\infty} a^n.$$

Now we have $\hat{n}$ in the limits of the sum, but n in the expression of the sum. We need $\hat{n}$ in **both**. So we replace the n in our expression.

$$\sum_{\hat{n}=0}^{\infty} a^{\hat{n}+N} = \sum \hat{n} = 0^{\infty} a^{\hat{n}} a^N$$

We can factor the a^N since it is common to every term in our sum.

$$\begin{aligned} \sum_{n=N}^{\infty} a^n &= a^N \sum_{\hat{n}=0}^{\infty} a^{\hat{n}} \\ \sum_{n=N}^{\infty} a^n &= \frac{a^N}{1-a} \end{aligned}$$

The last line comes from the fact that our sum involving $\hat{n}$ is exactly our identity.

We can even use this to find out an expression for a sum from 0 to N. This would be the sum from 0 to ∞ **minus** all the terms from $N+1$ to ∞.

$$\begin{aligned}\sum_{n=0}^{N} a^n &= \sum_{n=0}^{\infty} a^n - \sum_{n=N+1}^{\infty} a^n \\ &= \frac{1}{1-a} - \frac{a^{N+1}}{1-a} \\ &= \frac{1-a^{N+1}}{1-a}\end{aligned}$$

These formulae can be extremely useful when we are dealing with probabilities; particularly with the binomial distribution.

10.12 PROBABILITY AND PYTHON

Now that we have a theoretical underpinning, we can talk about how to actually use these ideas. Typically, these formulae are not meant to be used *by hand*, so to speak, at least not in modern times. The amount of data generated in almost any field is so overwhelmingly large that we do most of our probability calculations and statistics with computers. The important part is using our framework to make sure that when we present probabilities and statistics we are doing so under the correct assumptions and drawing correct conclusions.

For instance, heights of individuals in a certain species in a population may be normally distributed, but it would then be wrong to also assume the weights of the same individuals are also normally distributed. One study found that height and weight are related through a roughly square relationship[17],

$$w = ah^{2.11}$$

and if we look at the histograms of simulated heights and weights in figure 10.9, we see that while height is normally distributed, weight is not. This can lead to real problems as a lot of statistics depend on data being normally distributed. We will get into *why* a little bit in the next chapter.

For now, we will focus on realizing some of the ideas in this chapter in Python. To start, many of our common distributions are built-in in Python so that we don't need to create them from scratch. Table 10.2 gives the commands for common distributions.

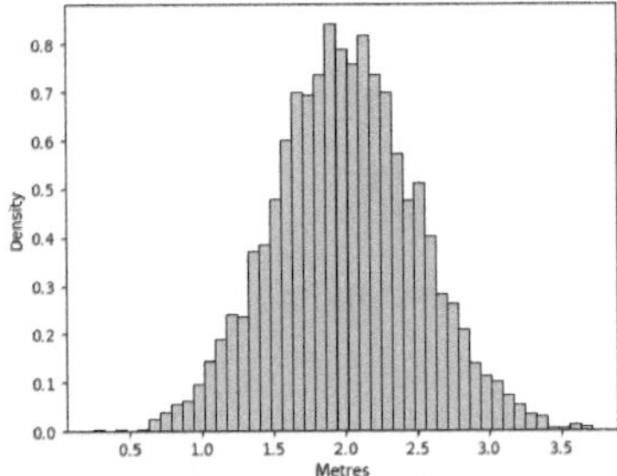

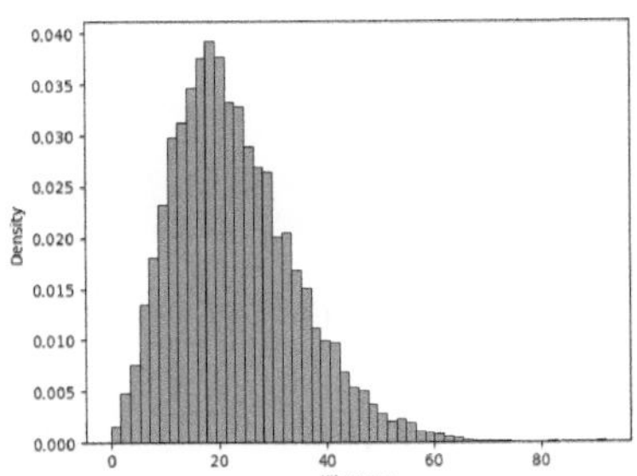

Figure 10.9: The left plot shows some normally distributed heights, and the weights on the right generated based on the allometric law above. Even though the heights are normally distributed, the weights most certainly are not.

Table 10.3: Python commands for creating probability density functions and cumulative density functions for common distributions.

Distribution Name	PDF/PMF command
Binomial	`stats.binom.pmf(k,n,p)`
Trinomial	`stats.multinomial.pdf([i,j],n,[p_A, p_B]`
Exponential	`stats.expon.pdf(x,scale = 1/lambda)`
Normal	`stats.norm.pdf(x,loc=mu,scale=sigma)`
Gamma	`stats.gamma.pdf(x,alpha,scale=1/beta)`
Beta	`stats.beta.pdf(x,alpha,scale=1/beta)`

Distribution Name	CDF
Binomial	`stats.binom.cdf(k,n,p)`
Trinomial	`stats.multinomial.cdf([i,j],n,[p_A,p_B])`
Exponential	`stats.expon.cdf(x,scale=1/lambda)`
Normal	`stats.norm.cdf(x,loc=mu,scale=sigma)`
Gamma	`stats.norm.cdf(x,alpha,scale=1/beta)`
Beta	`stats.beta.cdf(x,alpha,scale=1/beta)`

Empirical Probabilities & Conditional Probabilities

For instance, let's look at data set on video game movies. A sample of the data set is given in figure 10.10, and the data set itself is in the file repository as `videogamemovies.csv`. We of course first need to import the data set as usual using the command

```
vg_data = pd.read_csv(`videogamemovies.csv').
```

	Title	Release date	...	Distributor	Original game publisher
0	Super Mario Bros.	1993-05-28	...	Buena Vista Pictures Distribution	Nintendo
1	Double Dragon	1994-11-04	...	Gramercy Pictures	Technōs Japan
2	Street Fighter	1994-12-23	...	Universal Pictures	Capcom
3	Mortal Kombat	1995-08-18	...	New Line Cinema	Midway
4	Mortal Kombat: Annihilation	1997-11-21	...	New Line Cinema	Midway
5	Wing Commander	1999-03-12	...	20th Century Fox	Origin Systems
6	Lara Croft: Tomb Raider	2001-06-15	...	Paramount Pictures	Eidos
7	Resident Evil	2002-03-15	...	Sony Pictures Releasing	Capcom
8	Lara Croft: Tomb Raider – The Cradle of Life	2003-07-25	...	Paramount Pictures	Eidos
9	House of the Dead	2003-10-10	...	Artisan Entertainment	Sega Sammy
10	Resident Evil: Apocalypse	2004-09-10	...	Sony Pictures Releasing	Capcom

Figure 10.10: A sample of the data set on video game movie adaptations.

Estimating a probability for an event is as simple as counting the number of times our event occurs divided by the number of entries in our data set. For instance, if we want to calculate the probability of a video game movie adaptation that has a Metacritic score greater than 50, we first find all movies that satisfy our event condition

```
movies_over_50_meta = vg_data.loc[vg_data['Metacritic']>50].
```

We can then find the number of movies in this set and in the sample set by computing the respective lengths.

The binomial distribution can be used to determine the probability of seeing one success in n trials:

$$B(1; n, p) = np(1-p)^{n-1}.$$

In exactly one of these n cases, the one success happens during the nth trial. This means we can determine the probability of seeing $n-1$ failures before seeing a success as

$$B(n; 1, p) = np(1-p)^{n-1}.$$

We may then ask, what is the probability of *not* seeing a success before completing N trials? This is given by a cumulative probability, C_N.

$$P(n \geq N) = \sum_{n=N}^{\infty} p(1-p)^{n-1}$$

Since the p is common to all terms, we can factor it out.

$$P(n \geq N) = p \sum_{n=N}^{\infty} (1-p)^{n-1}$$

Now we can play the same game as above and make the substitutions $n = \hat{n} + N$ so that we can get our sum to look like our identity,

$$\begin{aligned}
P(n \geq N) &= p\sum_{\hat{n}=0}^{\infty}(1-p)^{\hat{n}+N-1} \\
&= p\sum_{\hat{n}=0}^{\infty}(1-p)^{\hat{n}}(1-p)^{N-1} \\
&= p(1-p)^{N-1}\sum_{\hat{n}=0}^{\infty}(1-p)^{\hat{n}} \\
&= p(1-p)^{N-1}\frac{1}{1-(1-p)}P(n \geq N) \qquad = (1-p)^{N-1}
\end{aligned}$$

which is exactly the same as the probability of seeing $N-1$ failures in a row and **not** seeing N failures in a row[60].

```
num_over_50 = len(movies_over_50_meta)
num_total = len(vg_data)
```

Recall that in our frequentist approach to probability, a probability is nothing more than a ratio; dividing these two lengths gives us a probability that a random video game adaptation has a Metacritic score over 50.

```
probability_over_50 = num_over_50/num_total
```

If our event, movies with a Metacritic score over 50, is called A, what we just did is compute $P(A)$ for this data.

What if we wanted to add a condition? As with our original dice-rolling example, a condition on a probability has the major effect of changing the size of the sample space. So, when dealing with data, what we need to do to add a condition is change the length we are dividing by. Say we want to compute the probability of a movie having a Metacritic score over 50 given that it also has a Rotten Tomatoes score over 50. If we call this second event B, we want to compute $P(A|B)$ from our data set. In this case we need to filter our data a little differently.

First, we filter by our condition, B[61]:

```
movies_over_50_rt = vg_data.loc[vg_data['Rotten Tomatoes']>50].
```

[60]Because the Nth trial is presumably the success we were looking for

[61]One way to remember this is that we *know* B is true, so we should take care of it first.

We then want to know how many movies *also* have a Metacritic score over 50. You may be tempted to copy and paste the line filter line we used before, but that would be a misstep. Our variable `movies_over_50_meta` contains movies that do *not* have a Rotten Tomatoes score over 50 and therefore do not satisfy our condition[62]. We apply the filter to our new sample space, those movies that have a Rotten Tomatoes score of over 50.

```
movies_over_50meta_and_50rt = movies_over_50_rt.loc[movies_over_50_rt['Metacritic']>50]
```

Now we can compute the lengths of these two data sets and divide them to find a conditional probability.

```
num_over_50_meta = len(movies_over_50meta_and_50rt)
num_over_50_rt = len(movies_over_50_rt)
prob_A_given_B = num_over_50_meta/num_over_50_rt
```

This computes $P(A|B)$.

Now, there would be no way for you to know this unless you are a connoisseur of video game movie adaptations, but even these probabilities are all conditional! The data set only contains live-action adaptations of video games, meaning any and all fully animated movies are excluded. Therefore, our $P(A)$ is actually $P(A|C)$, where C is the event that a movie is live action. Our probability $P(A|B)$ is actually $P(A|B \cap C)$, since this is a video game movie that has a Metacritic score over 50 *given that* the movie has a Rotten Tomatoes score over 50 *and* the movie is live action.

It bears repeating: almost every probability[63] is conditional in some way. Often times, we will drop the condition from our notation if it applies to everything in a data set, but this means we have to be extra clear in our statements what our sampling conditions were so that we don't overextend our results.

Empirical Probability Density Functions and Cumulative Density Functions

We saw in the last chapter how we can build a histogram of our data using seaborn. Generally, if we have discrete numerical data or categorical data, we would use

```
sns.histplot(x=data['column'], discrete=True)
```

[62]The word *also* in the first sentence in this paragraph is arguably the most important.

[63]Especially those coming from experimental data as in this example

to build a histogram. By default, this plot will give us the raw counts on the y-axis. We can turn this into a probability mass function by adding the condition that the heights of all the bars must sum to 1, which is a condition of a probability mass function. We do this with the option

```
sns.histplot(x=data['column'], discrete=True, stat='probability').
```

If we have a continuous variable, we know that the *total area* of the histogram must sum to 1. In this case we would use the command

```
sns.histplot(x=data['column'],discrete=False,stat='density').
```

This ensures that the total area of the histogram is 1 and that we indeed have an empirical probability density function for a continuous variable.

In either case, we may want to visualize the cumulative density function. This is easily done with the option `cumulative=True`. This gives us on the y-axis the probability that a random entry in the data set is the assigned value *or less*[64].

For our video game movie adaptation data set, we can look at the probability density function of Metacritic scores. While these are discrete, it might help to treat them as continuous for our purposes. We can plot the probability density function as

```
sns.histplot(x=vg_data['Worldwide box office'],
                              discrete=False,stat='density')
plt.show().
```

When we show this plot, we will see that the distribution looks exponential. Most of the values are close to zero. We could fit an exponential distribution to this data.

The `scipy.stats` package has all of our distributions built-in, but they are parameterized a little differently. The help files do let us know how to convert our parameterizations to those used by `scipy`. For instance, we can build an exponential distribution using the command

```
exp_dist = stats.expon.pdf(x_pts)
```

where `x_pts` is a series of numerical values on which to evaluate our distribution[65]. We can define a set of input points using

```
x_pts = np.linspace(0,5e8,10000).
```

[64]In other words, the value or anything on the x-axis to the left of the value

[65]Remember, probability distributions are just a special type of function and functions require an input.

This tells Python to take the interval from 0 to 500000000 and split it into 10000 equally spaced points. Then, `stats.expon.pdf(x_pts)` evaluates

$$E(x) = e^{-x}$$

at each value in `x_pts`.

You may have noticed that this is missing our parameter, λ, which is important for changing the mean of exponential distribution. By default, Python uses $\lambda = 1$. To use a different λ, we use the option

```
exp_dist = stats.expon(x_pts,scale=1/lambda).
```

How do we then determine the value of `lambda`? We know from our study of the exponential distribution that the mean can be given in terms of λ:

$$\mu_{exp} = \frac{1}{\lambda}$$

therefore,

$$\lambda = \frac{1}{\mu_{exp}}.$$

If we know the mean of our exponential distribution, we can find our parameter which fits the data!

Let's build our exponential distribution for the `worldwide box office` data for our data set:

```
box_office_expon = stats.expon.pdf(x_pts,
           scale = np.mean(vg_data['Worldwide box office'])).
```

Is this the best value to use for our parameter? Probably not. We discussed how we might find the *best* parameter by hand way back in Chapter 4. We will discuss this again, along with how to implement those ideas in Python in the next chapter. For now, we will use this crude estimate for our probability distribution.

If we plot this using a `seaborn` line plot on top of our histogram using

```
sns.lineplot(x=x_pts,y=box_office_expon),
```

we get something that looks like figure 10.11. We can see that this isn't an awful approximation to the data, even if it isn't the *best*.

What if we want the probability density as a specific point? We need only to give `stats.expon` a specific number instead of a bunch of numbers. If we want to know the probability density of 0, we would call

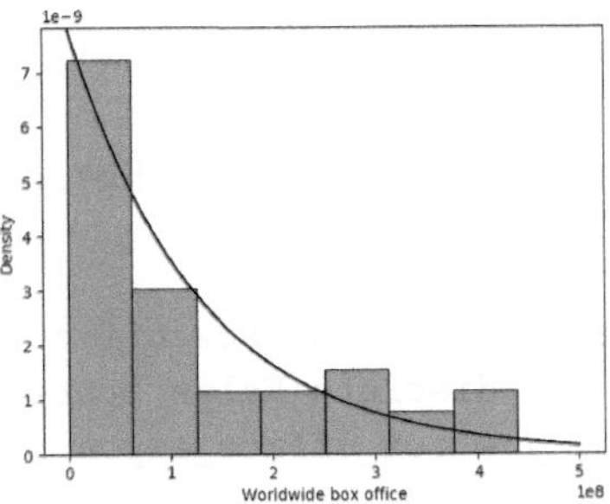

Figure 10.11: The distribution of worldwide box office revenue for our data set. The data looks exponential, so we fit an exponential distribution to it, and it doesn't look too bad. This isn't necessarily the best distribution or even the best exponential distribution, but it's a good start.

```
ww_box_0 = stats.expon.pdf(0,
                scale= np.mean(vg_data['Worldwide box office']).
```

The `scipy` stats package also has the ability to calculate the cumulative density function of any distribution it knows. For instance, to calculate the probability of a movie having a worldwide box office value of < 1000000, we would use the cumulative density function

```
ww_box_1m_cdf = stats.expon.cdf(1e6,
                scale= np.mean(vg_data['Worldwide box office']).
```

Due to the linearity of cumulative density functions, we could even determine the probability of a movie having worldwide box office value between, say, 1000000 and 2000000.

```
ww_box_1m_cdf = stats.expon.cdf(1e6,
                scale= np.mean(vg_data['Worldwide box office'])
ww_box_2m_cdf = stats.expon.cdf(2e6,
                scale= np.mean(vg_data['Worldwide box office'])
ww_bw_1m_2m = ww_box_2m_cdf - ww_box_1m_cdf
```

A question we might have is, *why did we choose an exponential distribution*? We brushed over this by saying the data *looks* exponential, so let's try it. This isn't necessarily a bad way to perform exploratory data analysis, but this doesn't explain any of the underlying mechanics that may lead to this behaviour. For that, we would need specialized knowledge in the economy of movies, how box office numbers are calculated, and if there are any confounding factors being hidden in λ. Models, experiments, and scientific rigour may help us to determine a distribution for data *a priori* or find an explanation as to why the data is distributed a certain way.

Let's increase the number of bins in our histogram and plot our same distribution on top of it.

```
sns.histplot(x=vg_data['Worldwide box office'],
                        discrete=False,stat='density',bins=100)
sns.lineplot(x=x_pts, y = box_office_expon)
```

Things start to look different[66], like in figure 10.12.

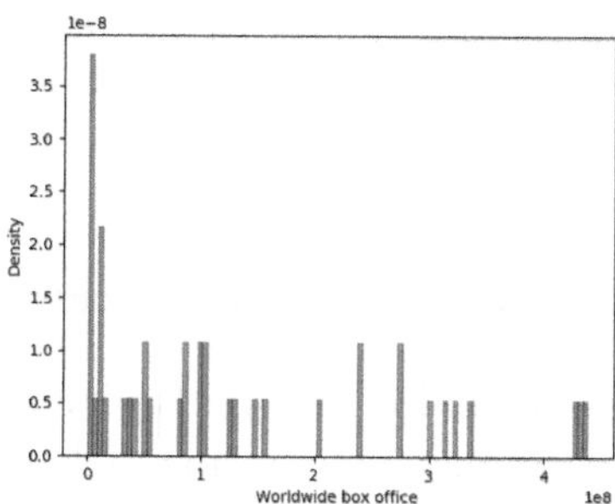

Figure 10.12: Using 100 bins in our histogram doesn't make much sense in this case, and we lose a lot of the pattern recognition we got with fewer bins. Choosing the right number of bins for your histogram is an important step in communicating the underlying information.

Where our exponential distribution used to look great when we used the default number of bins, it now seems to be underestimating. The key here is to remember that the probability density is **not** a probability. The probability, when we plot density, is the area of the bar. So while that first bar got taller, it also got thinner.

Generally, your modelled or fitted distribution should be independent of any scaling or changes in visualization, but it's important to be able to distinguish *artefacts* of changing bin size versus fundamental discrepancies in our model versus the data.

This leads to the natural question: *how many bins should a histogram have?*

[66] To say the least

Many people have asked this, and some have even attempted an answer. I'm partial to Scott's normal reference rule because it's quite simple and robust.

Definition 10.12.1. Scott's normal reference rule states that the best bin width, h, for a histogram is

$$h = \frac{3.49s}{n^{1/3}}$$

where s is the sample standard deviation and n is the number of data points. This means that the optimal number of bins is

$$b = \lceil \frac{M - m}{h} \rceil$$

where M is the maximum value in the data and m is the minimum value in the data. The symbols $\lceil \ \rceil$ represent the **ceiling function**; it means round up to the nearest integer.

Scott's rule fails pretty spectacularly when you have a large standard deviation and a very small number of points[67]. Tread carefully!

[67]Notice for our data set, where this is indeed the case, Scott's rule gives us almost half as many bins as Python's default computation.

10.13 PRACTICE PROBLEMS

Permutations and Combinations

1. Humans have 23 pairs of chromosomes, for a total of 46 chromosomes. Of these pairs, one copy comes from the mother and the other copy comes from the father. How many possible genetically different children are possible from two parents[68]?

2. There are 20 possible opening moves in chess[69]. If white moves first, then black moves, how many possible board configurations exist before white makes their second move?

3. A car vehicle identification number (VIN) is a 17-character serial number made up of digits one to nine and letters A to Z[70]. How many unique vehicles could be produced under this system?

4. I have salami, cheddar, sliced tomato, lettuce, banana peppers, proscuitto, and mustard, and I want to make the best possible sandwich in terms of flavour and mouthfeel. I have already determined that a good sandwich can't have more than four toppings, including those that have been doubled[71] up, and mustard must be on the top piece of bread or the bottom piece of bread[72]. How many sandwiches do I have to test in order to make an accurate assessment?

5. If the order doesn't matter because I plan on shovelling the food into my mouth like a monster, how many possible sandwiches can I make?

6. If you own 12 t-shirts, six sweaters, and three coats, how many ways can you layer three tops for a cold winter day if we **don't** consider what makes physical sense[73]?

7. How many options do you have if you must wear one t-shirt followed by one sweater followed by one coat[74]?

---

[68]This is a lower limit as there are all kinds of funny things that happen when creating gametes, which result in chromosomes sometimes mixing together!

[69]Each pawn can be moved either one or two spaces forward; each knight can be moved forward right or forward left.

[70]Capital only

[71]Or tripled, or quadrupled

[72]No one puts mustard between two toppings, except villains.

[73]In other words, for example, wearing a t-shirt over a coat is a perfectly acceptable thing to do or wearing three t-shirts is not non-sensical.

[74]Exactly as you would probably think to layer these clothes

8. If you were to pick three tops at random and commit to wearing them in the order you pick them[75], what is the probability that you pick a t-shirt, then a sweater, and then a coat?

Probability Notation

1. Consider a sample space, Ω, and a function P that acts on Ω. If

$$P(\Omega^C) = 0.00001$$

is P a probability?

2. Consider the following Venn diagram. Are events A and B mutually exclusive? What about A and C, or B and C?

3. Again referring to the Venn diagram in figure 10.13, what probabilities must we know in order to compute $P(A \cup B \cup C)$?

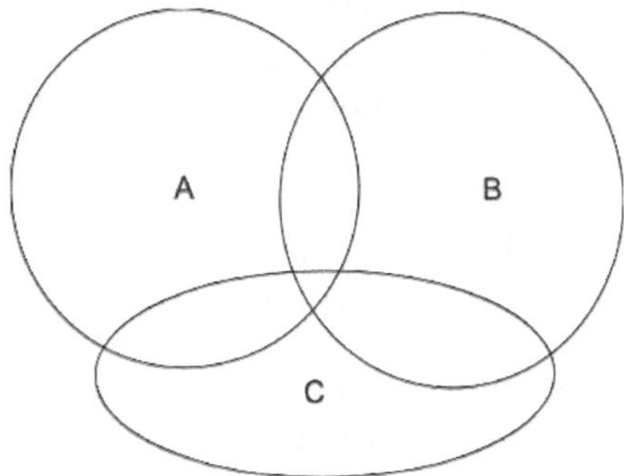

Figure 10.13: Diagram for questions 2 & 3

4. Consider the Venn diagram in figure 10.14. If $P(A) = 0.4$ and $P(B) = 0.2$, what is the probability of $(A \cup B)^C$?

5. Let's say we want to study the global age distribution of all humans at a specific time, say today[76]. If we define

$$\Omega = \{a | 0 \leq < a \leq 22\},$$

is this a proper sample space for our scenario? Why or why not?

---

75 Say in an early morning haze

76 Where today is literally whenever you're reading this

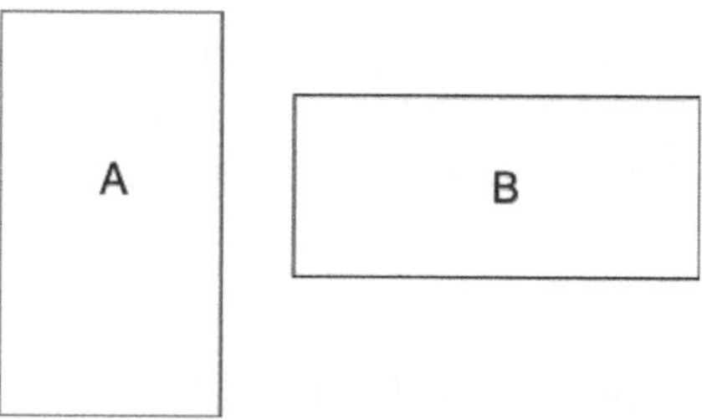

Figure 10.14: Diagram for question 4

6. Colours are sometimes represented digitally as RGB values, (R, G, B), which is a tuple of three integers ranging from 0 to 255. The first component represents how much red is in the colour, the second how much green is in the colour, and the third how much blue is in the colour. For instance, the colour pure red is given by (255,0,0) since it is all red, zero parts green, and zero parts blue. Black is given by $(0, 0, 0)$ since it is the absence of colour, and white is given by $(255, 255, 255)$ since it is the sum total of all colours. Write out, in set notation, the sample space of all digital colours, Ω.

7. Let's define event A as all colours which have equal parts red, green, and blue. How would you write this in set notation?

8. What is the probability of a digital colour chosen at random has equal parts red, green, and blue?

9. What is the probability that a colour has a non-zero red component?

10. What is the probability that a colour has a non-zero green component?

11. What is the probability that a colour has a non-zero red component **and** a non-zero green component?

12. What is the probability that a colour has a non-zero red component **or** a non-zero green component?

Conditional Probability & Bayes' Rule

1. We know from above that $P(A^C) = 1 - P(A)$. Does this hold true for conditional probabilities?

2. To start, show that if

$$P(A^C|B) = 1 - P(A|B)$$

then

$$P(A^C \cap B) = P(B) - P(A \cap B).$$

3. We can rearrange the above equation as

$$P(A^C \cap B) + P(A \cap B) = P(B).$$

If this is true, then our original statement, $P(A^C|B) = 1 - P(A|B)$, is true! Using a Venn diagram, show that the events $A^C \cap B$ and $A \cap B$ are mutually exclusive.

4. If $A^C \cap B$ and $A \cap B$ are mutually exclusive, show that $P(A^C \cap B) + P(A \cap B) = P(A^C \cap B \cup A \cap B)$.

5. Using all the above, argue that $P(A^C \cap B) + P(A \cap B) = P(B)$ is indeed true, and therefore $P(A^C|B) = 1 - P(A|B)$ is also true.

6. Consider the following events:

   A: a household with at least one dog

   B: a household with at least one cat

   In 2020, in Canada, it was reported that of all households in Canada, 58% of households have at least one cat or dog, 35% report to have at least one dog, and 38% report to have at least one cat.

   With this information, what is the probability that a household has one dog *and* one cat?

7. If we know a household has a cat, what is the probability that a dog also resides in the household?

8. If we know a household has a dog, what is the probability that a cat also resides in the household?

9. Consider the events:

   A: a roll of two dice has a sum of 7.

   B: one of the dice is showing a 4.

   What is the probability $P(A|B)$?

10. What is the probability $P(B|A)$?

11. What is the probability that one of the die does not show a 4 given that the sum of the two dice is 7?

12. Show that, at least in this example, $P(A|B) \neq 1 - P(A|B^C)$.

13. Why might we not expect $P(A|B) = 1 - P(A|B^C)$ to be true? Use our method from question 1. Where does it break down?

14. The true positive rate[77] for a chlamydia urine test is 99.3%. The probability of a random person in a population having chlamydia is roughly 5%. If we tested 2,500 students on a campus at random and 200 tests come back positive, what is the probability that any individual is infected with chlamydia given that they have tested positive[78]? In other words, what is the probability that a random individual does *not* have chlamydia given that they test positive?

15. Let's say we do another test on the same campus one year later, and we find that 900 out of our 2500 tests come back positive. Given that the true positive rate of the test is the same (the test hasn't changed) and there's no reason to believe the prevalence has increased[79]. What might we conclude happened? Should we order new tests?

16. PROBABILITY DISTRIBUTIONS

17. Apple's iOS requires a six-digit passcode made up of digits 0 to nine. After ten wrong guesses, the iPhone will permanently lock. If you are trying to guess a person's password randomly, what is the probability that the iPhone gets locked permanently?

18. The number of insurance claims made to a small insurance company on a given day is modelled by a Poisson distribution with parameter $\lambda = 12$. What is the probability of receiving exactly ten claims in a day?

19. What is the probability of receiving between eight and 12 claims in a day?

---

[77] The probability that you receive a positive test given that you indeed are infected

[78] How can we get a false-positive rate for our test from the above information?

[79] The diseases is in equilibrium, or an endemic state.

20. In one month, what is the probability of receiving between eight and 12 claims for 15 days in a month, assuming a month has 30 days?

21. In a criminal case, the prosecution presents DNA evidence from the scene of the crime as a match to the defendant. We can assume that our hypothetical court is founded on the principle that everyone is presumed innocent and must be proved guilty. We know the following facts:

    - There was DNA found at the crime scene.
    - The DNA matches the defendant.
    - The probability of a random DNA match is 1 in 4000000.
    - The prosecution tested the crime scene DNA against a database of 30000 people.

    What is the probability of finding at least one DNA match in a database of 30000 people?

22. Given a normal distribution of

$$N(x;\mu,\sigma) = \frac{2}{\sqrt{\pi}} e^{(x-\mu)^2/2\sigma^2}.$$

    can you write an approximate mathematical expression for the empirical rule using integrals? You, obviously, do not need to solve these exactly.

23. Consider the following table of data[80]

Participant ID	Sex	Participant ID	Sex	Participant ID	Sex
1	F	11	F	21	F
2	M	12	M	22	M
3	F	13	F	23	M
4	M	14	M	24	F
5	M	15	M	25	M
6	F	16	M	26	F
7	F	17	F	27	F
8	M	18	M	28	M
9	M	19	M	29	M
10	M	20	M	30	M

[80] This is self-identified sex data from a class list for Intro Data Science.

If we count F as a “success” and assume the sex of students is independent from one another, what is the most probable probability that a student is female?

24. In Canada, females make up roughly 50.4% of the population. Extrapolating from the above results, is the make-up of the Intro Data Science class representative of the population at large?

25. According to StatCan, females make up roughly 56.5% of the total post-secondary student body in Canada. Given the above results from your Beta distribution, argue why we can extrapolate from our sample to say that there is a large gender bias in mathematics[81]. How much more likely is our mean female frequency than a frequency of 0.565, given our data?

26. Using all 184 students in the class, we find that, experimentally, the proportion of females in the class is 0.359; how close is this to the most likely value given by the Beta distribution using a sample size of 30?

THE AIRPLANE PROBLEM

27. Let's say there is a small jet with ten seats in on it, and ten passengers who have each been assigned a seat. The passengers are lined up and ready to board the plane, but there is a problem. The first person to board forgot their seat[82]! Instead of trying to find a solution to their[83] problem, they just sit in a random seat and hope for the best. The first question is thus: what is the probability that they have sat in the correct seat?

28. Now the second person boards the plane and luckily remembers their assigned seat. One of the two things can happen now: either person 2 finds their seat unoccupied and sits where they are supposed to, or their seat is occupied by the entitled first passenger. In the case of the latter, person 2 picks a different seat at random and sits down. What is the probability that person 2 is sitting in their assigned seat?

[81] And, more generally, all of STEM

[82] Assume that on this magical plane the gate attendant takes your boarding pass and shreds it once you board.

[83] Given their behaviour, likely *his*

29. Now person 3 boards the plane, repeating their seat number to themselves as if it were a prayer. Person 3 follows the same rules as person 2: if their assigned seat is empty, they sit in it. If their assigned seat is occupied, they pick a seat at random and sit there. What is the probability that person 3 sits in their assigned seat?

30. Everyone boards the plane and follows these same rules. What is the probability that person 10 gets to sit in their assigned seat?

31. Extend this to a Boeing 747-8 which has a passenger capacity of 467 people. What is the probability of person 467 sitting in their assigned seat?

32. Make an argument why this rule set, even if extended to a stadium of $50,000$ people, always guarantees that the last person to enter has a 50% chance of sitting in their assigned seat.

33. What is the probability that at least one passenger is affected by the belligerent profligation of person 1? In other words, what is the probability that at least one passenger must change seats?

THE AIRPLANE PROBLEM, EXCEPT MORE JERKS

Let's say that it's not just the first person who enters the plane who decides to disregard decorum and our social contracts and sit wherever they want, but say the second person entering the plane does this as well. That is, even if the person 1 is *not* sitting in person 2's seat, person 2 will still choose where to sit randomly. Let's call these people jerks for lack of a more accurate word.

34. On a three-person private jet, where this is happening, what is the probability that person 3 sits in their assigned seat?

35. Let's say we still have two jerks (person 1 and person 2), but now we have four seats and four passengers. What is the probability that person 4 sits in their assigned seat?

36. Can you generalize this to get a probability that person N ends up in their chosen seat if there are k jerks[84]?

[84] And of course at least $k+1$ possible seats

THE POLYGRAPH TEST

37. A polygraph test intends to differentiate statements as true or false depending on a few biometric markers present while a subject is answering questions. In effect, the goal of a polygraph is to determine if someone is lying or not. It's been found that polygraph tests are fairly unreliable, with a false-positive rate of around 7%, and a false-negative rate of around 16.5%. Using Bayes' theorem and the law of total probability, can you determine the probability of someone lying given that they have a positive polygraph test as a function of $P(L)$, the probability of lying?

38. Using your above formula, what would the probability of lying, $P(L)$, need to be in order to be certain of lying given a positive test? In other words, what does $P(L)$ need to be to ensure that $P(L|+) \geq 0.99$? (We are 99% certain of lying given a positive test.).

39. Most people have no reason to lie. If the probability of lying is $P(L) = 0.12$, should we trust our polygraph test?

CHAPTER 11

Fitting Data

Now that we have some tools under our belt for describing data, we can start to think about predicting things *from* our data. It is here where our data and information begin to synthesize into knowledge.

I will give the caveat here that I am a mathematical modeller by trade and that I apply these skills to biological problems. My interpretation and methods when it comes to predictive data analysis is heavily driven by my motivating problems. I have developed a framework for exploratory predictive data analysis based on how I need and use it. That doesn't mean it's the only way to do things, and you may learn other techniques and other workflows from other places[1] that are just as good or better!

11.1 DEFINING RELATIONSHIPS

We often can see a *qualitative* relationship[2] just by plotting two things against each other; if we would like some *quantitative* information about the relationships, we must define the relationship mathematically.

To do this, we need to know about **functions**. Mathematical functions take an input x, manipulate it, and give back an output y. Usually, we write functions as

$$y = f(x).$$

In words, this means *y is related to x by way of f*, where f is a series of mathematical manipulations[3] that bring us from x to y.

[1]Or perhaps develop your own

[2]When X increases, Y increases; X and Y are unrelated. These are examples of qualitative relationships.

[3]Think addition, multiplication, exponents, etc.

DOI: 10.1201/9781003265405-11

Linear Relationships

The simplest possible relationship between two things is a linear relationship as such we have a lot of ways of describing linear relationships. Two things are linearly related to each other if

- They are *directly proportional* to each other.
- When plotted, a straight line can be drawn through all points.
- For any two points, the value of

$$\frac{y_k - y_i}{x_k - x_i}$$

  is constant.

We can sum all this up in the formula

$$y = mx + b.$$

The function in this case is $f(x) = mx + b$. This is the formula for a line. The *parameter* m is called the **slope** and b is called the **y-intercept**. The slope tells us the increase you can expect in y if you change x by 1. The y-intercept tells us what value of y you should expect when $x = 0$.

11.2 DATA AND LINES

So, from what we've seen so far, data does not often follow any kind of discernible curve. In 11.2, we can see the difference between perfectly linear data and more realistic, imperfect data[4].

When data is perfect, the line we should draw through it is obvious. When it's imperfect, there are many, equally valid lines we could draw like in figure 11.2. How do we determine which of these lines is the *best* at describing the linear relationship in the data?

That depends on our description of how the data relates to the line[5]. So we can describe the data points, x_i and y_i, as being related by the line, but then moved off it by an amount ε_i, an **error** term. We start on the line and then add or subtract an ε_i to get to the data point. Mathematically, this looks like

$$y_i = mx_i + b + \varepsilon_i$$

What this equation says is y *is related to* x *linearly, but with some error* ε.

[4] In truth, even this manufactured example is more perfect than most real data.

[5] And how we define *best*

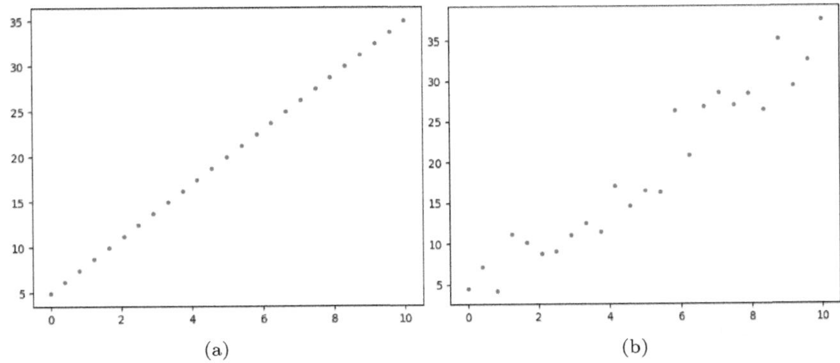

Figure 11.1: Perfectly linear data is obvious, but rare. Imperfect linear data is less rare, but leads to ambiguity.

Now we can answer the question, *which line is best?* We will define the **best fit line** as **the line which has the smallest total error**. In other words, we want to find an m and a b such that

$$E = \sum_{i=1}^{N} \varepsilon_i^2$$

is as small as possible. The line that does this is called the **best fit line in the least squares sense**. There are other definitions of *best.*

We could do this by hand or go into the mechanisms behind how it's done, but that's beyond our goal here[6].

Least Squares in Python

IN PYTHON, WE USE A PACKAGE CALLED `lmfit` to model data. It is not necessarily the quickest or easiest way to draw lines of best fit through data, but it's what I use and it is arguably the most robust set of tools.

For linear fitting, we will need to import some tools from `lmfit`:

```
import lmfit.Model as Model
```

BEFORE EVEN IMPORTING OUR DATA into Python, we need to talk about Python functions[7]. A function in Python is just like a mathematical function: it takes some inputs and gives us an output. If we would like to

[6] And is the primary goal of Chapter 4

[7] There's a brief discussion in the Appendix.

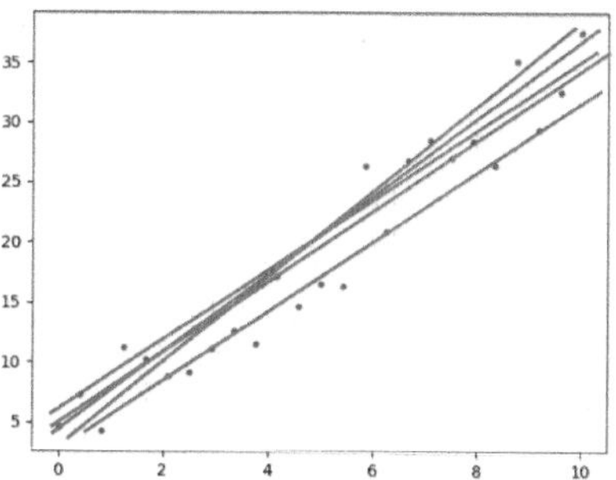

Figure 11.2: All of the lines on this plot are equally valid given the variation in the data. While they all look similar, if we try to predict outside this range, they will all start to give very different results. Only one of these can be considered the best line.

fit a mathematical function to data, it makes sense to make a Python function that corresponds to our mathematical function. In this case, our mathematical function is

$$y = mx + b$$

and in Python we will write

```
def linear_function(x,m,b):
                y = mx+b
                return y
```

The `return` keyword tells Python what the output is. This literally translates to *when I use the command* `linear_function` *with three inputs, calculate* $mx + b$*, and give me back* y. For instance, the command

```
linear_function(2,3,4)
```

will give back `10`.

LET'S WORK WITH THE DATA SET `heart.csv`[8]. This gives some data on different patients and measurements related to their cardiovascular health. If we create a scatterplot of age versus resting systolic blood pressure, we get figure 11.3. While the data doesn't have a strong relationship, there does seem to be a trend upward and to the right, enough of a trend that we may want to fit a line through the data.

[8]Found at https://www.kaggle.com/datasets/johnsmith88/heart-disease-dataset

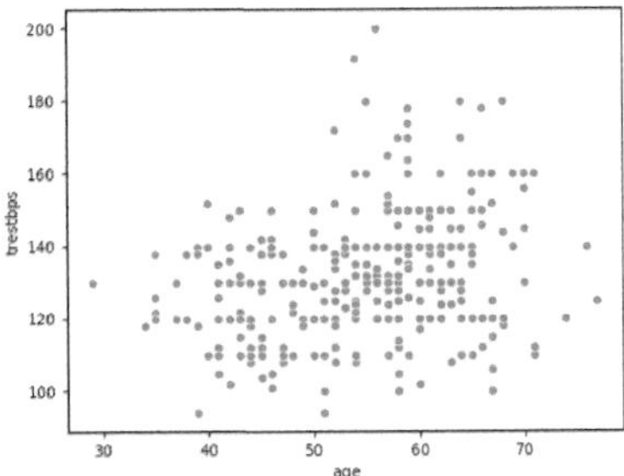

Figure 11.3: A graph of patient age vs. resting blood pressure.

To do so, first we need to define a model to use

```
linear_model = Model(linear_function,independent_vars=['x']).
```

This tells us to build a model[9] using the function `linear_function` and treating x as input data. Implicitly, this leaves m and b as parameters that we will fit.

Next, we actually perform the fit using

```
result =linear_model.fit(data['trestbps'],x=data['age'],m=1,b=1)
```

Let's break down this line:

- The `.fit` tells us Python that we intend to use our `linear_model` to fit some data.
- We give it our **output variable first**. In this case, it is the resting blood pressure. We do not specify a label of the form y= for our output.
- Next, we give it any input variables we have, specifying them by name. In this case we have only one, x. We still have to specify it by name.
- Next we are required to provide initial guesses for m and b. For simple, linear models, these initial guesses shouldn't affect the outcome so they can be anything.

We can then plot our fit with the command

```
sns.lineplot(x=data['age'], y = result.best_fit),
```

and we should get figure 11.4.

[9]For fitting

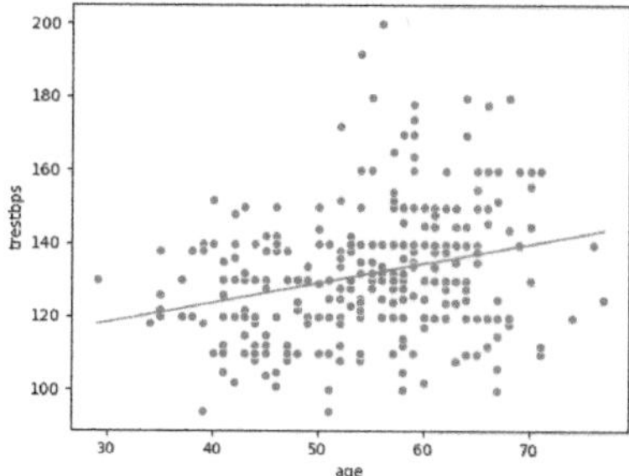

Figure 11.4: A graph of patient age vs. resting blood pressure with line of best fit.

```
[[Model]]
    Model(linear_function)
[[Fit Statistics]]
    # fitting method    = leastsq Method of fitting
    # function evals    = 6        "How long the fitting took"
    # data points       = 303      Number of points used
    # variables         = 2        Number of variables fit
    chi-square          = 85642.1720
    reduced chi-square  = 284.525489  Stuff we won't
    Akaike info crit    = 1714.19269  worry about
    Bayesian info crit  = 1721.62016
[[Variables]]
    m:  0.53944523 +/- 0.10687369 (19.81%) (init = 1) Value of fitted
    b:  102.296101 +/- 5.89058338 (5.76%) (init = 1)  parameters
[[Correlations]] (unreported correlations are < 0.100) with std. dev.
    C(m, b) = -0.986
```

Figure 11.5: The report on our fit.

We can also get some numerical data on the fit by using the command

```
print(result.fit_report()),
```

which gives us a table much like the one in figure 11.5. This tells us quite a bit of information like the fitting method used (least squares, as discussed above); `function evals` roughly translates to how long it took to fit[10], the number of data points used, the number of variables fit, a bunch of stuff we'll worry about later, and then the values of m and b. If we want to isolate the parameters of our model, we can use the commands

```
m_best = result.params['m'].value
b_best = result.params['b'].value
```

We can then use the slope and y-intercept as the numbers they are[11].

In this case, m is fit to be ≈ 0.54. This means that for every year older you get, on average your resting blood pressure increases by 0.54 mmHg. We see that b is ≈ 102, which means that if you're 0 years old,

[10] If this number gets very high, the fit may be bad!

[11] This will come in handy later

your blood pressure is 102 mmHG[12]. Is this a good predictive model? Probably not for ages below 30, but possibly for people in their 80s and 90s. Do m and b give us information we didn't have before? Yes. Can we maybe use this data if we were cardiologists? Maybe.

What if we want to fit a different function? We just have to make a few modifications! We could fit an exponential function to the same data. Exponentials come in the form

$$y = Ae^{bx}$$

where $e \approx 2.71$.

All we need to do is create a new function

```
def exp_func(x,A,b):
        y = A*np.exp(b*x)
        return y
```

then create a new model

```
exp_model = Model (exp_func,independent_vars=['x'])
```

and change the label on our guesses

```
result_exp=exp_model.fit(data['trestbps'],x=data['age'],A=1,b=1).
```

If you plot this, the fit will be terrible. This is clearly not a good model for this data, but what's important is that you can *change* the model.

We could try a quadratic model:

$$y = ax^2 + bx + c$$

We can modify our code to fit this model exactly as we did above for the exponential model[13]. Figure 11.6 shows this fit, it's not very different from our linear fit, but you see a bit of curvature.

Which of these fits is better? Which is more valid? Complication, just for the sake of complication, is usually frowned upon, but sometimes it's necessary! If only there was a way to start to determine which of these is better...

[12]Is this realistic?

[13]Don't forget that x^2 is x**2 in Python.

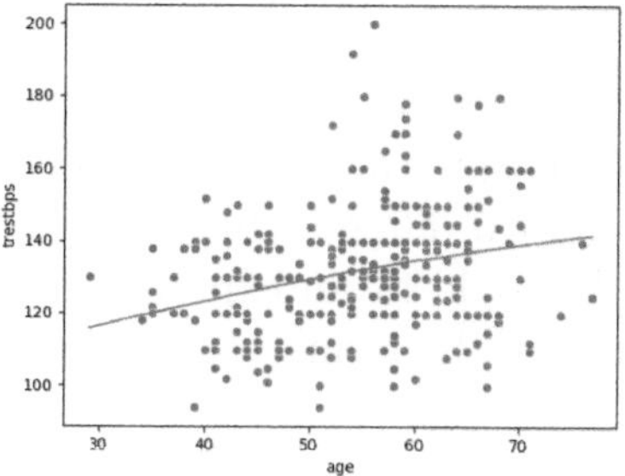

Figure 11.6: A quadratic fit for our data.

Goodness of fit

If you try to answer the question, *how do I know if my model is the best model?* by looking in books, online, or in academic literature, you will likely leave with less of an idea than when you started. There are many different ways to quantify the performance of one model over another, and many different strategies for choosing the *best*[14] model.

Generally, the best model is the one that has a strong theoretical underpinning. Mechanistic models[15] cannot only help us explain any patterns in data, but they can provide insight into *why* we are seeing certain behaviours and possibly what is not being considered and its possible effect on the observed relationship. If we have a model that fits the data perfectly, but does not have a valid interpretation or uses a mechanism that would undo a lot of established science, then it's probably not a good model. The latter caveat definitely has its exceptions.

The Solar System

As an example, consider our model of the solar system. Today, it would seem absolutely backwards to state that the earth is the centre of the universe, but the absurdity of the statement comes from how much we know about the space around us. Some of the first observations humans could make were

- The earth does not seem to move because we cannot feel it move.
- Everything seems to exist on a cycle of one day. The sun rises once per day, as does the moon, the stars, and the planets.

[14] Whatever that even means

[15] In my extremely biased opinion

Using just this information, the easiest, simplest conclusion is that the earth is stationary and everything is moving around it. Philosophically, it also *feels* nice: a universe with the earth at the centre makes us special and reinforces the fact that we are special and unique in the vastness of space. That's not to say that heliocentric models, which places the sun at the centre[16], were not posited[17]; they just didn't gather much traction because they could not easily explain, for instance, why we cannot feel the motion of the earth, nor how the daily cycle works.

Plato would be one of the first to formalize a model of the universe. He would place a spherical earth[18] at the centre of the universe, and the moon, sun, planets, and stars existed on spherical orbits around the earth.

Aristotle, a student of Plato, would expand upon this. More observation led to more data on the positions of planets in the sky. There was also the cycles of the moon that needed to be explained. Aristotle used dozens[19] of invisible spheres to explain the moving planets[20] and the phases of the moon.

This model started to break down as data accumulation grew. It still didn't quite explain all the planetary movement or how sometimes planets could disappear from the sky or become brighter or dimmer. It definitely didn't explain retrograde motion: the phenomenon where certain planets at certain times would appear to move backwards through the sky.

In the second century, Ptolemaeus had the idea that maybe it wasn't a series of sphere-like canvases surrounding earth, but maybe each individual planet, the moon, and the sun each existed on a circular path around the earth. The planets, the moon, and the sun also had an epicentre that existed on this circular path, but *outside* of the planet. The conclusion was then that each planet orbited its epicentre and each epicentre orbited the earth. The stars, still, remained fixed on a sphere at the edge of the universe.

People liked this idea. *Really* liked this idea. As more data came about, more precise measurements were made, and more circles on circles were added to Ptolemaeus' model to account for new observations. The geocentric model of the universe was becoming increasingly complex.

[16]And, surprisingly, the planets in the correct order

[17]Most notably by Aristarchus of Samos

[18]Yes, we have accepted as a people that the earth is round since about 400 BCE.

[19]Roughly 50

[20]Each planet was now on multiple spheres.

The assumptions started to break down in 1572 when Tycho Brahe discovered a new star in the sky[21]. In 1604, he discovered *another* new star. Perhaps the sphere of stars wasn't as fixed as humans had assumed.

It was around this same time Copernicus breathed new life into the heliocentric model of Aristarchus. His model that we were not *really* different than the other planets didn't sit well with a lot of people. While public opinions were deeply opposed to Copernicus and his model, it explained a lot of the observations with a far simpler system: the sun is central in the universe, and the earth along with the other planets orbit the sun. The moon orbits the earth. The stars, still fixed upon the backdrop.

It was Galileo who helped push heliocentric thinking to the forefront. Hearing descriptions of a new German invention that could magnify faraway things, Galileo built a telescope and started looking at the sky with it. Specifically Jupiter. The data he collected further tested the assumptions of the geocentric model of the universe. He saw that there were moons around Jupiter and that these moons moved in simple circles around the planet. It would take a lot of creative explanation as to why Jupiter's moons didn't orbit the earth the way everything else seemed to. Copernicus's model takes this into account by allowing planets to orbit stars and moons to orbit planets. Of course, things would not almost fully make sense until Kepler did away with the assumption that things move in circles[22], and replaced the orbits with ellipses. Instead of having circles upon circles and epicycles, we had simple, consistent rules for the orbiting of planets.

Newton would put the final nail in the coffin of a geocentric universe; using his models of gravity and motion[23] and extending these ideas to the motion of the planets. We still, mostly, use these models today; they are taught in almost any introductory physics course.

This was a long story, but I hope gives some perspective. Models are allowed to change; our understanding is allowed to change. Models are only as good as our observations and imaginations allow. The geocentric model of the universe was good enough for about 1500 years, and, frankly, for the lay person it would likely continue to be a fine explanation. The complexity was eventually overtaken by a simpler model that explained *more*. This is the philosophy that most use to define a *good model*. The model should

[21] In the constellation of Cassiopeia

[22] This assumption originally came about because the Greeks thought the circle was the perfect shape.

[23] Developed through observation and data

be the simplest possible model that explains the observations. Of course, Newton's models still couldn't explain everything, but we wouldn't know that for a few more hundred years. When observing the orbit of Mercury very precisely, one might notice a bit of a wobble that isn't accounted for with Newton's models. Einstein and his theory of relativity expanded upon Newton's model and managed to explain the precession of Mercury *and more*[24].

This story, I hope, imparts the flow of ideas. Observation leads us to ask how and why, which leads us to build models to explain the world we live in. Any model that can do so, like the geocentric model that worked for over millennia, is good enough. Eventually, when faced with an ever-increasing number of epicycles to explain what we see[25], we may want to rethink our model. Often, when translating the real world into the language of mathematics, I ask myself, *is it worth the epicycles?* as a reminder that sometimes we need to re-evaluate from first principles and challenge assumptions we thought were ironclad.

Standard Error of the Residual

There are many ways to quantify *simple* and *how well a model explains data*. Generally, we say a model is simple if it has fewer parameters. An exponential distribution is simpler than a gamma distribution because it has one parameter instead of two. A line is of the form

$$y = mx + b,$$

has two parameters, and is simpler than a quadratic model

$$y = a + bx + cx^2$$

which has three parameters.

Quantifying how well we explain the data can be done with something like a variance. Generally, we usually want to explain one piece of data using other pieces of data. We have input data X_i and output data Y_i. In our heart rate example above, our input is age and our output is resting heart rate. How well we explain the data is given by the difference between our model values, y_i, and our observed values, Y_i. If we have N data points and a model $f(x)$, then we can measure the error (also

[24]In fact, it is Einstein's special relativity that allows GPS satellites to work.

[25]I'm sure there is some formulation where the moons of Jupiter *appear* to be orbiting Jupiter, but it really is just epicycles upon epicycles of earthly orbits.

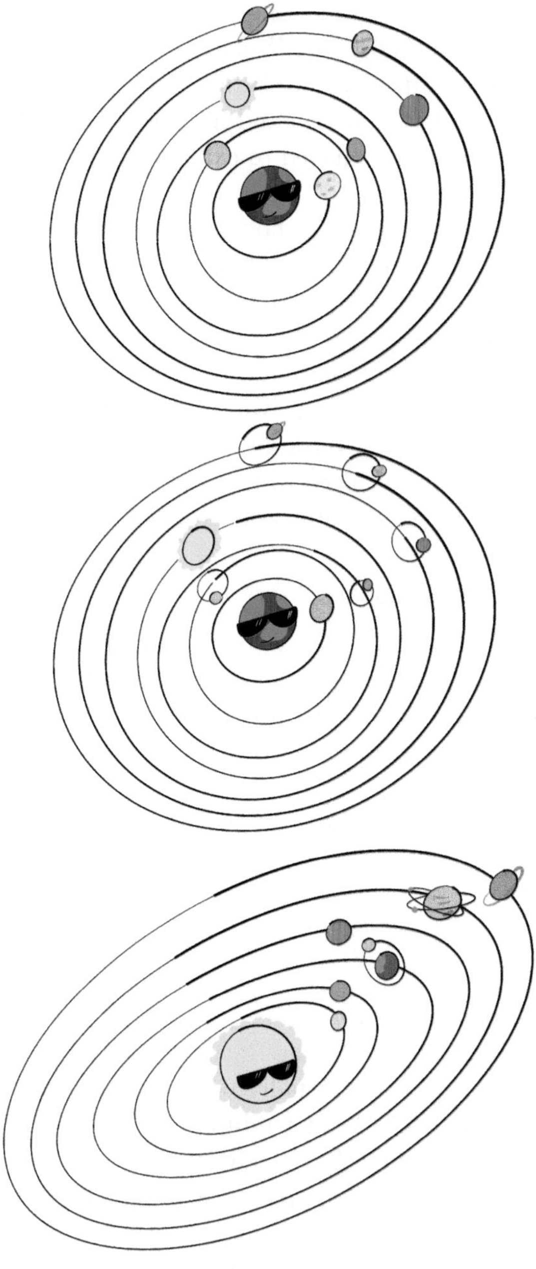

Figure 11.7: The evolution of models of the solar system: a visual dramatization in three parts

known as the residual), R, as

$$\begin{aligned} R &= \sum_{i=1}^{N} (Y_i - y_i)^2 \\ &= \sum_{i=1}^{N} (Y_i - f(X_i))^2 \end{aligned}$$

Of course, this number will be larger if we have more data points to compare since N will be bigger, so we should account for the number of data points. We also want to incorporate some notion of *simple.* We will subtract the number of parameters in our model, m, from the number of data points, N, and divide. This penalizes our model for having lots of parameters by making the denominator smaller[26]. The value $N - m$ is called the **degrees of freedom** of a model on a data set.

We are now in a place where we can define standard error.

Definition 11.2.1. The **residual standard error** of a model $f(x)$ explaining data (X_i, Y_i) is given by

$$S = \sqrt{\frac{\sum_{i=1}^{N} (Y_i - f(X_i))^2}{N - m}}$$

This gives the average distance from the model of a single data point *in units of* y. The smaller the standard error, the better the model explains the data.

A standard error of 0 means your model perfectly fits the data. This is less desirable than you might think and often leads to the data being high suspect of tampering.

We have three potential models for resting heart rate: linear, exponential, and quadratic. We can compute the standard error for each. Surprisingly, there isn't a built-in function for the standard error of the residual in Python, but that's why Python gives us the ability to write our own functions[27]! We will call our function `rste` for **r**esidual **st**andard **e**rror.

Like before, we need to tell Python we would like to define something new. These definitions should come below our import statements. Essentially, what we would like to do is convert our mathematical expression in

[26] This is my intuitive explanation; in truth the number of parameters is equal to the number of points you can fit *exactly*, so we discount those from the total number of points.

[27] See Appendix A for a brief summary on what functions do.

the definition into Python language so that the computer can understand what we mean by residual standard error. Our function requires three things in order to compute S: the degrees of freedom, the data points, $Y - i$, and function values evaluated at our X_i's.

```
def rste(Yi,fXi,dof):

        s = np.sum((Yi-fXi)**2)/dof

        s = np.sqrt(s)

        return s
```

While we've split the mathematical expression into two lines for clarity, if you examine closely, you will see that all the pieces are there.

Now to actually *find* the residual standard error, we have to *use* this function somewhere in our code just like we would any other Python command.

For our linear fit, we have parameters m and b, but they're hidden in our `result`. First thing we should do is extract them:

```
m = result.params['m'].value
b = result.params['b'].value
```

Now, we can generate $f(X_i)$ by putting these values into our linear function.

```
fX = linear_function(data['age'],m,b)
```

Our linear model has two parameters, m and b, and therefore our degrees of freedom are two fewer than the number of data points in our data set. Using this information, we can compute our residual standard error:

```
res_std_lin = rste(data['trestbps'],fX,len(data)-2)
```

We can work similarly for our exponential fit and our quadratic fit.

The standard error for all three models is roughly 17. This means on average our data is ± 17 bps away from our line. None of these models are great. Generally, the smaller the standard error of the residual, the better. Does this mean we should take this to be law? Always select the model that has the lowest residual standard error?

No. Your first and main line of defence when championing a model should be that it explains the data and it makes sense in the context of other specialized knowledge of your field. In this particular case, of any polynomial model, a cubic model of the form $f(x) = ax^3 + bx^2 + cx + d$ seems to have the lowest residual standard error. The highest residual standard error for polynomials of reasonable size is if we fit a constant function $f(x) = c$, which says that resting systolic blood pressure is independent of age and is roughly 131 mmHg. Normal[28] for an adult is 120 or less, but this data comes from hospital admissions, so we might expect elevated blood pressure for a number of reasons. Our specialized knowledge tells us that our fit of a constant function might be the best, despite the high residual standard error. Perhaps we're just looking at the wrong input.

Other information tells us that blood pressure tends to go up with age[29], so maybe our linear model has some bearing. The cubic model, should you choose to work through the example, seems to oscillate, and to date I have yet to find any evidence that blood pressure does this with age.

In the absence of any other information, the residual standard error can help you determine a direction in which to look, but as always *context* is the most important factor.

A VERY SPECIAL CASE

Many, many statistical tools you will go on to learn only really work if your residuals are normally distributed around 0. If they are not, it means that your model is missing something more than just random variations; usually some fundamental relationship between x and y. If your residuals are normally distributed around 0[30], then the residual standard error is exactly equivalent in Python to

```
S = np.std(data['trestbps']-fX,ddof=len(data)-M),
```

where M is the number of parameters in your model. In the case of our heart rate data, we can indeed use this because if we compute

```
res_mean = np.mean(data['trestbps']-fX),
```

we see that the mean of the residuals is 3×10^{-14} which, for all intents and purposes, we can treat as 0.

[28]For a very loose definition of normal

[29]Your arteries naturally get stiffer, which leads to higher pressure.

[30]Or at the very least have mean 0

11.3 DISTRIBUTION FITTING AND LIKELIHOOD

Let's look at the Age column of heart.csv as a histogram,

```
scotts = 3.49*np.std(data['age'])/(len(data)**(1/3))
sns.histplot(x=data['age'], binwidth = scotts).
```

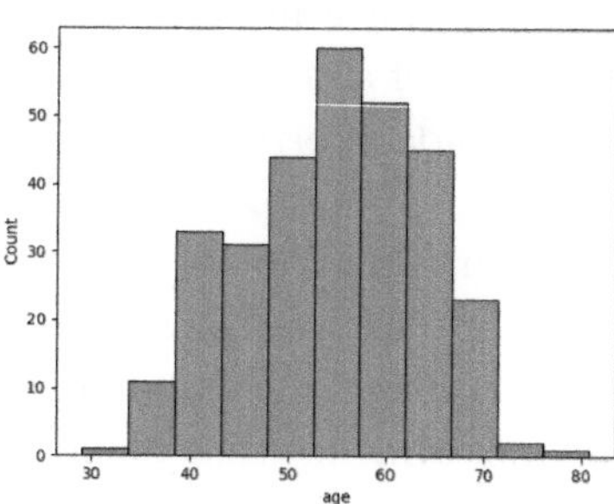

Figure 11.8: A histogram of ages from the public heart.csv dataset

This histogram is shown in figure 11.8. We can see the data is not quite normal. It seems skewed to the right a little bit. We may want to figure out what distribution we are sampling from. If this data was indeed all individuals who were admitted to the hospital, knowing the underlying age distribution of our data may tell us something about how many people of a particular age we should expect in a given time period in the hospital.

Since the data is skewed a little bit to the right, and in theory the ages can run from 0 to ∞[31], we can try to fit this to a gamma distribution. As a reminder, the gamma distribution is defined as

$$G(x; \theta, k) = \frac{1}{\Gamma(k)\theta^k} x^{k-1} e^{-x/\theta}.$$

Conceptually, we fit this the same way we do a function model to a relationship in data: we want to optimize *something*. The difference is in what we want to optimize.

Before fitting a histogram of data to a distribution, it's important to note some assumptions: the data in the histogram must be independent[32], and all the data must actually belong to the same underlying distribution[33].

[31] Ok, we can't have people who are infinity years old, but it's an approximation.

[32] One data point does not affect the others.

[33] We could not mix age data with a scaled version of systolic blood pressure say and make one histogram. The two values come from two different distributions.

In our case, we would like to fit a gamma distribution, which means we must prescribe two parameters: k and θ. We know the data we've seen, so if we pick a candidate k and θ pair, we can calculate the probability of k and θ being what we guessing them to be *given the data we've observed.* For our first age data point, X_1, this means the probability density is

$$p(k,\theta|X_1) = \frac{1}{\Gamma(k)\theta^k} X_1^{k-1} \mathrm{e}^{-X_1/\theta}$$

We don't only need to know the probability density of k and θ at X_1 though; we need to know the probability density for all of our data points *simultaneously* for a gamma distribution with k and θ[34]. Since the data points are all independent, this is just the product of all the probability densities:

$$L(k,\theta;X_1,X_2,\cdots,X_n) = \prod_{i=1}^{n} G(X_i;k,\theta)$$

This is called the likelihood of k and θ given our data. More generally, we have the following:

Definition 11.3.1. For n independent data points, X_i, all taken from the same distribution, we can define the likelihood of a probability distribution D with parameters $a_1, \cdots, a_m$ explaining the data to be

$$L(a_1,\cdots,a_m;X_1,\cdots,X_n) = \prod_{i=1}^{n} D(X_i;a_1,\cdots,a_m)$$

We define the best parameters as those that maximize the likelihood.

In common parlance, likelihood and probability are interchangeable. When we ask, *what is the probability of X happening?* we mean, *how likely am I to experience X?* Mathematically, they mean different things. A probability is a measure of how frequently a specific outcome will occur if we repeat an experiment enough times. A **likelihood** is a measure of fit of a model. It answers the question, *how likely were we to observe the data we have if the underlying probability distribution is P?*

[34] We need the probability density for X_1, X_2, X_3, etc.

You can almost think of it as the complement of probability.

- **Probability** answers the question *given a distribution, how likely is a particular outcome?*
- **Likelihood** answers the question *given an outcome, how likely is it that it came from a particular distribution?*

Again, we won't go into the details here, that's done in Chapter 4. Instead, we will work with `scipy` to let the computer maximize likelihood for us.

For any of our built-in distributions in `scipy.stats`, we can just call a `fit` method instead of `pdf` or `cdf` method like we did in Chapter 9.

If we would like to fit our age data from `heart.csv` to a gamma distribution, we need to only use the following:

```
k, loc, theta = stats.gamma.fit(data['age'],floc=0)
```

Generally, you will want to fix the `loc` parameter with `floc` since most of our mathematical expressions for distributions are not shifted from 0; the exception is the normal distribution where `loc` gives you the mean.

This is where, once again, help files and documentation are super important. There are 90 built-in distributions in Python. Those are just the 90 people bothered with. There are far too many of them to explain them all and some of them, like the normal pass parameters through `loc` and `scale`; the gamma passes the θ parameter through `scale` and `loc` should be 0; and others need both and have different numbers of parameters. It's important to read through the documentation for the particular distribution you need.

For instance, a distribution that we *didn't* talk about at all, but is quite useful is the Weibull distribution. We could also fit this to a Weibull distribution. The documentation for a Weibull distribution tells us that it is defined in Python as `weibull_min` and that it takes one parameter c, and λ is provided through the `scale`. We can fit our data to a Weibull distribution with

```
c,loc,lambda = stats.weibull_min.fit(data['age'],floc=0)
```

We can plot our both of these fits on top of our histogram using

```
x_pts = np.linspace(27, 83,1000)
sns.lineplot(x=x_pts,y=stats.gamma.pdf(x_pts,k,scale=theta))
sns.lineplot(x=x_pts,y=stats.weibull_min(x_pts,c,scale=lambda))
```

to get a figure like figure 11.9.

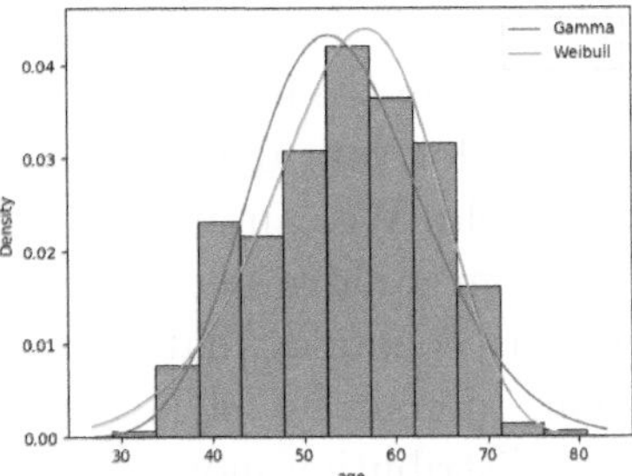

Figure 11.9: A comparison of a Gamme and Weibull distribution against the distribution of our age data.

Just by eye, it looks like the Weibull distribution might better match the data, but we can quantify this. Remember, we are trying to maximize the likelihood, and so if we compute the likelihood for each fit, we can see which number is larger.

```
print('The likelihood of the Weibull distribution is:',
    np.product(stats.weibull.pdf(data['age'],c,scale=lambda))
print('The likelihood of the Gamma distribution is:',
    np.product(stats.gamma.pdf(data['age'],k,scale=theta))
```

... and they're both zero. What gives? This is a quirk of computers. The probability density at any given point is going to be quite small, so if we have *a lot* of points, we are multiplying many, many small numbers. When we do this, the numbers get so small that the computer cannot distinguish them from 0.

For this reason, we usually compute the **log-likelihood**. It has all the same properties of the likelihood, but the log-likelihood is additive instead of multiplicative.

$$\mathcal{L}(a_1, \cdots, a_m; X_1, \cdots, X_n) = \sum_{i=1}^{n} \ln(D(X_i; a_1, \cdots, a_m))$$

If we instead compute this in Python, we get numbers we can actually compare.

```
print('The likelihood of the Weibull distribution is:',
    np.sum(np.ln(stats.weibull.pdf(data['age'],c,scale=lambda)))
print('The likelihood of the Gamma distribution is:',
    np.sum(np.ln(stats.gamma.pdf(data['age'],k,scale=theta)))
```

We see that indeed the Weibull distribution fits the data better than the gamma distribution.

This seems to answer the question, *which distribution is better for this data?* The question is a trick though, and one that I may have led you down on purpose so you can see how easy it is to fall into the trap. The best distribution to use for your data is the one supported by the science in your particular field. We can argue the Weibull distribution fits this age data better from a naive perspective, but there could be some underlying biology or economics that would eliminate a Weibull distribution as an option. In contrast, if nothing else is known, then identifying an underlying distribution motivates questions into *why* we are seeing that particular distribution, and this can lead to interesting routes of inquiry. In this case, the Weibull might be the distribution to go with because we should expect to skew right[35] because older people are more likely to need hospital treatment than younger people in general.

11.4 DUMMY VARIABLES

When dealing with mathematics, we generally need numbers that can be acted upon. This means that fitting and categorical variables create a slight impasse. We have to treat categorical variables in a special way, and that way is by creating **dummy variables**.

THE SIMPLEST ANALOGUE TO A DUMMY VARIABLE IS A CATEGORICAL VARIABLE THAT TAKES ONE OF THE TWO VALUES. As an example, let's look at data from a sleep survey in figure 11.10. If you focus on the

Sleep_Analysis

Age	Gender	meals/day	physical illness	screen time	bluelight filter	sleep direction	exercise	smoke/drink	beverage	sleep time
22	Male	two	no	2hrs	yes	west	sometimes	no	Tea	6.757500000000000
22	Female	three	no	3-4 hrs	no	south	no	no	Coffee	8.0
23	Male	three	no	3-4 hrs	no	south	no	no	Tea	8.0
23	Female	two	no	1-2 hrs	no	east	sometimes	no	Coffee	6.5
22	Male	three	no	more than 5	yes	east	sometimes	yes	Tea and Coffee both	6.0
22	Male	two	no	2-3 hrs	yes	west	sometimes	no	Tea	6.757500000000000
22	Male	four	no	1-2 hrs	yes	south	yes	no	none of the above	7.0
24	Female	three	yes	4-5 hrs	no	east	sometimes	no	Tea	6.5
24	Male	four	no	2-3 hrs	yes	east	yes	no	none of the above	8.0
22	Female	three	no	more than 5	yes	north	yes	yes	Tea and Coffee both	7.0
20	Female	three	no	0-1 hrs	yes	north	yes	no	Tea	6.5
[illegible]	Male	two	no	0-1 hrs	no	west	no	no	Tea	6.0
23	Female	three	no	4-5 hrs	no	west	sometimes	no	Tea	6.0
25	Prefer not to say	more than 5	no	more than 5	yes	east	yes	no	Coffee	1.5
23	Male	three	no	3-4 hrs	no	north	sometimes	no	Coffee	6.0
23	Female	two	no	2-3 hrs	yes	west	yes	no	none of the above	6.0
24	Male	three	no	more than 5	no	west	no	no	none of the above	7.0
24	Male	two	no	more than 5	yes	north	sometimes	no	none of the above	6.0

Figure 11.10: A data set from a sleep survey

`smoke/drink` column, you will notice it has one of the two responses: `yes` or `no`.

[35]Which the Weibull does; the gamma distribution we have actually skews a little left.

If we want to make a linear model that relates, say, whether you smoke and/or drink to the amount of sleep you get, we may make a model

$$y = mx + b$$

where y is `sleep time` and x is `smoke/drink`, but equations need numbers not yes or no. The simplest thing we can do is make the change.

$$\texttt{no} \to 0$$
$$\texttt{yes} \to 1$$

Now we have a number, and everything will work out! The difference is what m and b represent. If we make this substitution-

WAIT, *how* DO WE MAKE THIS SUBSTITUTION? We use a command called `pd.get_dummies` in the following way:

```
data_with_dummies = pd.get_dummies(data_set,
            columns=['column','names','to','make','dummies'])
```

We input the data set we are working on, and we use the `columns` keyword in order to specify which columns we would like to convert to dummy variables. When we use the command `get_dummies`, we get two values instead of one. The first new column, `smoke/drink_no`, treats `no` as 1, whereas as the second new column, `smoke/drink_yes` treats `yes` as 1. Notice these two columns give us the same information. We only really need one of them. We will use the column `smoke/drink_yes`.

Now, with this done, we can perform a linear regression! We set

```
X = data['smoke/drink_yes']
Y = data['sleep time']
```

Then we will once again build our model and fit it.

```
linear_model = Model(linear_function, independent_vars=['x'])
result = linear_model.fit(Y,x=X,m=1,b=1)
```

Doing this will give the values

$$m = 0.54$$
$$b = 6.71$$

	Age	Gender	meals/day	physical illness	screen time	bluelight filter	sleep direction	exercise	smoke/drink	beverage	sleep time
0	22	Male	two	no	2hrs	yes	west	sometimes	no	Tea	6.7575
1	22	Female	three	no	3-4 hrs	no	south	no	no	Coffee	8.0000
2	23	Male	three	no	3-4 hrs	no	south	no	no	Tea	8.0000
3	23	Female	two	no	1-2 hrs	no	east	sometimes	no	Coffee	6.5000
4	22	Male	three	no	more than 5	yes	east	sometimes	yes	Tea and Coffee both	6.0000
5	22	Male	two	no	2-3 hrs	yes	west	sometimes	no	Tea	6.7575
6	22	Male	four	no	1-2 hrs	yes	south	yes	no	none of the above	7.0000
7	24	Female	three	yes	4-5 hrs	no	east	sometimes	no	Tea	6.5000
8	24	Male	four	no	2-3 hrs	yes	east	yes	no	none of the above	8.0000
9	23	Female	three	no	more than 5	yes	north	yes	yes	Tea and Coffee both	7.0000
10	28	Female	three	no	0-1 hrs	yes	north	yes	no	Tea	6.5000

data = pd.get_dummies(data, columns = ['smoke/drink']

	Age	Gender	meals/day	physical illness	screen time	bluelight filter	sleep direction	exercise	beverage	sleep time	smoke/drink_no	smoke/drink_yes
0	22	Male	two	no	2hrs	yes	west	sometimes	Tea	6.7575	1	0
1	22	Female	three	no	3-4 hrs	no	south	no	Coffee	8.0000	1	0
2	23	Male	three	no	3-4 hrs	no	south	no	Tea	8.0000	1	0
3	23	Female	two	no	1-2 hrs	no	east	sometimes	Coffee	6.5000	1	0
4	22	Male	three	no	more than 5	yes	east	sometimes	Tea and Coffee both	6.0000	0	1
5	22	Male	two	no	2-3 hrs	yes	west	sometimes	Tea	6.7575	1	0
6	22	Male	four	no	1-2 hrs	yes	south	yes	none of the above	7.0000	1	0
7	24	Female	three	yes	4-5 hrs	no	east	sometimes	Tea	6.5000	1	0
8	24	Male	four	no	2-3 hrs	yes	east	yes	none of the above	8.0000	1	0
9	23	Female	three	no	more than 5	yes	north	yes	Tea and Coffee both	7.0000	0	1
10	28	Female	three	no	0-1 hrs	yes	north	yes	Tea	6.5000	1	0

Figure 11.11: When we use the command `get_dummies`, *we get two values instead of one. The first new column,* `smoke/drink_no`, *treats* `no` *as 1, whereas as the second new column,* `smoke/drink_yes` *treats* `yes` *as 1. Notice these two columns give us the same information. We only really need one of them.*

We can interpret these as follows:

- b gives the average for the 0 value of the dummy variable. In this case, b tells us on average those who do **not** smoke/drink sleep 6.71 hours.
- m gives us how much **more**[36] the average for the 1 value is. In this case, m tells us that those who smoke and/or drink sleep on average 0.54 more hours than those that do not. In other words, those that smoke/drink sleep on average 7.25 hours.

WHAT DO WE DO WHEN OUR CATEGORICAL VARIABLE HAS MORE THAN TWO CATEGORIES? Let's think about the column `excercise` in the sleep data. We may be tempted to substitute `no`, `sometimes`, and `yes` with 0, 1, and 2, but this just won't do. If we were to do such a thing, we are creating an order that otherwise doesn't exist. We are saying that a `yes` is worth two times as much as a `sometimes`, but this is completely arbitrary! We could use 0, 1, and 9, and then the implication is that `yes` is worth nine times as much as `sometimes`. The lesson here is that **a dummy variable cannot take on any value other than** 0 **or** 1. A thing either exists or it doesn't. There is no in between.

[36] Or less if the number is negative

How do we turn three things into 0 and 1 then? Well, let's make a dummy variable x_1 that maps like

$$\texttt{yes} \to 1$$
$$\texttt{anything else} \to 0$$

So if we see a `yes`, then $x_1 = 1$; if we see a `no` or a `sometimes`, then $x_1 = 0$. This differentiates yes from no/sometimes. How do we differentiate no and sometimes?

Let's make a dummy variable x_2 that maps like

$$\texttt{sometimes} \to 1$$
$$\texttt{anything else} \to 0$$

So if we see a `sometimes`, then $x_2 = 1$; if we see a `yes` or a `no`, then $x_2 = 0$.

Using x_1 **and** x_2, we can differentiate all three!

$$(x_1, x_2) = (1, 0) \to \texttt{yes}$$
$$(x_1, x_2) = (0, 1) \to \texttt{sometimes}$$
$$(x_1, x_2) = (0, 0) \to \texttt{no}$$

Notice that $(x_1, x_2) = (1, 1)$ is impossible.

We can then fit liner model using

$$y = mx_1 + nx_2 + b$$

In general, if our categorical variable has k categories, we need $k - 1$ dummy variables to exhaust all possibilities.

Let's look at the Titanic data set, a sample of which is given in figure 11.12. You will notice that even though the passenger class is a number, it is not a numerical variable. Someone in third class is not 1/3 the worth

titanic_data

PassengerId	Survived	Pclass	Name	Sex	Age	SibSp	Parch	Ticket	Fare	Cabin	Embarked
1	0	3	Braund, Mr. Owen Harris	male	22	1	0	A/5 21171	7.25		S
2	1	1	Cumings, Mrs. John Bradley (Florence Briggs Thayer)	female	38	1	0	PC 17599	71.2833	C85	C
3	1	3	Heikkinen, Miss. Laina	female	26	0	0	STON/O2. 3101282	7.925		S
4	1	1	Futrelle, Mrs. Jacques Heath (Lily May Peel)	female	35	1	0	113803	53.1	C123	S
5	0	3	Allen, Mr. William Henry	male	35	0	0	373450	8.05		S
6	0	3	Moran, Mr. James	male		0	0	330877	8.4583		Q
7	0	1	McCarthy, Mr. Timothy J	male	54	0	0	17463	51.8625	E46	S
8	0	3	Palsson, Master. Gosta Leonard	male	2	3	1	349909	21.075		S
9	1	3	Johnson, Mrs. Oscar W (Elisabeth Vilhelmina Berg)	female	27	0	2	347742	11.1333		S
10	1	2	Nasser, Mrs. Nicholas (Adele Achem)	female	14	1	0	237736	30.0708		C
11	1	3	Sandstrom, Miss. Marguerite Rut	female	4	1	1	PP 9549	16.7	G6	S
12	1	1	Bonnell, Miss. Elizabeth	female	58	0	0	113783	26.55	C103	S
13	0	3	Saundercock, Mr. William Henry	male	20	0	0	A/5. 2151	8.05		S
14	0	3	Andersson, Mr. Anders Johan	male	39	1	5	347082	31.275		S
15	0	3	Vestrom, Miss. Hulda Amanda Adolfina	female	14	0	0	350406	7.8542		S

Figure 11.12: A sample of the Titanic data set.

of someone in first class. These values do not scale properly. If we were to try to use linear regression while assuming that Pclass is numerical, we will get a terrible fit. By using

```
t_data = pd.get_dummies(titanic_data, columns=['Pclass'])
```

we can convert this column into dummy variables. We always get one more column than we need. In this case we will get Pclass_1, Pclass_2, and Pclass_3. Just like above, we only need the first two of these columns to determine the behaviour.

Now, there aren't too many columns in our Titanic data set that are numerical so we don't have many choices when it comes to fitting. You will notice that there is some data in the Age column that is missing. This makes it a prime candidate for modelling because with a model we can estimate the missing ages.

The first thing we need to do is create two data sets: one of known data and one of unknown data.

```
t_data_known = t_data.loc[t_data['Age'].notna()]
t_data_unknown = t_data.loc[t_data['Age'].isna()]
```

Now we run into a new problem. If we want to predict age by passenger class, we need *two* input variables into our function, but so far we have only dealt with one. This is a simple change.

```
def fitting_function(x,w,m,n,b):
        y = m*x+n*w+b
        return y
```

Now we are fitting to two inputs: x is going to be the Pclass_1 dummy variable, and w is going to be the Pclass_2 dummy variable. From here it is just two slight modifications.

- First, we must add w to our list of independent variables when building our model.

  ```
  linear_model = Model(fitting_function,
                                  independent_vars=['x','w'])
  ```

- Then we specify which column goes in which variable when we fit

  ```
  Y = t_data_known['Age']
  X=t_data_known['Pclass_1']
  W = t_data_known['Pclasas_2']
  result = linear_model.fit(Y, X, W, m = 1, b = 1, n = 1)
  ```

When we look at our `fit_report()`, we see

```
[[Variables]]
    m:  13.0928211 +/- 1.22172469 (9.33%) (init = 1)
    n:  4.73701034 +/- 1.25148345 (26.42%) (init = 1)
    b:  25.1406197 +/- 0.71635977 (2.85%) (init = 1)
```

b represents the average age of passengers in third class (because `Pclass_3` corresponds to two zeros), which the model predicts is around 25. n is the average *increase* in passenger age if you are in second class. Our model predicts that, on average, second class passengers were 4.74 years older than third class passengers. Finally, m tells us how much older on average first class passengers were compared to third class passengers. On average, first class passengers were 13 years older than third class passengers. What can we glean from this new found information? Mainly, wealth has never belonged to the young.

WE CAN ALSO MIX AND MATCH NUMERICAL AND DUMMY VARIABLES AS INPUTS. There is no reason why we can't. The order of interpretation is numerical first, then categorical. For instance, if we were to add the `Fare` column to our model so that we are now fitting `Age` to

$$y = mx + nw + b + kz$$

where z is our `Fare` column, then our interpretation of parameters changes slightly.

- k is the average age increase of a person per dollar spent on fare **of someone in third class**.
- m is the average increase in age of someone in first class versus third class **after age-related effects are accounted for**.
- n and b have the same interpretation as m.

The values are

```
m:   16.9743325 +/- 1.49426536 (8.80%) (init = 1)
n:   5.16509768 +/- 1.23944551 (24.00%) (init = 1)
k: -0.05193898 +/- 0.01180129 (22.72%) (init = 1)
b:   25.8277431 +/- 0.72430754 (2.80%) (init = 1).
```

Notice that in general, the lower your fare, the more likely you were to be older, but this number is so small that it is almost negligible. We see that passenger class has a far bigger effect on age than fare price.

We can then predict the ages in our `t_data_unknown` data with the command

```
predictions = result.eval(x=X, w = W, z = Z).
```

We can then augment our unknown data set with these predictions by creating a new column.

```
t_data_unknown['predicted_ages'] = predictions
```

11.5 LOGISTIC REGRESSION

One of the limitations of linear and non-linear regression is that it only works on **numerical data**; if we want to predict a categorical variable, we have to use different techniques.

Off the bat, we should understand that basic logistic regression works on **binary categories**; *i.e.* a categorical variable that takes one of two values[37].

WE WILL CONTINUE WITH OUR TITANIC DATA AS A WORKING EXAMPLE. Let's first try to look at survivorship vs. age and sex. Survivorship is a good variable to test logistic regression on, as a person either survived (1) or didn't (0). If we look at figure 11.13 we can see that we can calculate a probability of `survived` being 1 at each age by counting dots in the 1 swarm and dividing by the total dots at that age. For instance, in the first column of points (at age 0) we see a total of 14 points, of which 12 are in the 1 category. Therefore $P(1|age = 0) \approx 0.857$. That is, the probability of survival given the reported age is 0 is 0.857. On the other end of the spectrum, we have $P(1|age = 80) = 1$ since we only have one data point for this age. It is this probability that we want to fit.

WE DO THIS USING A LOGISTIC FUNCTION. A logistic function is a function that runs between 0 and 1. Mathematically, it looks like

$$P = \frac{1}{1 + e^{-(b_1 x1 + b_2 x2 + \cdots + b_M xM}}$$

where the values $x1, x2, \cdots, xM$ are M different input variables, which we are using to determine the probabilities for our categorical variable.

[37]Think *is a thing* or *not a thing*.

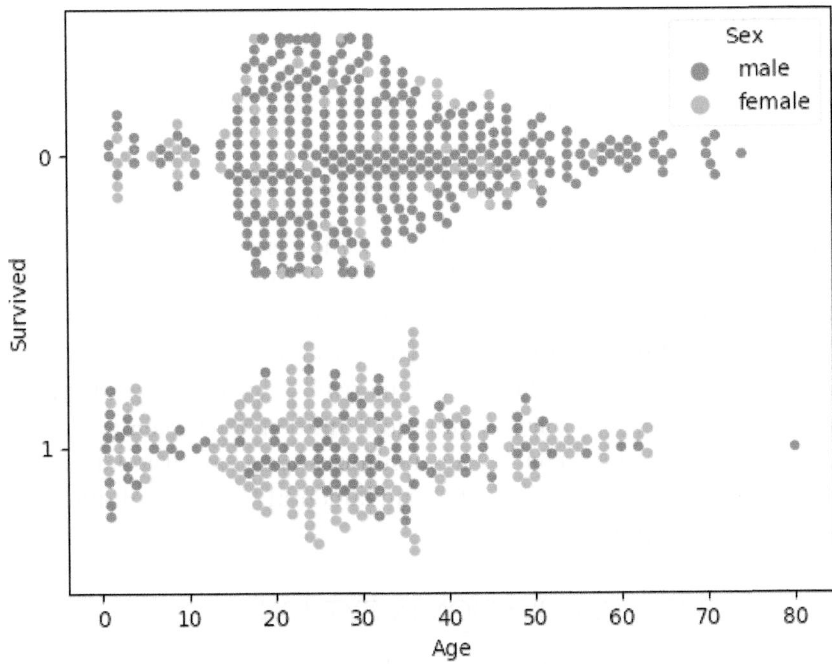

Figure 11.13: A swarm plot of survivorship vs. age, with sex information given in colour. Comparing the two swarms, we see that survivors were overwhelmingly female. We also see that more children survived than not (there are more dots in the 1 *swarm for low ages than in the* 0 *swarm). We can also see that more people in their 20s and 30s did not survive (more dots in the* 0 *swarm in this region than dots in the* 1 *swarm).*

This is where conditional probabilities come in. For each data point, we can compute the probability that $Y = 0$ or $Y = 1$, given some b values and our input data, x's,

$$P(Y = 1|X, B) = \frac{1}{1 + \mathrm{e}^{-(b_1 x1 + b_2 x2 + \cdots + b_M xM}}$$

where X is shorthand for all $x1, x2, \cdots, xM$ and B is shorthand for all the $b1, b2 \cdots, bM$. Then, by the nature of only having two options, we get

$$P(Y = 0|X, B) = 1 - \frac{1}{1 + \mathrm{e}^{-(b_1 x1 + b_2 x2 + \cdots + b_M xM}}$$

NEXT WE CAN ASK, *what is the probability of getting y% 1's for a given X and B?* This is where the binomial distribution comes in.

$$p(y|X,B) = \left(\frac{1}{1+e^{-(b_1x1+b_2x2+\cdots+b_MxM}}\right)^y \left(1-\frac{1}{1+e^{-(b_1x1+b_2x2+\cdots+b_MxM}}\right)^{1-y}$$

This tells us the probability of getting a y fraction of 1's when we have particular values for X and B. This follows a binomial distribution, and when given for a single value of y like this, we call it a **Bernoulli trial**.

This is the function we want to fit to the data, but we don't necessarily want to minimize the error[38]; instead we will try to **maximize likelihood**. This is the biggest practical difference between (non)linear and logistic regression. The former minimizes error, and the latter maximizes likelihood. This is due to the fundamental, conceptual differences used in our interpretations.

11.6 LOGISTIC REGRESSION IN PYTHON

Thankfully, we don't need to do any of this as a computer takes care of it for us! In fact, we don't even have to switch our two categories to 0 and 1[39].

Unfortunately, `lmfit` is not the best package for logistic fitting. We can make it work, but it's more involved than it's worth for this class. Instead, we will use a package called `sklearn`. We import

```
from sklearn.linear_model import LogisticRegression.
```

We then create a logistic model

```
logistic_model = LogisticRegression(max_iter=1000)
```

here; `max_iter` is an option for telling the model how many times it should run before giving up. The default is 100, which usually isn't enough.

Now we do the usual thing: define which columns we want as inputs and which we want as an output.

```
t_data = t_data.dropna()
X = t_data[['Pclass_1','Pclass_2','Fare','Age']]
Y = t_data['Survived']
```

[38]Mostly, because there is no error term!

[39]As long as there are *only* two categories – we will touch on this later.

Notice that when using sklearn we can put *all* input columns into one variable X. Then we can do a logistic regression with one line.

```
result = logistic_model.fit(X,Y)
```

Now, we have a bit of an issue. The survivorship data for this data set is complete. We know the survival status for everyone in the data set, but the Titanic had 2200 passengers, of which only 890 are present in our data set, so we can't predict things for passengers we know nothing about. Even if we could, how confident can we be in our model?

What we can do is use *most* of our data to build a logistic model and then use the remainder of the data as a test. In data science lingo, we will create two data sets from our data: a **training set** and a **test set**. The **training set** is used to build the regression, the **test set** is used as if we didn't know the output column, and we predict it.

With sklearn we can do this automatically. First, we have to import another thing

```
from sklearn.model_selection import train_test_split.
```

At this point, it might be tempting to import skleran in its entirety, but sklearn is very large; it's usually not ever a good idea.

Now, **before** we fit the model, we split our data into two.

```
X_train, X_test, Y_train, Y_test = train_test_split(X, Y,
                  test_size=0.33, random_state=42
                  )
```

This creates a random training and test set from our data. The test_size says what percentage of the data to use for testing, and random_state is optional. If we put a number in here, we will produce the *same* training and test sets every time we run the code; if we leave it, out we will get different sets every time.

Once we have these sets, we use the training set to build a model.

```
result_train = logistic_model.fit(X_train, Y_train)
```

Now, we can predict values from our test set.

```
predictions = result_train.predict(X_test)
```

We can compare our predictions to the actual `Y_test` data to see how accurate our model is. We can do this manually, or we can import one more thing

```
from sklearn.metrics import accuracy_score,
```

and then we can get an accuracy score for our model by using

```
accuracy = accuracy_score(predictions,Y_test)
```

We see that our model is about 64% accurate, better than chance but not by much.

You will notice we also didn't include what the plot showed us to be the most relevant factor: sex. If we include this by

- Creating a dummy column for `sex`
- Modifying our X to include a dummy column for `sex`
- Rebuilding the model and re-running the model

we can see our accuracy increases to $\approx 69\%$.

11.7 ITERATED LOGISTIC REGRESSION

Iterated logistic regression is something we would likely never do, but I think conceptually it can help us. The next logical question is *how might we predict a categorical variable that has more than two categories?*

Just like we did when creating dummy variables, a categorical variable with N categories is just like $N-1$ categorical variables with two categories. So we can **iterate** our regression by looking at one category, determining whether the data is in that category or not, then looking at the second category and doing the same, and continuing for $N-1$ times.

Let's look again at our sleep data (figure 11.10) as a working example. The column `beverage` has four categories: `Coffee`, `Tea`, `Coffee and Tea both`, and `none of the above`. If we create four dummy variables, we will get four new columns[40]: `beverage_Coffee`, `beverage_Tea`, `beverage_Coffee and Tea both`, and `beverage_none of the above`.

[40] Remember, we only need three.

We can use the other columns to try and build a logistic regression using beverage_Coffee as our output. When we do this, we will get a binary categorization: the model will categorize things as Coffee (1) or notCoffee(0).

```
result = logistic_model.fit(X,Y)
```

where X is all the appropriate columns used for input and Y is the column beverage_Coffee.

Now let's augment our data set with this result.

```
sleep_data['CoffeeOrNot'] = result.predict(X)
```

Now, we theoretically know which rows are Coffee (1) and which are notCoffee (0). We need to determine what the notCoffee's actually are. So we should filter our data set to only have notCoffee.

```
sleep_data_notcoffee=sleep_data.loc[sleep_data['CoffeeOrNot']==0]
```

Now we can perform a logistic regression on sleep_data_notcoffee but using beverage_Tea as the output. This will then give us a prediction of 1 if it is Tea or 0 if it's notTea. Since this data set was created from just the rows that are notCoffee, we know the 0's are notCoffee and notTea.

So we do the same thing:

- Augment sleep_data_notcoffee with a column called "TeaOrNot."
- Filter the data set for 0's to get a smaller data set called sleep_data_notcoffee_or_tea.
- Use beverage_Coffee and Tea both as an output and perform logistic regression.

We can summarize this whole process with a diagram like in figure 11.14

Keep in mind that each of these models is independent, and therefore the accuracy scores multiply. This has the detrimental effect accuracy decreasing as categories increase. Also, the order in which we implement the models may change the outcome! There is a better way, and this discussion of iterated logistic regression is mostly because it is conceptually similar to the superior random forest classifier.

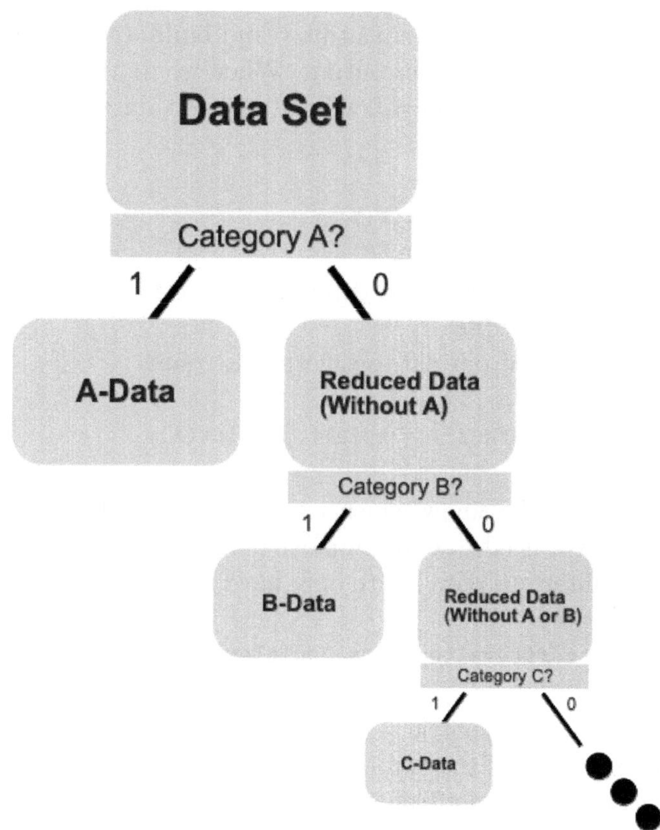

Figure 11.14: We can iterate over many logistic regression models to deal with cases when there are more than two categories.

11.8 RANDOM FOREST CLASSIFICATION

Iterated logistic regression would require us to create $N - 1$ models for N categories in our output data. Moreover, these models would be independent and run in sequence; this means that for each class our predictive accuracy would decrease for every new category we have. This makes iterated logistic regression somewhat undesirable for classifying more than binary categories.

Why did we spend time learning about an undesirable method? The concepts introduced in learning logistic regression and extending to iterated logistic regression are transferrable to other methods, which are more powerful and more robust classification algorithms. One of these classes of algorithms is the decision tree. Decision trees use different specific algorithms to take our input data and narrow down possible categories for specific rows. In the logistic regression, we are using a linear combination of all attributes and then splitting on a probability modelled by a logistic function. The algorithms for decision trees let us split more finely and allow for more accurate results.

A heuristic example of a decision tree is the game 20 questions. One player thinks of a thing, and the other person has 20 yes/no questions in order to narrow down and guess the thing. The next question we ask is likely to be related to the answer to the question before. By accounting for different input data (the answers), we can split the set of all possible things into smaller and smaller subsets and, hopefully, eventually narrow ourselves down to a guess between one or two things.

Decision trees are very sensitive to the first question we ask. Sometimes if we start with the wrong question, it can throw off the whole process and lead to very inaccurate results. Overfitting is also often an issue with decision trees; a decision tree may find a spurious relationship that is an artefact of our sampling and not a feature of the population.

To combat this, we allow for *many* trees to be created, each starting from a random attribute and a random subset of our data. If we do this enough times, then we wash out many of the issues that creep up with using individual trees. This is called creating a random forest[41]. With many different decisions trees each coming to its own independent answer, we then take a majority rule approach to determine the final result.

In Python, creating a random forest is very easy. First, we import the classifier from `sklearn`.

```
from sklearn.ensemble import RandomForestClassifier
```

We then create the classifier much the same way we created a logistic regression

```
our_forest = RandomForestClassifier(n_estimators=100)
```

where the option `n_estimators` tells Python how many trees we would like in our forest. We can then fit some training data `X_train` and `Y_train` with the forest with

```
result = our_forest.fit(X_train,Y_train)
```

and again we can use

```
predictions = result.predict(X_test)
```

[41]Because a forest is made of trees

Table 11.1: Every evening, my dog lets me know it's time for a walk by moving to patiently sitting by the door and letting out a single whine. This table is a month's worth of when that whine happens.

Day	Walkies Time	Day	Walkies Time	Day	Walkies Time
1	7:00	11	7:02	21	7:01
2	6:58	12	7:12	22	7:08
3	6:48	13	7:08	23	6:58
4	7:10	14	7:01	24	7:04
5	7:08	15	7:00	25	6:59
6	7:06	16	6:57	26	7:02
7	7:00	17	7:02	27	6:45
8	7:01	18	7:00	28	7:20
9	6:55	19	7:02	29	7:05
10	6:59	20	7:00	30	7:07

to predict classifications for our test data, `X_test`. From here, all our same metric commands, like accuracy and the confusion matrix, work just the same.

11.9 BOOTSTRAPPING AND CONFIDENCE INTERVALS

Bootstrapping

Maybe you have thought about it already, but if you haven't, I will pose and then discuss a question. Data is necessarily noisy; samples are ideally random; how can we be sure that we are seeing a *real* relationship and not some artefact of randomness? For instance, look carefully at the following table of data. My dog has a pretty particular personality. Every evening when he decides he is ready for a walk, he will sit by the door and let out one single whine. For one month, I recorded when this whine happened. We can rewrite this time to something we can plot by defining the minutes from $7:00$; we can call it the time to first whine (TTFW) Figure 11.15 shows the TTFW as a function of day. Clearly, there is no real relationship between these values. My dog likes to go out around $7:00$ every day; he's just bad at telling time.

Yet, if I only measured, say, once a week, I might get a plot like figure 11.16, and this relationship looks positive. A random sample from random days like in figure 11.17 may even make the relationship look negative.

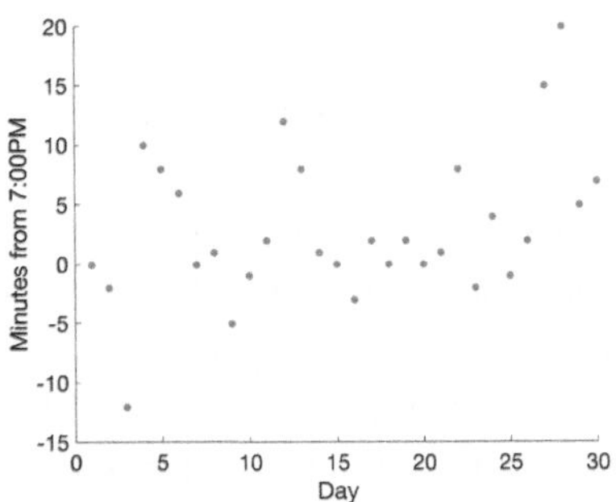

Figure 11.15: Data visualization for Table 11.1.

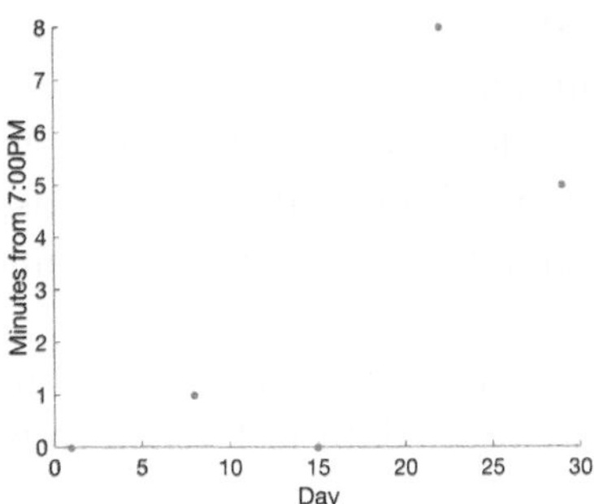

Figure 11.16: Measuring walk time once a week gives a different trend in the data.

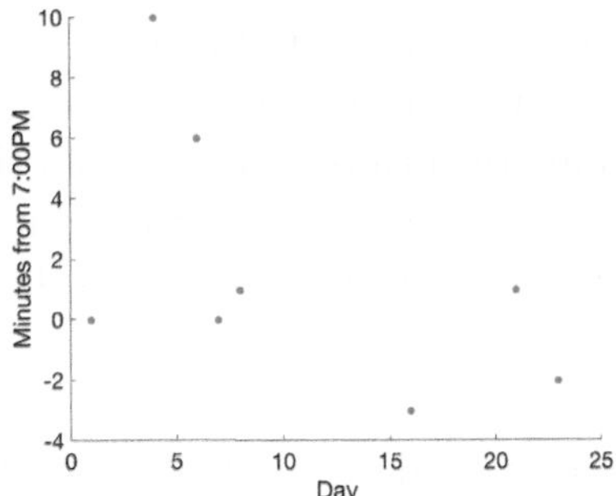

Figure 11.17: Measuring walk time randomly can show a pattern that isn't actually present.

While this example is silly, it shows that the very act of sampling randomly may lead to the observation of relationships that aren't actually there. Likewise, we may miss relationships in data because of random variability in measurement. How might we be able to parse out what is real and what isn't?

First, we need to have a hypothesis. What kind of relationship are we expecting? This is where specialized knowledge comes in, and where the real legwork of mathematical modelling is. Using all known knowledge, along with reason and structure, is how we build good, robust models that can give us deep insights. There is only so much to be done with drawing lines through things.

The technique of taking different sets of samples is called bootstrapping. It is a way to simulate repeating experiments without having to repeat experiments. It works best when we have a lot of data points to choose from. Bootstrapping works on the simple premise that the distribution of measurements we get from resampling will match the population distribution for the measurement. While computationally more intensive than many data analysis methods, it is generally quite robust and does not require any assumptions on underlying distributions of data[42].

Resampling with replacement is easy to do in Python. First, let's turn our data table into some data in Python to play with.

```
x = np.linspace(1,30,30)
y = np.array([0,-2,-12,10,8,6,0,1,-5,-1,2,12,
  8,1,0,-3,2,0,2,0,1,8,-2,4,-1,2,-15,20,5,7])
```

These are two columns of data, without the actual DataFrame structure around it[43], but we can still use them in the same way. We just refer to them by their variable names instead of column headers.

To start, let's say we want to use bootstrapping to try and estimate the mean number of minutes before or after 7 pm that my dog likes his walk. We are going to create many, many means using resampled data; we should have somewhere to store these means. Let's create a variable.

```
walk_means = []
```

What we've done here is created an empty *list* variable. A list in Python is a data structure we can use to store multiple pieces of information.

[42]The most common is that the quantity we're looking for and/or residuals must be normally distributed.

[43]Two **vectors**, cf. Ch. 3

Anything can go into a list. In some sense, a list is an abstraction of a column of data. We will also want to create a variable for the number of times we wish to resample. We have 30 spots to fill, with 30 different values, and we will allow repetition. Since these data must be ordered in the histogram, there are a maximum of

$$M = {}_{30}R_{30} \approx 5 \times 10^{16}$$

where R is the *combination with replacement* formula from Chapter 9.

This is obviously too many samples to take, so we can limit ourselves to a few hundred or few thousand. Let's try 10000. We will create a `for` loop that will do the following:

1. Take random points from our `y` array, with replacement, such that the new sample array is the same size as the old one.

2. Compute the mean of the new random sample.

3. Store this mean in our list of means.

We want to do these three steps over and over again a fixed number of times; hence the `for` loop.

In Python, this will look like

```
for i in range(0,10000):
        Y_samples = random.choices(Y,k=len(Y))
        sample_mean = np.mean(Y_samples)
        walk_means.append(sample_mean)
```

This makes use of a package we haven't seen in earnest yet: `random`. The `random` package in Python gives us access to tools for generating random[44] numbers. After this loop runs, our variable `walk_means` will be populated with 1000 different means which together give us a *distribution* of possible means of our data set. If we plot a histogram of the density,

```
scott = 3.49*np.std(walk_means)/(len(Y)**(1/3))
sns.histplot(x = walk_means,bin_width = scott, stat = 'density')
```

we will get something like figure 11.18. Note that yours won't be *exactly* the same because of the random nature of bootstrapping, but it should be close. This histogram is an approximation of the probability distribution that the mean follows. It should, in this case[45], look mostly normal.

[44] They're not quite random, but they're close enough.

[45] And in most cases

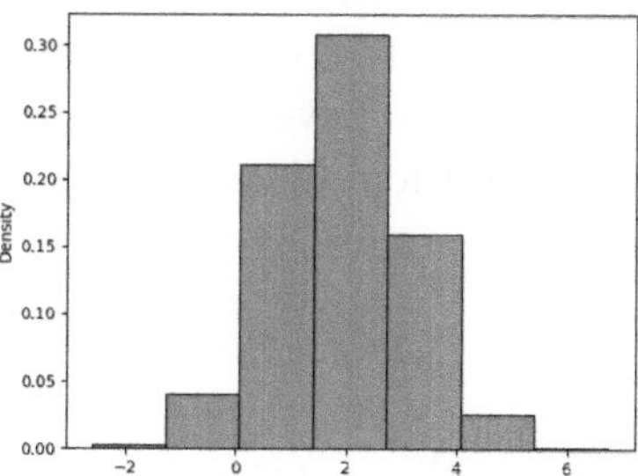

Figure 11.18: A histogram of the average walk time; since our data isn't perfect, neither is our average measurement.

We see that the mean number of minutes away from 7:00 PM is approximately 1.9 with a standard deviation of approximately 1.19. We can calculate this and the standard deviation with

```
est_mean = np.mean(walk_means)
est_std = np.std(walk_means)
```

We can similarly use bootstrapping to get a better estimate of the slope, m, and y-intercept, b, of our linear model. The steps we'd like to carry out in the bootstrap are similar to the ones for the mean, but we need to take care that we keep our x and y data paired:

1. Take random (x, y) pairs from our sample set, with replacement such that the new number of pairs is the same size as the original
2. Use least squares to fit the slope and y-intercept to the newly sampled data
3. Store the slope and y-intercept

Firstly, we will need to create two new lists so that we can store our slopes and our y-intercepts.

```
m_array = []
b_array = []
```

Assuming that we have a `fitting_function` already defined as a line with a slope and y-intercept, we make a model, as we did before.

```
linear_model = Model(fitting_function, independent_vars = ['x'])
```

Now we *iterate* the above steps, once again, in a for loop.

```
for i in range(0,10000):
        samples = random.choices(range(0,len(Y)),k=30)
        X_samples = X[samples]
        Y_samples = Y[samples]
        result = linear_model.fit(Y_samples,X_samples,m=1,b=1)
        m_array.append(result.params['m'].value)
        b_array.append(result.params['b'].value)
```

We can then find the mean m value (≈ 0.09) and the mean b value (≈ 0.5) along with the standard deviations: 0.17 and 2.6 for m and b, respectively. When we plot the histograms for `m_array` and `b_array`, we get approximations of the probability distribution that these values follow.

Confidence Intervals

Another thing we can compute with bootstrapping is called a **confidence interval**. A confidence interval is exactly what it sounds like: a range around a parameter or statistics that quantifies our certainty in our estimate. The $M\%$ confidence interval is an interval within which we expect $M\%$ of new measurements of the parameter or statistic to fall if we were to redo the experiment from scratch or uncover new data. You will most often see confidence intervals given as 95% confidence intervals. These are values between which we expect 95% of our measurements to fall.

The beauty of bootstrapping is that we can read off confidence intervals pretty directly. We need a lower limit, x_L, and upper limit, x_U, of our probability density function such that

$$\int_{x_L}^{x_M} P(x)\mathrm{d}x = 0.95$$

Unfortunately, there is no one and done command in Python for getting this value, but that doesn't mean it can't be done. One of the reasons we are using Python is that it is incredibly robust and amenable to being manipulated and added to. Since we know the number of samples we have, 10000, we can first sort our data.

```
mean_sorted = np.sort(mean_array)
```

We can then find the number that appears 5% of the way into our data and 5% from the end of our data.

```
val_95 = int(0.025*10000)
```

We use 10000 here since we have 10000 bootstrapped samples; keeping in mind that we need `ints` to access elements, we convert to an int. We then get the values 2.5% from the start of our sorted data and 2.5% from the end of our sorted data[46].

```
low_95 = mean_sorted[val_95]
high_95 = mean_sorted[10000 - val_95]
```

If we do this, we should get values `low_95 = -0.04` and `high_95 = 4.3`; keep in mind yours might not be exactly the same because of the element of randomness in the process. We can then say that on average my dog likes to go out on average 1.9 minutes after 7:00 PM, but this value ranges from 0.04 minutes *before* 7:00 PM to 4.9 minutes *after* 7:00 PM. If we repeat had infinite samples, we could be fairly confidence that the true mean value falls somewhere in this range.

How is this different than what the standard deviation tells us? Remember that while the distribution for the mean *looks* normal, there's no guarantee that a parameter will always be distributed normally. If the parameter is distributed normally, then the 95% confidence interval will correspond to exactly 1.96 standard deviations[47]. If our data is *not* normally distributed, then the standard deviation tells us where *at least* 50% of our data lies. In this case, the confidence intervals determined from bootstrapping are far more robust and precise than the standard deviation. Note further that the upper and lower bounds of our confidence intervals do not need to be symmetric, so we can see which direction the asymmetry in our distribution goes!

The beauty of bootstrapping and confidence intervals is that we can use them on any statistic or metric we would like to compute about a data set. If we look at the confidence interval for the slope of our linear model, we will see that while the mean slope is $m = 0.09$, the 95% confidence interval is between -0.25 and 0.42. Notice that this interval includes 0, the slope we expected from our previous knowledge and from looking at the histogram; the idea that my dog cannot tell time and is not progressively creeping his walk later and later.

In both of these cases, the 95% confidence intervals are relatively large compared to the estimated mean, but that is because our original sample size is extremely small. As we get more data, as our sample size grows, our confidence intervals will begin to converge on a smaller and

[46]So that in total we have the middle 95% of our data.

[47]This is the slightly more exact value that corresponds to the second condition in our empirical rule.

smaller region giving us a better estimate of the parameter or statistic we are trying to estimate.

Bootstrapping is, in a way, a *brute force* method of estimating statistics and parameters. It requires time and a lot of computational power. This is the trade-off for methods that are extremely versatile and work with minimal assumptions, but that doesn't mean that there are *no* assumptions.

Assumptions underlying Bootstrapping

- The sample should represent the population. If we want to say something about dogs in general, we probably need more than just my dog. My dog is not representative of all dogs; he's not even representative of all mutts.
- Each subsample should be independent of the other samples.

The first of these assumptions is the easiest to break; bootstrapping is simple and powerful, and we may *think* we can extract more information than is actually present in our data and we may over-extend our conclusions. It's important to note that **bootstrapping cannot give us any more information than in our original sample**. No amount of increasing the number of resamples is going to narrow our confidence intervals or drastically change our estimated values. There are only two ways to do that: better models which use more assumptions to provide more "soft data"[48], or more samples.

Of course, if we have other information or if we are willing to make other assumptions, we can do more with less.

Hypothesis Testing and p-values

Going back to our example of the judicial system, we saw that there are two abstract concepts at play in a criminal trial[49]: there is the truth, whether a person is guilty or not, and there is what the evidence says is true. When the evidence and the truth agree, we can sleep easy knowing that justice was done. We saw many examples of how manipulating data, statistics, and people's intuition can lead to evidence and the truth being in disagreement.

[48]Context-specific information that we know through other means that we can apply to our analysis.

[49]Or any trial for that matter

Sometimes this doesn't even have to be as intentional as the examples we gave make it out to be. Sometimes, the evidence can point us in a certain direction because of the observations we *didn't* make, or perhaps we just happened to see only rare pieces of evidence of guilt and completely ignore some very obvious evidence in favour of innocence.

For instance, many criminal cases before the advent of DNA testing would rely on evidence that was far less definitive and far more circumstantial. All of this evidence may be extremely convincing, when looked at together, but technological limitations meant that one piece of exonerating evidence could not be observed.

We saw that for one piece of evidence, no matter how convincing that evidence may be against a guilty party, $P(E|G)$[50], this does not necessarily translate to any certainty that the evidence points to guilt, $P(E|G) \neq P(G|E)$. Bayes' theorem gave us a way to take the probabilities of seeing evidence and turning it into a probability, or level of confidence, in our guilty hypothesis.

For many pieces of evidence, we can iterate over Bayes' theorem to get an estimate of confidence we could have in our entire case. This, in a sense, is an example of what we call hypothesis testing.

Definition 11.9.1. Hypothesis testing is a procedure used to test whether a given set of evidence supports a particular hypothesis.

Is that vague? Yes, because hypothesis testing, generally, is extremely dependent on the context of your data, your hypothesis, and even the experimental design that generated your data.

The general algorithm is as follows:

- State a hypothesis.
  - For our example: my dog is slowly drifting his walks later and later as time goes on.
- State an alternative or null hypothesis.
  - My dog is **not** drifting his walks at all.
- Decide on a test statistic.
  - We will use the t-test; but there are many.

[50]Where E is the event of seeing the evidence and G is the event that the defendant is guilty

- Be sure that your data set is compatible with the assumptions underlying your test statistic.
    - For example, the t-test assumes that your data is roughly normal; that the mean and standard deviation are independent of one another. If this is *not* the case, the t-test would be inappropriate.
- Compute your test statistic.
- Report what's called the p-value.

There are a lot of terms in there that are probably unfamiliar. We will go through a hypothesis testing example, defining them along the way to try and build our intuition.

Let us go back to our calculation of the mean time in which my dog prefers his walks. Our bootstrapping method shows that the mean is roughly 1.9 minutes after 7:00 PM, but that the mean likely exists somewhere between 0.04 minutes before 7:00 PM and 4.9 minutes after 7:00 PM. If we compute the mean of our total sample, the mean is 1.93, in agreement with our bootstrapped estimate.

Let's form a hypothesis: MY DOG PREFERS HIS WALKS TO START 2 MINUTES AFTER 7:00 PM.

The null hypothesis is the statement that there is no difference or no relationship between the values in your hypothesis: THERE IS NO DIFFERENCE BETWEEN WHEN MY DOG PREFERS HIS WALKS AND 7:00 PM[51].

Since we don't know any test statistics, we can't really *decide* on one. We will use the t-test along with the t-statistic, as it is conceptually the easiest.

Definition 11.9.2. The t-statistic is defined, generally, as

$$t = \frac{X - X_0}{\nu}$$

where X is the relevant statistic or parameter determined by the data, X_0 is the relevant statistic or parameter expected in the null hypothesis, and ν is the standard error, sometimes called a **scaling factor**. Its form depends on the context.

[51] This is an awkward way to phrase this, but it is to highlight the format of a null hypothesis. The way you would say it to not get (or at least get fewer) weird looks at parties would be *my dog prefers his walks at exactly 7:00 PM.*

In the event that the distribution X comes from is normally distributed; the t-statistic will follow a t-distribution. We will speak more on this later; file it in the back of your mind[52].

Our null hypothesis is that there is no difference between 7:00 PM and when my dog likes to start his walk. What would this mean in terms of the ideal mean of my data set? It would be that, on average, the difference between 7:00 PM and the start time of the walk is 0. Therefore, we can conclude that $X_0 = 0$ for this particular situation with this particular null hypothesis.

The mean of the data, as we've seen, is $X = 1.93$. The scaling factor ν, in this case where we are only using one column of data, is given by

$$\nu = \frac{\hat{s}}{\sqrt{N}}$$

where N is the sample size and s is the sample standard deviation.

We could do this by hand, but Python is all set up to do this for us. Since the data we would like is in an array `Y`, we can use the command

```
tstat, pvalue = stats.ttest_1samp(Y,0).
```

This says perform a one-sample t-test, so named because we only have one column of data, on the data Y with the null hypothesis mean of 0. Keep in mind that by default `ttest_1samp` only computes the t-statistic for the mean of your data Y.

This will give us back two values. If our data was initially normally distributed, then it is guaranteed that the values of t are t-distributed[53]. This is just a different probability distribution, one that we did not discuss because it doesn't show up in raw data all that often. In our case, we get

```
tstat = 1.59.
```

This corresponds to a particular probability density on the t-distribution.

Recall that probability distributions don't really tell us too much, but an area will tell us a probability. We can ask, *what is the probability that the t-statistic falls between* -1.59 *and* 1.59*?* We take the positive and the negative because it shouldn't make a difference if we compute t as above or as

$$t_{neg} = \frac{X_0 - X}{\nu}$$

[52]Not too far back though

[53]Yes, it sounds circular.

The p-value is the probability that our t-statistic does *not* fall between -1.59 and 1.59. It is the measure of confidence we can have that our sample mean disagrees with the null hypothesis. In this case, there is a $\approx 12\%$ chance that our true mean is 0, and our value of 1.9 is just an artefact of our small sample size, variation in measurements, and other unconsidered factors.

In terms of conditional probabilities, we can say the probability that we observe a mean of 1.93 *given that the actual mean is* 0 is ≈ 0.122.

$$P(1.93|0) = 0.122$$

Notice this tells us *nothing* about our actual hypothesis; it doesn't even tell us that the mean is actually 0 instead of 2 because Bayes' theorem shows that conditional probabilities are not symmetric. The probability that the true value of the mean is 0 given that our sample mean is 1.93 is given by

$$P(0|1.93) = P(1.93|0)\frac{P(0)}{P(1.93)}$$

something we can't know without $P(0)$ and $P(1.93)$.

11.10 T-STATISTICS

A t-statistic always has the form

$$t = \frac{M - M_0}{\nu}$$

but what ν is depends on what we are comparing. This is guaranteed to follow a t-distribution in M_0 if M was sampled from a normal distribution.

Single sample t-test

Above, we used what is called a single-sample t-test. When we are comparing a single summary statistic to a fixed value, we proceed as above with ν being the quotient of the standard deviation and sample size:

$$\nu = \frac{s}{\sqrt{N}}$$

Slope of a regression line

When we perform a fit, we get a single value for the slope of our line[54]. How then might we calculate the standard error of the slope if we only have one value?

Since the slope is a function of our Y data and our X data, it stands to reason that the standard error of the slope would be related to the standard error of our data X and Y. Since the slope is the quotient of Y and X[55], the standard error can be written the same way

$$\nu = \frac{1}{\sqrt{N-1}}\sqrt{\frac{\sum_{i=1}^{N}(Y_i - \bar{Y})^2}{\sum_{i=1}^{N}(X_i - \bar{X})^2}}$$

where $\bar{X}$ and $\bar{Y}$ are the means of X and Y, respectively. In this case, our measurement, M, is the slope of our regression line, m. If we are comparing to the null hypothesis, then $M_0 = 0$. In other words, the null hypothesis is that there is no linear relationship between X and Y.

Comparing two data sets

What if we have two sets of data measuring the same thing, but collected, or measured in different ways, in different places, or at different times? For example, let's look at two columns of data of honey production in a given year in two different states.

Over the 15 years of the study, California produced on average 23.17 million pounds of honey and Florida produced on average 16.47 million pounds of honey. We might ask, *are these means as different as they seem, or is this difference the product of imperfect measurement?* In other words, if we were to take more measurements over more years, would these means get closer together or stay far apart? The null hypothesis would be that the means are *not* different, and any difference we see is the cause of randomness and imperfect or limited measurements.

We can use a t-test to determine the probability of our null hypothesis being true or not, provided the means of our data would be normally distributed[56].

[54] Unless we bootstrap, in which case we can get a distribution and can calculate the standard error from that distribution directly.

[55] Rise over run

[56] For the sake of the example, we will assume they are.

Table 11.2: Honey production in millions of pounds per year in California and Florida

Year	CA	FL
1998	37.35	22.54
1999	27.90	23.26
2000	30.80	24.36
2001	28.06	22.00
2002	23.50	20.46
2003	32.16	14.91
2004	17.55	20.09
2005	30.00	13.76
2006	19.76	13.77
2007	13.60	11.36
2008	18.36	11.85
2009	11.72	11.56
2010	27.47	13.8
2011	17.76	10.98
2012	11.55	12.35

The t-test for comparing two independent summary statistics is given by

$$t = \sqrt{\frac{X_1 - X_2}{\nu_1 + \nu_2}}$$

where

$$\nu_1 = \frac{s_1^2}{n_1}$$
$$\nu_2 = \frac{s_2^2}{n_2}$$

X_1 is the summary statistic from the first data set, X_2 is the summary statistic from the second data set, s_1 and s_2 are the sample standard deviations of the first and second data set, respectively, and n_1 and n_2 are the sample sizes of the first and second data set.

In Python, we use the command

```
tstat = stats.ttest_ind(data1, data2, equal_var = False)
```

to compute the t-statistic. By default the option `equal_var` is set to `True`. This means that Python will assume that the variance of both samples should be the same. Here, we set it to false because there is no reason to believe that they are or should be the same.

Computing this, we see that the t-statistic has a value of 2.74, and the corresponding p-value is $p = 0.0115$. We can probably safely conclude that California on average has higher honey production than Florida. The probability that we observe this particular different in the means *given that the means are not different at all* is roughly 0.01.

11.11 THE DICHOTOMOUS NATURE OF p-VALUES

Without going into a deep history lesson, somewhere along the way people started grouping p-values into two classes: significant and not significant. Generally, when reading scientific studies you may see that if the calculated p-value for a given hypothesis is $p < 0.05$ or $p < 0.01$, then the authors will claim their hypothesis significant that it is unequivocally true[57]. If the p-value is $p > 0.05$ or $p > 0.01$, they may claim the relationship they hypothesized is not significant.

As the Jedi would tell us, dealing in absolutes is rarely a good idea. The dichotomous use of p-values has always been problematic and has always drawn critique. On a purely practical level, it seems foolish to dismiss what could be an interesting result because a single number tells you too. Not only that, but years and years of people hearing the word not significant or significant has bastardized the intent.

A p-value has nothing to do with actual significance. When we frame the definition in terms of a conditional probability, we can clearly see that a p-value only tells us if we are on the right track or not. Another way to think about the p-value is *in our bootstrapping resamples, how many are likely to produce a mean of* 0*?* A high probability that what you measured agrees with the null hypothesis does *not* mean the null hypothesis is true. It does *not* mean that your hypothesis is false. It *does* mean that you cannot be very confident either way. It means *more investigation is needed.*

Similarly, an extremely small p-value doesn't mean that your hypothesized relationship is true, and it doesn't necessarily mean that the null hypothesis is false. It only means that the observations you made are highly unlikely under the null hypothesis. Therefore, it is more likely than not that there is *something* going on[58].

---

[57] I'm being *slightly* hyperbolic for the sake of argument.

[58] Not necessarily what you have hypothesized is going on

Figure 11.19: p-values are not gospel and must be treated with the same careful, context-sensitive interpretation as any other conditional probability, no matter who is preaching otherwise.

Much of this significance business as it is often misunderstood arises from the following slight logic:

1. The p-value suggests that our observations are extremely unlikely under our proposed null hypothesis.
2. Since the null hypothesis is unlikely, here is an alternative hypothesis that explains the data we have observed.
3. Since these observations are unlikely under the null hypothesis, they must be likely under the alternative hypothesis.
4. Our alternative hypothesis is highly likely to be correct.

Another way this gets misunderstood is equating an insignificant p-value with an insignificant study.

By reporting confidence levels, estimates, sample sizes, sample collection methods, and just on the whole being extremely transparent with the whole scientific process, we can do better. There is nothing inherently wrong with calculating or stating p-values; they can be extraordinarily useful. There is something wrong with dismissing data, results, and hypotheses based on an arbitrary cut-off of an oft manipulated number[59]. In short, while hypothesis testing can inform us whether or not we should trust what our data is telling us, the binary accepting or rejecting

[59]Sometimes it's strategically removing outliers, or transforming data in ways that break assumptions of your test statistic.

leads to false-positive, accepting a null hypothesis that is not true, and false-negative, rejecting a null hypothesis that *is* true, errors[60].

Moreover, if data is presented in a clear, open, transparent way, then data will find life beyond the original experiment it is designed for. For instance, our resting systolic blood pressure data was taken from patients as they entered a hospital. If we can mirror the same conditions while taking new data on systolic blood pressure, we can use *both* sets of data to get even clearer results.

[60] Where as if we just report what we do, we can avoid the problem all together. Sometimes, this is the best course of action.

APPENDIX A

A Crash Course in Python

You can think of Python as a Rosetta Stone for translating the thoughts and language of humans to those of computers. The golden rule here is that computers are notoriously literal[1], and so we have to be very careful when telling them what to do[2].

When we communicate with a computer through the Python programming language, we must follow the *syntactic* rules of Python and the *semantic* rules of a computer. Programs are, in essence, a series of commands for the computer to follow. The types of programs we will be writing are *procedural* in nature, meaning they are, for the most part, a series of commands that are executed in order from top to bottom[3].

How do we write commands in Python that a computer can understand? We use the language rules of Python. I've had many students in the past, particularly in my data science course, tell me, "*I'm no good with computers* or *I'm bad at math, so I can't do this programming stuff.*" The good news, firstly, is that you don't need to be good at math in order to be a really good programmer or even computer scientist. In fact, learning a programming language has a lot more in common with learning a natural language[]! The even gooder news is that you've all already mastered at minimum one language[4] and so you're well on your way to being good programmers! In terms of not being good with computers,

[1]Think Data from ST:TNG or Sheldon from The Big Bang Theory or those people that say things like "It's not illegal, so I'm within my rights to do this terrible thing" as if morality and legality are interchangeable. I'm ranting now.

[2]Like wishing on a monkey's paw

[3]There are a few exceptions to this, but we'll deal with them in time.

[4]Many readers have probably mastered even more, which is just amazing.

DOI: 10.1201/9781003265405-A

many people seem to hold computers in a hallowed state. It's really no different than dealing with a person, if said person doesn't talk back, is highly organized, can sometimes be opaque, and is a stickler for rules[5].

WITH THAT OUT OF THE WAY, LET'S TALK ABOUT THE BUILDING BLOCKS OF A COMPUTER PROGRAM. Computer programs are made up of **variables**, **functions**, and **keywords**. Together, these create **statements** which are then run by the program.

A.1 VARIABLES

Variables are extremely similar to those used in math (which are discussed in almost too much detail in Chapter 1). We can assign values to variables and then refer to the variable name when we need to use it. For instance in Python, if I were to type

```
x = 5
print(x),
```

I will see `5` printed in my terminal. If 5 isn't your favourite number, you can replace the value assigned to `x` so that your favourite number is printed to the screen.

When assigning values to variables, there are *three* main types of values that we will use in this course:

- **Integers:** these are whole numbers, *e.g.* $\cdots, -4, -3, -2, -1, 0, 1, 2, 3, \cdots$.
- **Float:** these are decimal numbers, *e.g.* $3.14159, 10.2$. Note that 1 is an integer but 1.0 is a float[6].
- **String:** these are what we use to represent words or symbols literally. Strings are always kept between a set of single[7] quotation marks. The object `"This is a sentence"` is a string, as is `"Let my people go!". Note that verb`"5"+ and 5 are different. The former is just a symbol, some scratches on paper[8] that have no meaning. The latter, the integer 5, is a symbol that represents our concept of five. In this sense, `'5'+'5'` is `'55'` and `5+5` is `10`.

[5] If you haven't dealt with someone like this in your life yet, just you wait

[6] even though they are functionally the same mathematically

[7] Or double

[8] Or pixels on a screen

We use variables in computing for the same reasons we may use them in math: we might not always know what value needs to go in, or we may want to do represent multiple values all at once.

A.II KEYWORDS

Variables can have almost any name we wish. The almost is important as there are some words that have very special meaning in Python which we cannot overwrite. These are known as **keywords** and are used to execute very specific behaviours in Python. Keywords exist in all computer languages, and while they may look different, their functions are largely the same.

A.III CONDITIONALS

The first set of keywords deals with **conditional statements**. Sometimes, when programming, we want something to happen only when something else is true. To create this branching process, we use conditionals. In Python, the main conditional keywords are `if`, `elif`, and `else`.

Conditionals are used in conjunction with **tests**. A test is a statement that can be `True` or `False`. For instance,

```
print(786521485 > 1}
```

will evaluate to `True` because 786521485 is indeed much bigger than 1.

To test if two things are equal, we use ==. We would like to use = to test if two things are equal, but as we saw above the = sign is used to *set* two things equal to each other; therefore we require a different symbol to *check* if two things are equal. That symbol, again, is ==.

As an example, we can check the truth value of `'dog'=='cat'` by printing the result of this test to the screen.

```
print('dog'=='cat')
```

will print the word `False` to the screen because the word[9] *dog* is definitely not *cat*.

Any code that we want to be executed if the test in our `if` statement is true must be **indented**[10], like in the following example:

[9] And the being

[10] Either with a tab or four spaces

```
if x < 2:
        print('x is less than 2')
elif x > 2:
        print('x is bigger than 2')
else:
        print('x is 2')
```

Here, we use two tests to check if x is greater than or less than 2 and our program will behave differently as a result. We don't need a third test to conclude equality; why might that be?

A.IV LOOPS

Sometimes we want to repeat the same commands many times, or on slightly different variables. In this case, we have **loops** which allow us to run code multiple times without retyping[11] the code multiple times. Loops come in two flavours: `for` loops are used when we know exactly how many times we want to repeat something[12], and `while` loops are used when we **don't** know how many times code is to be repeated[13].

When we know how many times we need to loop through stuff, we use a `for` loop. There are two ways to loop through things in Python. We can either loop over the items themselves

```
a_list_of_things = ['item 1', 'item 2', 'item 3', item 4']

for item in a_list_of_things:
        print(item)
```

or we can loop over the *positions* in a list[14]

```
a_list_of_things = ['item 1', 'item 2', 'item 3', item 4']

for i in range(0,4):
        print(i)
        print(a_list_of_things[i])
```

Both code snippets above are equivalent.

---

[11] Or copy/pasting

[12] Think about serving dinner to a table. If there are six guests, you must repeat the action of serving six times.

[13] As an example, think of searching for something you lost. You don't know how many places you'll have to search or how long you will spend searching. All you know is that you must search until the thing is found.

[14] Noting that the first item is in position 0

If we *don't* know how many times we have to loop over something, we wait for a *condition* to be true. In this case, we use a `while` loop.

```
x = 2
h = 0.1
while x < 50:
        x = (x-1)/h
```

In this case, unlike with the `for` loop, the number of times we run through the loop can change depending on the values of `x` and `h`. With the values set as is, we will only run through the loop two times[15]. This is a bad loop, because if we choose x or h badly, the code will run forever!

A.V IMPORT

Python can do lots of things. Everything it can do requires its own sets of commands. These new commands often come in **packages**. It is estimated that the number of available Python packages is around 5GB and growing[]. When you download Python, it comes with the necessary commands for building any command you need[16]. Luckily, we need not write everything from scratch. Many[17] things have been implemented in Python far better and more efficiently than we could hope to do. New commands, algorithms, and ideas are implemented in **packages**.

You can think of the whole of Python, with all its packages and programs as a giant library. When you sit down in a library to write a research paper[18], you bring with you your writing tools: a computer, a notebook, a pen, and a smattering of books. As you plan your paper, you realize there are things missing. Instead of using your primary sources to recreate studies that have already been done, and conclusions that have already been reached, you might think to go into the stacks and find books that will help you fill in your gaps. In Python, these other books are called packages[19] and are used to expand and enhance the functionality of Python. Everything in these packages can be recreated using the tools that are loaded up when you start Python, but this is unnecessarily inefficient and time consuming. If someone else has already

[15] The first time sets x to 10, and the second time will set it to 90, which is bigger than 50.

[16] This is called Turing complete, but that is the subject of a computer science class

[17] Understatement

[18] Or essay, or your super amazing novel, or a confession of your love to that person that you see everywhere, but have never spoken to, but you just *know* you're meant to be

[19] Or libraries

created a set of tools and commands that we need, we can just `import` that code into our program and use them as we see fit.

We install new packages from an online repository using the `pip` command. The most common way to do this is through a command line tool and typing[20]

```
pip3 install PACKAGE_NAME
```

We can then tell Python that we're going to be using this new package and all the things it contains by typing

```
import PACKAGE_NAME
```

at the top of our Python file.

The import command comes with some tools to make our coding more efficient or to distinguish between two packages that may have the same name. For instance, it's generally accepted that the package `pandas`, which we will use for all of our data processing, is shortened to `pd`. To do this, we use the `as` keyword:

```
import pandas as pd
```

After we type this, we need only to type `pd` to refer to the pandas library.

Occasionally, we also would like to only import specific pieces of a package. In that case there are two ways to do so. We can either

```
import PACKAGE_NAME.tool as tool
```

to extract the tool command from our package or we can write

```
from PACKAGE_NAME import tool
```

Either way, we then need only to type `tool` to use the specified tool. For reasons we can't get into, the **first** way of doing this is usually better.

A practical example is the use of the package `matplotlib` for visualizing data. This package is large and can do a lot of things. We often don't require the entire package, but only the piece of it called `pyplot`. So, when we import this package, we will use

```
import matplotlib.pyplot as plt
```

[20]This is highly operating system specific. This command works on macOS and Linux, but you may need a different command for Windows.

As well, scipy contains a lot of information and we often only need small pieces of it. Throughout this book, we make use of the `stats` package included in scipy. When we do, we import.

```
import scipy.stats as stats
```

Sometimes, we use the `from···import` when we want to import multiple things from a package without importing the whole thing. For instance,

```
from scipy import stats, linalg, constants
```

will import all three of the tools listed.

When referring to commands and tools that have been imported, we use **dot notation** to signal that a command belongs to a certain package. Some examples include

```
pd.read_csv()
stats.mean()
constants.pi
```

The first command says we are using `read_csv` from `pandas`[21]. The second command says from `stats` we would like to call the command `mean()`, and the third command says we would like to use `pi` from our imported `constants`.

A.VI FUNCTIONS

A function is just about identical to a mathematical function[22]. A function is a named set of rules that takes some variables as inputs and gives back an output. In Python[23], it is best practice to only have a single output to a function. You may have as many inputs as you need though.

In Python, a function is a word[24] followed by a set of `()`. The inputs go inside the `()`. Unlike math, sometimes Python functions don't have any inputs at all!

In just this appendix, we have already seen a function. The `print` statement above is a function. Its purpose is to take an input and print it to the terminal. This is an example of a function that has no output; it has an input, an action that it executes, and it gives you back nothing.

[21]Which we call pd for convenience, remember

[22]It's not hard to see where the association of math and computer science may have come from, but they are distinct for a reason.

[23]And generally any computer language

[24]The function name

In math, we create functions by writing equations.

$$CowLife(grass, milk) = 2^{milk} * grass + 36$$

is a function with a name, $CowLife$, which takes two inputs, $grass$ and $milk$, and returns some number. In Python, we define a function using the `def` keyword. The skeleton of a Python function looks like the following:

```
def function_name(input1, input2, input3):

        output = input1 + input2

        return output
```

This function has the name `function_name` and three inputs, `input1`, `input2`, and `input3`, and returns `output` which is the sum of `input1` and `input2`.

We could, in fact, recreate our $CowLife$ function exactly in Python.

```
def CowLife(grass, milk):

        out_variable = 2**(milk)*grass + 36

        return out_variable
```

If we would like to store a *particular* output for a *particular* set of input for this function, we set it equal to a variable. For instance,

```
cow23 = CowLife(2,3)
```

would then store the number `48` in the *variable* `cow23`.

Often times, we use functions when we want to reuse code with different, arbitrary inputs. For instance, the `print()` statement that we often use to see things on the screen is a function. Its purpose is to take an input[25] and print it to the screen.

As an example that we would use fairly often in this course, we may want to define a linear function

$$y = mx + b$$

[25]Our input could be a string, integer, float, or even a function.

where we can specify an input x, but also specify a different slope and y-intercept. In Python, this would look like

```
def linear_function(x,m,b):
        y = m*x+b

        return y
```

We can then find the output for a given input, slope, and y-intercept. For instance,

```
linear_function(2,2,2)
```

will return 6.

For the sake of the sanity of future you[26], it is best practice to put any function definitions at the top of your code (under any import commands).

A.VII A SIMPLE PYTHON PROGRAM

Putting all the above together, we can write a very simple program that takes our `basicdatafile.csv` and computes the mean of a column.

```
import numpy as np
import pandas as pd

def get_column(data_set,column_name):

        column = data_set[column_name]

        return column

data = pd.read_csv('basicdatafile.csv')

m1_col = get_column(data,'Measurement1')

mean_m1 = np.mean(m1_col)

print(mean_m1)
```

[26]I know in my case, future me doesn't get enough love. Future me always suffers for the convenience and comfort of present me. Treat future you better than I treat future me.

In this short code, we have

1. Imported two packages in their entirety: pandas and numpy and given them easier to work with names

2. Defined a function which takes a data set and a string and gives us back a particular column from our data set

3. The first line of code to get executed[27] is the `pd.read_csv` command. It is telling Python to look in the pandas library and read the data from our csv file into a variable called `data`.

4. The next line *calls* our `get_column` function with a *particular* data set, verb+data+, and a particular column name, `'Measurement1'`. This column is then stored in a variable called `m1_col`

5. We then use numpy to calculate the mean of this column and store that result in `mean_m1`.

6. Finally, we print this value to the screen so we can see it.

If you would like to see this in action, you can type it into your own Python file. If you would like to see the contents of any variable, you can always add `print()` statements at any point in the code after you have defined a variable. For instance, `print(data)` will show us the data set on the screen.

[27] i.e. do something

Bibliography

[1] Russell L Ackoff. From data to wisdom. *Journal of Applied Systems Analysis*, 16(1):3–9, 1989.

[2] Thomas Alerstam, Mikael Rosén, Johan Bäckman, Per G P Ericson, and Olof Hellgren. Flight speeds among bird species: Allometric and phylogenetic effects. *PLoS Biology*, 5(8):e197, 2007.

[3] Timm Bruch. Vaccinated hospital patients outpace the unvaccinated, but it doesn't mean the shots don't work: Experts. *CTV News Calgary*, 2022.

[4] Statistics Canada. Table 13-10-0415-01, live births, by month, 2020.

[5] US Supreme Court. People v. collins, 68 cal.2d 319, 1968.

[6] M Davidian and DM Giltinan. Nonlinear models for repeated measurement data. Chapman Hall. *CRC Monographs on Statistics and Applied Probability*, 1995.

[7] Jenne De Koster, Miel Hostens, Mieke Van Eetvelde, Kristof Hermans, Sander Moerman, Hannes Bogaert, Elke Depreester, Wim Van Den Broeck, and Geert Opsomer. Insulin response of the glucose and fatty acid metabolism in dry dairy cows across a range of body condition scores. *Journal of Dairy Science*, 98(7):4580–4592, 2015.

[8] Josh Dehaas. 10 things I wish I'd known in my first year of university. *Maclean's*, 2013.

[9] Environment and Climate Change Canada. Glossary - climate. https://climate.weather.gc.ca/glossary_e.html.

[10] HENRY ERIKSSON, KURT SVÅRDSUDD, BO LARSSON, LENNART WELIN, LARS-OLOF OHLSON, and LARS WILHELMSEN. Body temperature in general population samples: The study of men born in 1913 and 1923. *Acta Medica Scandinavica*, 217(4):347–352, 1985.

[11] Michael O Finkelstein and William B Fairley. A bayesian approach to identification evidence. *Harvard Law Review*, 83(9):489–517, 1970.

[12] Pierluigi Gambetti, Qingzhong Kong, Wenquan Zou, Piero Parchi, and Shu G Chen. Sporadic and familial cjd: Classification and characterisation. *British Medical Bulletin*, 66(1):213–239, 2003.

[13] Sakshi Goyal. Credit card customers, 2018.

[14] Penelope Green. The audacity of taupe. *The New York Times*, 2010.

[15] Andrew Hacker. *The Math Myth: And Other STEM Delusions*. New York: New Press, 2010.

[16] Stephen Heard. Do biology students need calculus?, 2016.

[17] Karoline Hood, Jacob Ashcraft, Krista Watts, Sangmo Hong, Woong Choi, Steven B Heymsfield, Rajesh K Gautam, and Diana Thomas. Allometric scaling of weight to height and resulting body mass index thresholds in two asian populations. *Nutrition & Diabetes*, 9(1):1–7, 2019.

[18] Saravanakumar Jeevanandam and Prathibha K Muthu. 2d: 4d ratio and its implications in medicine. *Journal of Clinical and Diagnostic Research: JCDR*, 10(12):CM01, 2016.

[19] John Kruschke. *Doing Bayesian Data Analysis: A Tutorial with R, JAGS, and Stan*. Cambridge, Massachusetts: Academic Press, 2014.

[20] Doug Linder. The trial of orenthal james simpson, 2000.

[21] Supreme Court of Nevada. Brown v. state, 1997.

[22] Government of Ontario. CovidâĂŚ19 (coronavirus) in ontario, 2022. Accessed: 2021-01-30.

[23] Philip Sadler and Gerhard Sonnert. The path to college calculus: The impact of high school mathematics coursework. *Journal for Research in Mathematics Education*, 49(3):292–329, 2018.

[24] Knut Schmidt-Nielsen and Schmidt-Nielsen Knut. *Scaling: Why is Animal Size So Important?* Cambridge, Massachusetts: Cambridge University Press, 1984.

[25] Derek J Smith, Alan S Lapedes, Jan C De Jong, Theo M Bestebroer, Guus F Rimmelzwaan, Albert DME Osterhaus, and Ron AM Fouchier. Mapping the antigenic and genetic evolution of influenza virus. *Science*, 305(5682):371–376, 2004.

[26] Valerie Strauss. Why kids hate school: Subject by subject. *The Washington Post*, 2012.

[27] M Vázquez, P Fagiolino, M Ibarra, and L Magallanes. Safety assessment of efavirenz after a single-dose bioequivalence study: A trend to correlate central nervous system effect and plasma concentration. *International Journal of Pharmacy*, 5:46–52, 2015.

Index